Racial Science in Hitler's New Europe, 1938–1945

CRITICAL STUDIES IN THE HISTORY OF ANTHROPOLOGY

Series Editors: *Regna Darnell* | *Stephen O. Murray*

RACIAL SCIENCE IN HITLER'S NEW EUROPE, 1938–1945

Edited by Anton Weiss-Wendt and Rory Yeomans

University of Nebraska Press : Lincoln and London

Manufactured in the United States of America
♾

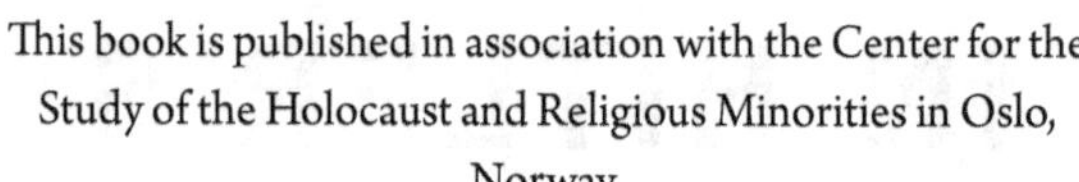

This book is published in association with the Center for the Study of the Holocaust and Religious Minorities in Oslo, Norway.

Library of Congress Cataloging-in-Publication Data
Racial science in Hitler's new Europe, 1938–1945 / edited by Anton Weiss-Wendt and Rory Yeomans.
pages cm. — (Critical studies in the history of anthropology)
Includes bibliographical references and index.
ISBN 978-0-8032-4507-5 (pbk.: alk. paper) 1. Physical anthropology—Europe—History—20th century. 2. Racism in anthropology—Europe—History—20th century. 3. National socialism and medicine—Europe—History—20th century. 4. National socialism and science—Europe—History—20th century. 5. Racism in medicine—Europe—History—20th century. 6. Eugenics—Europe—History—20th century. 7. Race—Research—Germany—History—20th century. 8. Germany—Politics and government—1933–1945. 9. Germany—Race relations. I. Weiss-Wendt, Anton, 1973– II. Yeomans, Rory.
GN50.45.E85R33 2013
305.800943—dc23 2012050639

Set in Arno Pro by Laura Wellington.
Designed by Ashley Muehlbauer.

In Memoriam:

Karin Isakova
Elisabeth Weiss-Wendt
Jeanne Margaret Pearson
Margaret Henderson Macdonald

Contents

Series Editors' Introduction

REGNA DARNELL AND STEPHEN O. MURRAY

Anton Weiss-Wendt, Rory Yeomans, and the contributors to this collection of essays explore the complex story of how eugenics and race as opposed to culture and class became the touchstones of German anthropological science during the Second World War (1939–45 in Europe). Nazi science placed remarkable value on anthropological justification for its policies of genocide, and the discipline, not only in Germany, still struggles with its complicity. These issues are usually framed for English readers in terms of Anglo-American responses to Nazi and Holocaust literatures, from outside what are judged to be the unambiguously deplorable misuses of science in the service of ideology and pragmatic politics. An ocean of separation from Europe, at least at the time, allowed American protestations of innocence, despite its own considerable development of eugenics and scientific racism.

This collection presents detailed case studies of how racial science was adapted to local conditions in a wide variety of European nation-states within the Nazi sphere of influence. The internal variability and cultural specificity of these cases have hitherto been invisible, especially outside Europe, because we have inherited a simplistic and unnuanced narrative of the Holocaust and the science underlying it.

Hitler's racial science was by no means confined to Germany, although its intellectual underpinnings arose from the international hegemony of the German research universities. These essays demonstrate how satellite states acted as active participants in defining racial science in relation to local agendas. In no case was the national ideology coextensive with that of Nazi Germany; yet in each country local issues found fertile ground for the pursuit of national identity and autonomy based on German racial science as translated into local terms. Motives, negotiations, bureaucratic regimes, and outcomes varied sharply in Denmark, Italy, Austria, Latvia, Estonia, Hungary, and Romania. Contemporary scholars from each of these national traditions often come to the comparative task with an anthropological or ethnographic mindset. By contrast, agency at the periphery is the common problem that unites this set of essays into a sustained critique of the arbitrary restrictions of scholarship that is confined by the national boundaries and official languages of particular nation-states. Europe in Hitler's time, as in our own, displayed a fluidity of communication—political, social, economic, and cultural—that renders any exclusively national analysis incomplete.

Race is also a quintessential American problem, one deeply embedded in the subconscious of American anthropology. During the years leading up to World War II, North American anthropology explored the autonomy but inextricability of race, language, and culture as ways of classifying the diversity of humankind. It was Franz Boas, himself an early product of the very German higher education system that produced Hitler's race theory, who deconstructed the typology of race and demonstrated the plasticity of racial types. His 1911 *The Mind of Primitive Man* remains a trenchant critique of scientific racism and eugenics, in part because it insists on seeing human culture and human biology as sides of a single coin. In any case, Boas's anthropology was directed to questions of immigrant assimilability rather than of permanent minorities that threatened the ostensible homogeneity of majority populations in European nation-states where the German Romantic tradition supposed a "genius" of one and only one folk (*Volk*) per nation. Eugenics rather than genocide was the permanent solution envisioned by scientific racism.

The questions opened up by this comparative analysis frame antisemitism and genocide in local terms that varied across European nation-states. The broad effect of Nazi racial science, then, is perhaps best understood as providing a focus point for multiple smoldering resentments based on ethnicity and race; these were played out radically under a regime that privileged the "othering" of some groups, to the point where they were not fully human, and lent the stamp of science to violent intolerance. Demography, history, internal diversity of region, ethnicity, and local circumstance all determined particular outcomes within the umbrella of Nazi racial science.

Acknowledgments

This volume traces its origins to the conference "Racial Science in Hitler's New Europe," which was held at the Center for the Study of the Holocaust and Religious Minorities in Oslo, Norway, in October 2009. The conference was cosponsored by the center and the University of Oslo. We would like to take this opportunity to thank all those at the center in Oslo who contributed toward the conference (listed alphabetically): Kari Amdam, Georg Andreas Broch, Terje Emberland, Odd Bjørn Fure, Ann Elisabeth Mellbye, Ewa Maria Mork, and Maria Rosvoll. A special thank-you goes to Jorunn Selm Fure of the University of Oslo, who enthusiastically embraced the idea of a conference and actively participated in all the stages of its planning.

Abbreviations

CULANO	Commissie tot Uitzending van Landbouwers naar Oost-Europa (Commission for the Employment of Farmers in Eastern Europe)
DNSAP	Danmarks Nationalsocialistiske Arbejder Parti (Denmark's National Socialist Workers' Party)
ERR	Einsatzstab Reichsleiter Rosenberg (Operations Staff Rosenberg)
ERÜ	Eesti Rahva Ühisabi (Estonian Relief Agency)
HSSPF	Höhere SS-und Polizeiführer (Higher SS and police leader)
KWI-A	Kaiser Wilhelm Institut für Anthropologie, menschliche Erblehre und Eugenik (Kaiser Wilhelm Institute for Human Heredity Sciences and Eugenics, Berlin)
MOVE	Magyar Országos Véderő Egyesület (Hungarian National Defense Association)
NOC	Nederlandse Oost Compagnie (Dutch East Company)
NSB	Nationaal-Socialistische Beweging (Dutch Nazi Party)
NSDAP	Nationalsozialistische Deutsche Arbeiter Partei (National Socialist German Workers' Party [Nazi Party])
NSDAP-N	Nationalsozialistische Deutsche Arbeiter Partei-Nordschleswig (National Socialist German Workers' Party, North Schleswig Branch)
RAVSIGUR	Ravnateljstvo za javni red i sigurnosti (Directorate for Public Order and Security, Independent State of Croatia)
REM	Reichserziehungsministerium (Reich Ministry of Education)

RFSS	Reichsführer-SS (head of the SS [Heinrich Himmler])
RKF	Reichskommissar für die Festigung deutschen Volkstums (Reich commissioner for the strengthening of Germandom [Heinrich Himmler])
RKO	Reichskommissariat Ostland (Reich Commissariat Ostland)
RKU	Reichskommissariat Ukraine (Reich Commissariat Ukraine)
RMO	Reichsministerium für die besetzten Ostgebiete (Reich Ministry for the Occupied Soviet Territories)
RSHA	Reichssicherheitshauptamt (Reich Security Main Office)
RuSHA	Rasse-und Siedlungshauptamt (Race and Settlement Main Office)
SA	Sturmabteilung
SK	Sonderkommando (subunit of the mobile killing units of the SS [Einsatzgruppen])
SS	Schutzstaffel (Nazi elite force)
TDSPI	Tautas dzīvā spēka pētīšanas instituts (Institute for Research on National Vitality, Latvia)
TDSVDK	Tautas dzīvā spēka veicināšanas darba kopa (Working Group for the Advancement of National Vitality, Latvia)
TU	Tartu University, Estonia

Racial Science in Hitler's New Europe, 1938–1945

Introduction

The Holocaust and Historiographical Debates on Racial Science

ANTON WEISS-WENDT AND RORY YEOMANS

It is now thirty years since the publication of Bernt Hagtvet, Jan Petter Myklebust, and Stein Ugelvik Larsen's *Who Were the Fascists?*[1] As stated in their introduction, one of the objectives of the book was the creation of an international network of scholars interested in the social history of fascism. Much has changed during the past three decades, both in scholarship and in the wider world. The dominant scholarly discourse under which many of the East European contributors to the book operated—Marxist historiography—has vanished along with the Communist countries to which they once belonged. Simultaneously, social history has been superseded by cultural history as the dominant tool for the study of totalitarian regimes. Nonetheless, the history of Nazi Germany and specifically the Holocaust is one field of research that has demonstrated continuous scholarly interest in modernization.[2]

Zygmunt Bauman was one of the first scholars to argue that the Holocaust—and by extension the racial theories that underpinned it—was "genocide with a purpose." Eradicating populations, he contended, was not an end in itself but a grand vision of a better and different kind of society. For Bauman, "modern genocide is an element of social engineering, meant to bring about a social order conforming to the design of the perfect society." In his now-famous metaphor of landscaping the human garden, physical destruction appears as a necessary chore of weeding, which can be framed as a creative process. Thus, Bauman wrote, "All visions of society-as-garden define parts of the social habitat as human weeds. Like all other weeds, they must be segregated, contained, prevented from spreading, removed and kept outside the society boundaries; if all these prove insufficient, they must be exterminated."[3]

The issue of modernity has defined the study of society in Nazi Germany from the end of the 1960s onward. While some earlier studies attempted to place the social history of the Third Reich firmly within the context of racial politics, others examined aspects of modernization and everyday life such as consumerism, leisure, tourism, and architecture divorced from the Nazis' racial agenda.[4] Among the sternest critics of the new social history were Michael Burleigh and Wolfgang Wippermann. Making a distinction between Nazi racial policy as "reactionary" and its social policy as "progressive" was deeply problematic, in their opinion; indeed, both racial and social policy were symbiotically linked, simultaneously modern and profoundly antimodern. According to Burleigh and Wippermann, race was to supplant class as the primary binding principle in a society with growing cleavages. The Nazis sought to create a racial state by means of modern social policies. Therefore, racial and social policy had to be studied as "an indivisible whole."[5]

One of the major controversies in the study of the Third Reich and the Nazi regime concerned the evolution of the Holocaust. Specifically, scholars probed the extent to which the Holocaust was the result of deliberate policies by the Nazi leadership from the late 1930s onward and/or how far it reflected a range of external and internal pressures.[6] While the "intentionalists" were largely political and diplomatic historians who focused overwhelmingly on the personality and ideology of Hitler and suggested a top-down model of Nazi rule, "functionalists" were often social and institutional historians who interpreted the Nazi regime in polycratic terms. The latter argued that the Holocaust was driven by improvisation and the internal struggle for power and therefore came about as the result of pressure from below rather than arbitrary decisions from above.[7] One of the most influential, if controversial, examples of the functionalist interpretation of the Holocaust was *Vordenker der Vernichtung* (translated into English as *Architects of Annihilation*), by Götz Aly and Susanne Heim, published in 1991. Aly and Heim insisted that the explanation for the Holocaust was to be found not in *völkisch* (racial) ideas or academic racial treatises but in the utopian economic, industrial, agricultural, and social programs devised by a new generation of ambitious young agronomists, policy analysts, economists, and social planners in the service of the Nazi Party since the late 1930s. Without the input of these young technocrats, they argued, the campaign against the Jews would not have escalated into industrialized mass murder but likely would have remained at the level of pogroms and massacres. Combining ideas about economic rationalization and social engineering, they linked the genesis of the Final Solution to the push for *Lebensraum* (living space) and the attempt to create an empire in Eastern Europe. According to Aly and Heim, the Jews were systematically murdered because economic planners considered them to be an obstacle to the transformation of the rural East European population into a modern, urban middle class, which would constitute a support base for Hitler's New Europe.[8]

Aly and Heim contended that elements of everyday routine such as mass tourism and Volkswagen cars were part of the same process that led to genocide. In their view, the Holocaust belonged to the same idea of remaking the world through economic restructuring, the decimation of classes and groups, and working toward the realization of a "modern technocracy." Model landscapes complete with motorways, railway lines, canal projects, and integrated economic and transport systems were as much a part of the landscape of the Holocaust as the barbed wire, watchtowers, and gas chambers of Dachau and Auschwitz.[9]

To advance their thesis, Aly and Heim used, among others, the example of scientific advisers in the General Government for the Occupied Polish Territories and the administration of the Warsaw ghetto. Having determined that the ghetto economy was unsustainable without substantial financial support, those advisers suggested to Governor General Hans Frank that the ghetto and its inhabitants should be liquidated. The removal of the ghetto, they advised, would enable the rationalization and modernization of the Polish economy. Christopher Browning, however, demonstrated the flaws in Aly and Heim's thesis. When arguing for the destruction of the Warsaw ghetto, the General Government's economic advisers (the "productionists") sought to keep the Jewish population alive in order to fuel economic production. This line of argumentation contradicted the wishes of racial experts and much of the occupation authorities, who demanded the destruction of the ghetto population on racial and ideological grounds. As Browning has pointed out, rather than a utilitarian tool of economic rationalization, the Holocaust was economically ruinous for the new Nazi European empire, draining manpower and resources.[10]

Nevertheless, Browning conceded that Aly and Heim had demonstrated the significance of social utopia and modernization, in addition to social mobility and ambition, in the genesis of the Holocaust. Furthermore, he sided with both scholars in emphasizing that Nazi planning agencies were marked by conflict, disagreement, factionalism, and competing agendas as far as the treatment of the Jews was concerned.[11] Historians such as Burleigh and Wippermann agreed that, by their very nature, Nazi power structures were diffuse and that the regime encountered pressure from working-class constituencies, especially on issues related to social mobility. However, while insisting that local dynamics had induced the policy of *Lebensraum* in the East and the Holocaust, Burleigh and Wippermann argued that social pressures could not be separated from the racial agenda.[12] To underscore this nexus, this volume explores the development of racial science under Nazi rule from a broader perspective. In the main, the essays demonstrate that while the Final Solution was ostensibly informed by racial and national aspirations, these ideals in turn were affected by wider societal processes that played an important role in the creation of a utopian society and in building consensus.

Positive Eugenics into Racial Science: Defining the Terms

Racial theory, as it came to be known, developed during the nineteenth century through a collective effort of European scientists and thinkers. In a tumultuous period of social and political change, the offer of finding immutable characteristics in humans gave extra credence to experimental science. The determination to classify all flora and fauna in the world inevitably led to the idea of redefining the place of humanity in nature. It appeared only natural that the great variety of cultures, languages, and physical features had to be explained. One of the most striking ideas advanced by Charles Darwin in his *On the Origin of Species by Means of Natural Selection* (1859) was that of struggle for survival. The term *survival of the fittest* was further popularized by English social scientist and philosopher Herbert Spencer. Arthur de Gobineau readily conflated linguistic families with racial types to conclude that mixing of races led to degeneration, gradually decreasing the quality of blood. In his pioneering *Essay on the Inequality of the Human Race* (1853–55), Gobineau cast race as the primary moving force of world history.

The word *eugenics*, coined by the British statistician Francis Galton in 1883, denoted selective breeding both for promoting favored characteristics and for eradicating features deemed harmful. Eugenics effectively merged anthropology, Darwinism, and medicine into something German scientists later termed "racial and social biology." The German eugenics movement emerged in the late nineteenth century in response to the supposed "degeneration of the human species." The German term *Rassehygiene* (racial hygiene) had a broader scope than the English word *eugenics* and loosely meant the hereditary improvement of a population or all of humanity.

As difficult to capture is the meaning of the word *Rasse*, which, according to German eugenicist Alfred Ploetz, signified any interbreeding human population that, over the course of generations, demonstrated similar physical and mental traits. Whereas Ploetz was the first to start using the term *Rassehygiene*, Wilhelm Schallmayer was the first German scholar to publish a treatise on eugenics, in 1891. Schallmayer introduced the cost-benefit analysis theory, which later came to dominate the race hygiene movement. He saw a direct correlation between the biological vitality of the nation and the scope of state power. Neglect of the hereditary fitness of the population would allegedly have a negative impact on politics and could eventually result in the downfall of the state, according to Schallmayer. German zoologist Friedrich Ratzel interpreted Darwin's theory of evolution as the violent struggle between species for territory. The book that he published in 1904 referred to it as *Lebensraum*.[13] That same year the first journal ever dedicated to eugenics, *Archiv für Rassen- und Gesellschaftsbiologie* (Journal of racial and social biology), was founded in Germany by Ploetz. Not coincidentally, eugenics and imperialism have exercised a mutually beneficial relationship. Imperialism provided rich material for eugenics, which supplanted a scientific legitimation for the domination of the "lesser races." By 1900 racial thinking had become a "science." The First World War, marked by the brutalization of warfare and dehumanization of the enemy, only intensified the tendency to think in racial and national terms.[14] The goal now was no longer preservation of race, but its improvement. Concern shifted from the health of the individual to improving the hereditary fitness of the human race. Consequently, racial progress came to be increasingly interpreted in negative terms as an aversion to racial decline.

Criminal anthropology applied elements of racial theory in practice, making a bridge from physical appearance to mental abilities and to habitual criminal tendencies. Coined in 1888 by lawyer Franz von Liszt, the term *Kriminalbiologie* (criminal biology) suggested that crime was a manifestation of innate characteristics of the offender rather than a reflection of social environment. Although criminal biology predated the Nazis, it acquired scientific currency only after 1933. In effect, criminal law theory was transformed into a concept that was meant to strengthen the vitality of the German people through eliminating "harmful elements" and more vigorous implementation of "racial-hygienic measures." From then on, criminal biology became the preserve of psychologists and medical professionals who focused in their research not on the offense but on the nature of the offenders. Anthropological characteristics of individuals were supposedly enough to determine their predisposition to crime. As a result, a criminal act made an offender a lasting danger to society, someone who could not be resocialized.[15]

Typically, racial scientists came from the medical profession, traditionally seen as the guardian of the health of the nation. In the late nineteenth century academic physicians received much social esteem and, by extension, political importance. Otherwise, German eugenicists were far more heterogeneous than has been assumed in literature. Their scientific interests, personal beliefs, and political allegiances often predated the Nazis. All but a few German eugenicists accepted the superiority of the white race, but so did most of their colleagues abroad. Their ideas of increasing the number of Germany's "fitter" elements and eliminating the masses of the "unfit" were not dissimilar to those of other Western eugenics movements. Hitler's seizure of power, however, instantly placed them in the service of the Third Reich. Eventually, over 90 percent of German anthropologists and eugenicists joined the Nazi Party.[16] The ideal of a healthier, more productive, and therefore more powerful nation echoed Nazi calls for national revival. As Sheila F. Weiss has emphasized, "Eugenics embodied a technocratic, managerial logic—the idea that power was a product of the rational management of population."[17]

Scholars have noted a remarkable consistency between the views of German anthropologists and Nazi officials. The Nazi seizure of power eliminated the very possibility of any other but a racist interpretation of eugenics. Anthropologists helped to shape Nazi racial policies, either directly or indirectly, while the Third Reich invested significant resources in research that was expected to perfect the human race. Indeed, by the late 1930s the goals and activities of professional eugenicists came to closely resemble the rhetoric and racial policies of the Nazi

Party. One of the fundamental Nazi principles—the immutability of human genetic material—came straight from the annals of racial science. However, once racial science had been proclaimed the guarantor of the people's welfare, and by extension the state's welfare, it inevitably became the subject matter of political decision making. The blatant manifestation of that link was the appointment of Heinrich Himmler as Reich commissioner for the strengthening of Germandom (Reichskommissar für die Festigung deutschen Volkstums, or RKF) in October 1939, tasked with supervising the population transfer that followed (*Umsiedlung*).[18] Eventually Himmler gained control of virtually every institution involved in both racial policy and organized terror.

Through the blending of the concepts of people and race, the Nazis developed the centerpiece of their ideology—the *Volksgemeinschaft* (the community of the people). The existence of the *völkisch* community, however, was predicated on the stigmatization of the "enemies of the people," who were to be excluded, expelled, or annihilated. The Decree on the Protection of the People and the State from February 1933 marked the first attempt at restructuring the relationship between subjects and state. This and subsequent legal acts widened the spheres of private life in which the law could be applied.[19] However, the *Volksgemeinschaft* could not be attained through gratification, compliance, terror, and legislation alone. It was a transformative, political process that encompassed the whole of society. Seen from this perspective, the persecution of German Jews was instrumental in destroying civil society and the constitutional state.[20] Viewed through the prism of race, society was composed not of individuals or social classes but of the *Völker* (races) that should, with the help of biology and genetics, be segregated into "valuable" and "less valuable." Those defined outside the *Volksgemeinschaft* did not have the right to enter the community of the chosen few and therefore supposedly aimed at destroying it from within. Jews in the first place were stigmatized simultaneously as a foreign race and *Gemeinschaftsfremde* (alien to the society).[21] In equal measure, the preservation of the homogeneity of the German nation "required" the physical separation of Gypsies (Roma). By merging the concepts of *Volksgemeinschaft* and *Kulturkampf*, Nazis evolved the idea of *Volkstumskampf*, that is, racial struggle.

Nazi ideology developed out of the twin concepts of *Rasse und Raum* (race and space) and *Blut und Boden* (blood and soil), which encompassed antisemitism and Nordic supremacy on the one hand and eugenics and pronatalism on the other. Resultant policies were supposed to improve the human stock by favoring the most "advanced" racial type, the Aryan or Nordic race. Nazi ideologists viewed the pursuit of living space as part of an inevitable racial struggle for existence that was driving human evolution forward. The primacy of that, ultimate, goal had never been in question, despite the oscillation of Nazi policies due to pragmatic considerations.[22] Euthanasia and forced sterilization in Nazi Germany evolved into a program of mass murder, defined by a deadly mixture of ideology and cost efficiency.

Problematic from the viewpoint of definition, the intricate connection between positive and negative eugenics came to eventually haunt the practitioners of the new discipline. The Nazi agenda further blurred the border between the two. The ban on smoking and extensive cancer research in Nazi Germany, as Robert Proctor has shown, were supposed to benefit only the few select "Aryans" to the exclusion of many, to empower the race rather than improve the health of the individual.[23] The nutritional intake and health of babies and their mothers of "good blood" in Germany proper, as Götz Aly has vividly demonstrated, were conditioned upon deliberate food withdrawal from the populations of subjugated Eastern Europe.[24] In retrospect, almost any hygienic measure in Hitler's New Europe that could be loosely evaluated as "positive" had inbuilt negative consequences. Heralded by Hitler's regime as "racial war," the Nazi attack on the Soviet Union in 1941 turned eugenics into an ultimate weapon of destruction on behalf of the Aryan race. It was the image of the superiorly built and mentally attuned "Aryans" defending the white European civilization against the devious, degenerate barbarians from the east that became the epitome of racial science.

Nazi Racial Science and Organic Nationalism in East Central Europe

Racial doctrine is grounded in existential fear, as illustrated in the European colonial experience in Africa in the nineteenth and early twentieth century. The further European settlers segregated the native populations the more they feared that their own religion and culture were under threat of extinction. Colonial conquest promised instant riches as well as cultural and genetic demise. Colonial discourse builds a bridge from German imperial adventures, which had commonly resulted in massacre, to Nazi racial policies in occupied Europe. Ironically, Hitler was probably the first politician who started speaking of a "New

Europe." His vision for the continent, however, was drastically different from that of most radical thinkers and scientists of his time. In his programmatic book, *Mein Kampf*, Hitler infamously stated that the future of Germany lay not in southern and western Africa but in Eastern Europe. Together with the doctrinal transfer came the old fears of the barbarians' revenge.

Rather than the tribal areas of Africa, however, Germans marched through countries with newly won national identities. In fact, in most of the new East and Central European states the nation-building process was still underway. Diaspora minorities, which on average constituted over 10 percent of the total population, were universally viewed as a potential fifth column. It was not only nationalist politicians who wanted to push them over the border or, alternatively, extend the state borders, notwithstanding the obvious contradiction. Throughout Europe, the references to culture were as frequent during the Nazi occupation as they were during the previous period of independence. Coincidentally, in accordance with the existing conventions, the terms *English race* or *Estonian race* had been used interchangeably with *English nation* or *Estonian nation* throughout the 1920s. In the scholarly writings and popular media of the period alike, the words *race*, *ethnicity*, and *nationality* often meant one and the same thing. Whenever it was conceived of as the struggle for survival, the conflict took on a military, cultural, or even metaphysical form. Although the term *Kulturkampf*, which was coined by anthropologist and leading left liberal Rudolf Virchow, originally denoted Bismarck's policy toward the Catholic Church, its meaning expanded dramatically. By stressing "natural" difference based on language and culture, organic nationalism lent credence to anthropology. Conversely, hereditary fitness supposedly ensured the long-term survival of a nation and the alleged superior cultural traditions that it embodied. Within this discourse, the Aryan race emerged as the "cultural race" par excellence.

The German Society for Racial Hygiene (Deutsche Gesellschaft für Rassenhygiene) was replicated in the years leading to and immediately after the First World War across East Central Europe. Thus, similar institutions were established in Austria, Czechoslovakia, Hungary, Poland, Romania, Estonia, and Bulgaria, although the specific national context and social and medical practices in those countries did not differ significantly. That was the conclusion of Swiss eugenicist Marie Thérèse Nisot in her comparative study of eugenics published in 1926. Throughout interwar Europe, racial anthropologists and eugenicists enjoyed the status of a constructive force contributing toward the creation of the modern nation-state, since their research advanced the program of national regeneration. Dependent on the state for funding, eugenics movements entered a mutually beneficial symbiosis with the state. Regardless of the preferred political structure—a liberal democracy, a peasant state, or a corporatist state—organic nationalism ruled supreme in interwar Europe, while the ethnic majority embodied the idea of a nation-state. Hence the biological laws of heredity promoted by eugenics and racial nationalism seemed to offer the most appealing definition of the state. Eugenics claimed the guardianship of the "biological capital" of the nation, a "healthy body politic" required for a strong nation-state. At the same time, eugenicists promoted a program of national regeneration that would stamp out the proliferation of the "genetically inferior." Notably, when referring to the "unfit" they meant not only people with disabilities but sometimes also individuals of different ethnic origin. Thus national belonging was redefined in biological terms.[25]

From the late 1920s onward eugenics in East Central Europe increasingly looked to German racial hygiene for inspiration. For some countries, the Nazi sterilization law of 1933 served as both a model and the affirmation of the eugenics movement's vitality. The dominance of the German academic tradition was the reason why the majority of East European eugenicists had been educated in Germany and Austria.[26] Cost-benefit analysis, motivated by the need to reduce the national welfare budget, provided a further incentive for scholars outside Germany to look for a connection between race hygiene and various forms of rationalization. By the mid-1930s the ideological underpinning of the eugenics movement became even more pronounced. The papal 1930 encyclical against the eugenics law and Stalin's censure of eugenic research in the Soviet Union in 1936–37 propelled Nazi Germany to the center of "racial science."[27] The onset of authoritarianism in East Central Europe strengthened state monopoly and promoted corporatism. The subsequent discourse advanced the idea of a planned economy in relation to health policy, introduced in the interests of a nation in possession of high culture. As in the late Weimar period in Germany, eugenics brought with it the promise of economic efficiency and cultural aptitude. In short, it appeared to be a scientific means of solving social and political problems.

What has largely gone unnoticed are the hundreds of scientists and belletrists in the occupied countries of Europe who worked selflessly to implement Hitler's racial plans, although for the benefit of their own countries. Since the end of the First World War eugenicists and racial nationalists in East Central

Europe had been debating the issue of ethnic minorities, suggesting various solutions, from birth control and sterilization to population transfer. Scientists relied on anthropometry to establish the "racial origin" of various ethnic groups, particularly within the multiethnic context of East Central Europe, and thus their proper place within the national community. The perception of the nation as a racial community—as opposed to a political community—became dominant whenever culture was made the formative element thereof. *Culture*, however, not only rhymed with *nature* but often came to replace it. The concept of culture was instrumental in asserting primordial ties within a community. Simultaneously, it advanced the romantic notion of a heroic past with a double emphasis on struggle for freedom and conquest. Coined in 1839, the term *ethnography* was the confluence of natural and human sciences, meant as a tool in the search for biological and historical origins.[28]

The focus on Nazi Germany has superimposed a notion that racial policies were enforced on the countries occupied or dominated by Hitler's armies. The campaign of mass murder, ingrained in the concept of Aryan superiority, perpetrated by the Nazis in subjugated East Central Europe made any suggestion of indigenous agency appear exceptional and/or exaggerated. Although not totally incorrect, this view seems to ignore local dynamics firmly rooted in the national histories of the great many states that came into existence in the wake of the First World War. In fact, statehood and historical continuity proved to be two potent factors leading to the establishment and promotion of "racial science" locally. Not coincidentally, war served as a consolidating factor for the field of eugenics. The First World War made eugenicists change their perception of warfare as a natural selection process that benefitted the fittest. It simultaneously invited military analogies and contributed to the creation of a truly international eugenics movement. The Nazi conception of racial war recast the Second World War as an exercise in fundamental science carried out by violent means.[29] In spite of the reality of military occupation or political dependence, semiofficially the Nazi occupation authorities not only tolerated but actually endorsed local academic and scientific research that might have coincidentally advanced Nazi geopolitical goals. The idea of splitting the so-called eastern populations into as many parts as possible (as emphasized by the head of the German Schutzstaffel [SS] and the police, Heinrich Himmler, in his position paper submitted to Hitler in May 1940) was fully compatible with promoting scholarship that could have inadvertently accelerated that process.[30]

Suddenly, racial hierarchies that had separated white Europeans from the rest of the world now were being used to redraw boundaries within Europe, Hitler's New Europe. Anxious to carve a piece from the emaciated body of the continent caught in the struggle between "good" and "evil"—the idea pressed home by Nazi propaganda—many occupied European countries developed their own programs for national renewal. This proved to be one idea that united politicians, intellectuals, and scientists on both sides of the ideological divide—those who chose to collaborate with the occupation regime and those who did not. However little trust they placed in Nazi assurances of a brighter future, one thing was clear: democratic, multiethnic states were finished. The promise of racial doctrine was too hard to ignore and too easy to follow. In effect, the occupied eastern territories, as they became known in Nazi parlance, sometimes engaged in a self-destructive process. They thought they were laying the foundation for a sovereign or independent state but in fact they were helping to build the Thousand-Year Reich, grounded in racial superiority.

The American Connection

In order to better comprehend the uneasy relationship between Nazi racial science and its numerous variations across occupied Europe, it may be useful to briefly look at the interaction between American and German eugenicists prior to the outbreak of the Second World War. Historiographically, in the past thirty or so years understanding of the latter issue has undergone substantial revision, exemplified by the studies of Daniel J. Kevles, Stefan Kühl, and Edwin Black.[31] From the simple admission that a handful of American scientists, who nonetheless represented the radical-right fringe, had exercised some influence on German racial hygiene, the discussion moved to an examination of the comprehensive exchange of ideas and eugenic policies between the United States and Nazi Germany and to the provocative conclusion that American eugenics paved the way for the Holocaust. Indeed, this connection can be effectively established within the broader context of international eugenics.

American and German scientists played a principal role in establishing the international eugenics movement. The First World War and the vehement opposition of French and Belgian members of the International Federation of Eugenic Organizations prevented their German counterparts from rejoining the fold until 1927. By that time the United States had emerged as an absolute leader in eugenic legislation. Since the adoption of the first-ever law on sterilization in Indiana in 1907, thirty-two more American states had followed suit by 1930. The Immigration Act of 1924, which discriminated against arrivals from Southern and Eastern Europe to the benefit of those from Northern Europe, received a boost in the form of a scientific rationale supplied by American eugenicists. Adolf Hitler was among those who positively commented on "advancement" in population control in the United States.[32] Nazi racial scientists carefully studied American state legislation before introducing their own sterilization law in 1933. They were quick to note that, unlike in the United States, the German law extended to the whole of the country and served as a preventive measure rather than punishment against criminal offenders.

Scientific cooperation between the United States and Nazi Germany went both ways. Several leading American scholars, most notably the head of the Eugenics Record Office in Cold Spring Harbor, Charles B. Davenport, and his deputy, Harry H. Laughlin, came dangerously close to endorsing Nazi racial policies. Even before Hitler conceived of the idea of a New Europe, Laughlin proposed establishing a world government based on eugenic principles. Davenport fended off criticism by making a distinction between politics and science; Laughlin, meanwhile, proudly accepted an honorary doctorate from Heidelberg University in 1936 yet decided it would be better for his career not to attend the award ceremony in person. American foundations such as the Rockefeller Institute continued sponsoring racial research in Germany (but also throughout Central, Eastern, and Southeastern Europe) after 1933. Mutual praise periodically appeared in professional journals on both sides of the Atlantic throughout the 1930s; a few American eugenicists went to Germany on a study tour as late as 1940. The Japanese attack on Pearl Harbor and the subsequent U.S. entry into the war against Nazi Germany in December 1941, however, put an end to any form of cooperation between eugenicists in the two countries. Laughlin was forced to retire (Davenport had already retired, in 1934) and his laboratory was closed, ceasing its use as a conduit of Nazi racial propaganda.[33]

The American contribution attests to the truly global appeal of eugenics in the interwar years. Much as the Darwinian devaluing of human life should not be regarded as proto-Nazi—as Richard Weikart has insisted—the call for a "biological revolution" was not confined to the Nazis.[34] Along with the Americans Davenport, Laughlin, Lothrop Stoddard, Clarence Campbell, Madison Grant, and others, Nazi eugenic know-how received enduring support from the Norwegian Jon Alfred Mjøen, the Swede Herman Lundborg, the British Cora B. S. Hodson, and many other scholars of international repute who hailed racial research as the science of the future. The first-ever international meeting of eugenicists, in Dresden in 1911, brought together scholars from eight different countries: Germany, the United States, Great Britain, Austria, Czechoslovakia, the Netherlands, Sweden, and Denmark. By the end of the 1930s the eugenics network had expanded to include five times as many countries and dominions: France, Italy, Spain, Belgium, Switzerland, Norway, Finland, Estonia, Latvia, Hungary, Poland, Mexico, Cuba, Columbia, Guatemala, Venezuela, San Salvador, Uruguay, Chile, Brazil, Panama, Peru, Argentina, the Dominican Republic, Siam, Japan, Australia, New Zealand, Canada, South Africa, and India. A dozen or so countries that had passed sterilization laws in the late 1920s and 1930s took cues from both the United States and Nazi Germany. André Pichot went as far as to argue that Germany would most likely have implemented sterilization legislation regardless of Hitler's coming to power in 1933.[35] In fact, the international reputation of eugenics proved very important to the Nazi regime, especially in its early days. Characteristic is a Nazi poster from 1936 that features an "Aryan" family of three holding a shield inscribed with Germany's 1933 Law for the Protection of Genetically Diseased Offspring. The heading "We Stand Not Alone" is illustrated by flags of the nations that had already enacted sterilization legislation.[36] Conversely, it can be effectively argued that if it had not been for Nazi Germany, the international eugenics movement would not have been able to project its influence indefinitely; by the early 1930s population studies and genetics, built on a more solid scientific basis, increasingly put eugenics on the defensive.

The preeminence of Nazi Germany in the field of racial science has prompted Stefan Kühl to pose the following, two-pronged question: why relatively many eugenicists, specifically in America, supported Nazi racial policies and why so few opposed them. Even though none of the protagonists could have known that the discipline of eugenics would ultimately pave the way for mass murder and genocide, Kühl's conclusion remains partially valid for this volume as well. Originally the eugenics movement was meant to promote the national cause. Thus the Nazis did not have the prerogative on the policy of race improvement; national, political, and scientific peculiarities shaped perceptions of eugenics in individual countries during the interwar period. At the same time, Nazi scientists had, for the most part, succeeded in ensuring their approach dominated the international eugenics movement. The stronger the Nazi regime, the more circumspect became international criticism of its policies and the more effectively it could deflect that criticism. Many eugenicists viewed Nazi policies as the triumphant embodiment of their own scientific and political goals. With regard to the Nazi sterilization law, many eugenicists specifically pointed out its comprehensive scope and scientific foundation. Social conservatism, augmented by the antidemocratic tendency among eugenicists, as it had transpired in eugenic literature, resonated with the resolute implementation of the law in Nazi Germany. The ill-defined correlation between positive and negative eugenics enabled advocates of racial science always to strike a discursive balance between the two. As late as 1942, for example, the American geneticist T. U. H. Ellinger in the *Journal of Heredity* explained away the persecution of Jews in Germany as a "large-scale breeding project."[37] The relationship between science and politics in the programmatic function of eugenics made it equally easy to impress the argument depending on the circumstances. In short, what was ideally supposed to make racial science advance human progress in reality reintroduced barbarism; the ambiguity surrounding eugenics' principles opened up to interpretations that were effectively utilized by the Nazi regime to advance its destructive visions. Along the way, the eugenics movement's ideas helped to legitimate various nationalist ideologies in Europe, both inside and outside Germany, before and after the Nazis began building their racial empire. Among the critics of Nazi eugenics were socialist eugenicists, particularly from Great Britain, France, the Netherlands, and the United States, as well as scholars who had fled persecution in Germany. Yet reform eugenicists never came to resemble anything like a common front against the Nazi racial project.

Redefining the Agenda for the Study of Racial Science

Recent decades have seen renewed interest in the history of racial science and eugenics, especially during the Second World War. Indeed, numerous studies have been published on the history of eugenics in Nazi Germany, the campaign of euthanasia against people with disabilities in particular. Despite that, there has been relatively little comparative analysis of Nazi racial policies in the occupied territories, especially in East Central Europe. The traditional literature on the Nazi occupation has tended to paint a conventional picture of compliant and collaborative regimes obediently doing the bidding of the Nazi rulers.

More often than not, many of the regimes and states that collaborated with the Nazis have been portrayed as possessing an ideology and outlook rooted in Nazi ideas of race and nation that had no resonance in national culture and therefore enjoyed little support among the local population. At the same time, scholars have often advanced a top-down analysis of racial science and racial policy in the occupied territories, focusing on decision makers and supreme rulers, to the disregard of the diverse means by which these ideas were received by the general population, interpreted by local and national ideologues, or implemented by bureaucrats at the lower level. Hence the lack of comparative analysis of the role that idealism and social mobility might have played in the implementation of a racial agenda under Nazi occupation.

The thirteen chapters that make up this volume pursue three main avenues of inquiry. First, they explore the connections between racial science in Hitler's Germany and similar ideas and intellectual trends pursued in the Nazi-occupied or Nazi-dominated countries of Europe by establishing numerous links at the level of scientific exchange, ideological borrowings, institutional or individual collaboration, and policy making. Second, they probe the continuity between scientific developments in various parts of Europe before and during the Second World War. Indeed, both positive and negative eugenics were given even greater priority under the conditions of Nazi occupation. In this context, to determine what changed may be as important as to establish what remained the same.[38] Finally, they trace the pursuit of a racial scientific agenda in each of the cases under discussion to peculiarities of national history and culture. Undoubtedly, the collective reading of history and its projection into Hitler's future New Europe helped to mold racial science into a respectable field of studies, no matter what the long-term consequences might be. Racialization—the construction of race—intentionally invited the recasting of national history in racial terms and overemphasized its own cultural mission. Despite the temptation to attribute the appeal of racial science solely to the rise of the Nazi Party in Germany, it enjoyed much wider currency throughout Europe than many scholars have been willing to admit. Rather, the Nazis capitalized on preexisting sentiments and pseudoscientific interpretations to promote their program of radical restructuring of the ethnic map of Europe.

This volume advances several theses that challenge the traditional interpretations of racial science and racial politics as applied throughout German-dominated Europe. First, racial science was a prominent feature of population policies throughout interwar Europe, including the territories that were later occupied by or allied with Nazi Germany. Efforts to define and manage various population groups—ethnic minorities in particular—in biological terms stemmed from the local context. Due to the sheer amount of resources and credence invested in negative eugenics in Nazi Germany, however, it was only natural for scientists and nationalist politicians from other parts of Europe to look to their German colleagues for guidance and inspiration. Although their objectives and means were not identical, the existential fear that fueled racial science permeated the entire European continent. Therefore, it proved relatively easy for the Nazis to ensure the cooperation of racial scientists (as well as ethnologists, linguists, archeologists—scholars who worked toward preservation of the national heritage) in the occupied territories.

Second, Nazi divide-and-rule policy extended into the academic and scientific disciplines. The "racial value" assigned by the Nazis to any given national group defined the limits within which the latter could operate, including the pursuit of an independent research agenda. Creating hierarchies within hierarchies may represent by far the biggest paradox of German rule in Eastern Europe. Looking from a contemporary perspective, most unexpected was that the local elites genuinely believed in the promise of race as an escape route from the multifaceted crises facing their nations. To find another ethnic group that could be categorized as racially inferior to one's own was both emotionally liberating and politically promising. The idea of the ethnically based transformation of society expunged the nightmare of degradation and emasculation. It simultaneously fed into foundation myths and personified the onward march of modernization. Most important, the drive for racial purity appeared to many intellectuals as the epitome of originality and the avant-garde, a notion encouraged by the Nazis. Racial science, organic nationalism, and radical policies came to reinforce each other in a pursuit of a purer, or rather purified, society.

Third, local support for eugenic and racial ideas was a multifaceted as well as multicausal phenomenon. While ambition, career aspirations, and a desire for social mobility certainly motivated many young scientists and intellectuals to embrace Nazi eugenic and racial scientific ideas, idealism and utopianism were often as important factors. Contrary to traditional narratives of eugenics and racial science as an end in itself, these disciplines also contributed to the radical restructuring of society through social engineering. The cumulative radicalization of racial policy in Nazi Germany, as well as in the satellite states, was a result of the influence and ideas not only of leading racial ideologues but also of lower-ranking bureaucrats, intellectuals, and policy analysts. While the idea of polycratic rule has long been recognized as part of the power structure in Nazi Germany, it applies in equal measure to Nazi satellites during the Second World War.

Fourth, because power relationships often remained in a state of flux, radical regimes were frequently able to use the Nazi occupation to further their own short- and long-term goals, sometimes divergent from those of the Nazis. While the Nazi occupation authorities used their power and influence to achieve their own political, strategic, racial, and economic goals, they were forced to work with regimes and societies that had very different agendas. Otherwise, there was only limited resistance to the racial ideas and policies of the Nazi occupation authorities. Indeed, opposition to Nazi racial, biological, and eugenic ideas was not always synonymous with "moderate" views on race and nation. Paradoxically, many of the most extreme advocates of racial purification and transformation were also among the strongest opponents of Nazi euthanasia and eugenics.

Fifth, as long as we are dealing with the problem of motivation, it makes little sense to divide the diverse group of academics, thinkers, scientists, officials, and politicians—all those who ever expressed their opinions on the merits of racial science—into set categories based on their reading of eugenics. The messy administrative structure and the sheer ethnic, geographic, social, and political diversity of Hitler's New Europe prevented the emergence of specific schools of thought with a mass following. Besides, the conditions of Nazi occupation and the war of extermination made the very notion of *pure* science obsolete. Particularly in the case of racial studies, which had traditionally aspired to policymaking, its practitioners came to view it literally as *applied* science. Theirs was an experiment, an experiment in existential survival, and as such had countless variations. The term *kämpfende Verwaltung*—a kind of political fighting administration exemplifying a proactive approach to governing—that Michael Wildt has used with regard to the Reich Security Main Office (Reichssicherheitshauptamt, or RSHA), equally well captures the self-imposed mission of the racial scientists.[39]

This divergence prompts the question of whether it makes sense to talk about racial *science* at all. With hindsight, *scientific racism* would be the most accurate term to use, whereas to account for all of its too-numerous variations, *racial studies* may appear an even better construct. One could go one step further and declare the race-related exploits pseudoscience, which they effectively were. However, such linguistic precision may prove counterproductive when it comes to answering the question of why the study of race and the practical application of its findings enjoyed such a broad appeal across Nazi-occupied Europe. Without buying into euphemistic language, it is important to recognize that for practitioners of racial science what they did professionally certainly constituted science. It is this, nearly universal, belief that experimental science could, and should, correct perceived imbalances in humanity that guided their action. Thus, to better understand their motivation we have to look at the larger picture and, wherever possible, through their eyes. The diverse use of terms such as *race, racial,* and *racism* by the contributors to this volume reflects the heterogeneity of national experience under Nazi occupation and the divergent conceptions of eugenics that informed those experiences.

The peculiarity of racial science as a body of knowledge was its intricate connection to politics and policymaking. From the outset, academics and scientists engaged in eugenic theorizing and research, seeing themselves as the agents of social change. As one of the most influential German racial scientists, Lothar Loeffler, remarked in 1934, eugenicists should not hesitate to draw political conclusions from their research.[40] In other words, new scientific knowledge legitimized the political program of the eugenicists (more often than not built around the perceived threat of racial mixing). In the age of the nation-state, however, apart from a few American foundations that had supported that kind of research, for it to have any long-term impact, eugenics could be implemented only from above. Initially perceived as a handicap, and occasionally a point of criticism, the link between racial science and state power reached a whole new dimension with the emergence of Nazi Germany in 1933. As several scholars have pointed out, the major appeal of the Nazi regime to eugenic circles in North America and Western Europe lay less in its particular brand of racial science than in the might of a state that had unequivocally championed its cause—indeed, the state that was built on racial principles. Even the starkest critics of the Nazi regime among reform eugenicists, as Stefan Kühl has contended, attacked only its discriminatory treatment of minorities, specifically the Jews, but never questioned the premise of race improvement per se.[41] In short, they clung to the ephemeral notion that negative and positive eugenics were two worlds apart that never intersected.[42] In consequence, this made it possible for them to reconcile with certain elements of Nazi racial policy without claiming any intellectual affinity to Nazi racial ideology—the paradox that underlined the pursuit of racial science in Hitler's New Europe. In retrospect, racial *science* was successful only as a policy.

The connection between eugenic research and its political implementation justifies, in our opinion, approaching racial science during the Second World War as a complex project not limited exclusively to professional scientists and academics who built their career within this particular field. As intended by its practitioners, racial science truly became a guiding principle backed by the might of the Nazi state and various indigenous agencies. It stopped being the realm of the select few, confined to their writings and conference papers, and became at once theory and ideology, policy and reality. Inevitably, individual contributors, and the volume as a whole, grapple with the issue of ethics. Yet this collection of essays goes beyond the dilemma of applying scientific findings extracted by coercive means—inflicting pain or suffering in the name of knowledge, universally perceived as the common good.[43] The case studies discussed in this book relate the pursuit of experimental science to, among others, national foundation myths, the status of minorities, nationalist rhetoric, border disputes, and local and international politics. Predictably, this multilayered interrelation took on a new quality during the Nazi occupation. However local context might have shaped it, as a comprehensive project racial science had a cumulative effect of life redistribution. Willing or not, all those who took part in that project advanced its ultimate objective of race improvement.

As with many other aspects of Nazi ideology, the concept of "rational selection" entailed a contradiction. In order to build a modern, racial state, Nazi social engineers recast the asocial individuals created by industrialization in biological terms. However, the seemingly symbiotic relationship between processes of modernization and eugenics was less evident in the European states that came under German control. In most cases eugenic and racial thinking was driven more by concerns about national survival than a grand utopian vision comparable to the one envisaged by the Nazis. Similar to Robert N. Proctor's observation regarding the earlier eugenics movement in Germany, its ideology in other parts of Europe was less racialist than it was nationalist or meritocratic. The gradual takeover of the German racial hygiene movement by the radical right, much like the entire bureaucratic apparatus in Germany, was not necessarily replicated in other parts of Europe. Whenever individual scholars sided with their German counterparts in advocating a racial-biological resolution, a majority of the population rarely became a part of the pseudoscientific discourse, let alone had time to internalize it. The fact that the institutionalization occurred in the occupied countries only haphazardly further widened the gap between theory and praxis.

As a broader objective, by advancing a comparative and interdisciplinary analysis of racial science and eugenics this volume aims to better integrate East Central Europe into the mainstream of Holocaust research and the history of the Second World War. Despite a number of fine monographs dealing with the history of racial science in interwar East Central Europe, and quite a few case studies of the Holocaust published in the past decades, we are lacking a synthetic work encompassing the entire Nazi-dominated continent. By scrutinizing the conventional narrative of both Nazi racial science and occupation policies during the Second World War, this volume surveys a range of countries and ideologies whose relationship to the discussion of the Third Reich and Nazi racial science has remained peripheral at best.

Designing the Perfect Society: Racial Science, Resistance, and Social Mobility

The essays in this volume strike a balance between functionalist and intentionalist interpretations of the Holocaust and racial science. Several essays expose the factionalism and discord at the heart of many of the Nazi agencies dedicated to the implementation of racial politics in the new empire. A number of contributions highlight the tensions and conflicts accompanying policy making that existed in Nazi agencies such as the Racial Office of the Race and Settlement Main Office of the SS (Rasse-und Siedlungshauptamt, or RuSHA) and the Reich commissioner for the strengthening of Germandom. At the same time, they illustrate the extent to which many of the disputes—routinely framed in the language of scientific integrity—served to conceal another, ambition-driven agenda of racial theorists, scientists, academics, and bureaucrats to remake the Nazi empire in their own image. Furthermore, they demonstrate the willingness of academics and scientists to work with Nazi agencies intimately engaged in the implementation of the Final Solution for what they considered to be the greater good of humankind.

In his chapter Thomas Mayer advances a comparative analysis of three Austrian universities—Graz, Vienna, and Innsbruck—tracing the evolution of racial and eugenic studies in each. Mayer looks at how professional and institutional ambitions and frustrations affected the relationship between Nazi occupation and racial science. He compares the earlier attempts by racial scientists, anthropologists, and eugenicists to establish a eugenic center in Austria and the opportunities that opened up following the 1938 Anschluss. To realize their ambitions, however, Austrian racial scientists had to compromise their professional integrity by providing scientific support and intellectual legitimization to the Nazi campaigns of euthanasia and sterilization and pathological examinations of prisoners of war.

Like the pioneering work of Aly and Heim, a number of the contributions in this collection demonstrate the symbiosis of racial and nonracial agendas in the formulation and implementation of racial policy. Thus, Isabel Heinemann examines the SS policies of racial categorization, forced population transfer, and mass murder in the German-occupied East. Heinemann argues that despite divergent dynamics the main factor driving mass expulsions of populations and genocide was racial homogenization in the nascent Nazi empire. While the brutal population policies in the East certainly widened the scope of genocide by advancing the racial categorization of a significant proportion of the local population, it was more than just a response to war, Heinemann argues. Moreover, many of the racial theories that the Nazis implemented during their colonization of Eastern Europe had been part of the mainstream discourse in German racial science and anthropology long before 1939. Nonetheless, Heinemann contends that another important factor driving Nazi racial policies in the East was the attempt to establish a socially and economically viable empire populated by a new elite of ethnic Germans and German settlers from the Third Reich.

As with most ambitious colonial projects throughout history, from the beginning the Nazi racial science program was marked by contestation and resistance on the one hand and support and collaboration on the other, from within as well as from outside. Sometimes the ideas and plans advanced by Nazi agencies and individual officials faced opposition or even active resistance, even among the most ideologically committed party activists and members of the SS. At other times, Nazi agencies were forced to subvert the very ethical basis of Nazism for the greater good of racial science and the perceived survival of the German "race." Amy Carney illustrates these conflicting tendencies in her examination of Himmler's policy to increase the number of children born to

SS men, thereby enlarging the size of the SS "family" and preserving the Aryan race. This racial imperative prompted several ideological innovations, some of which violated the very ethics on which the organization had been based, and therefore faced opposition from certain Nazi agencies and individual SS officers. According to Carney, the need to increase the stock of "good blood" and win the race war superseded all other concerns: while racial principles could never be negotiated the Nazi moral revolution frequently could.

In his detailed discussion of the existential conflicts within the SS, Wolfgang Bialas deals with the tension between ideology and reality, looking specifically at the role of Nazi ethics and the Nazi moral order in the prosecution of mass murder. Bialas examines the often-conflicting origins, elements, and manifestations of Nazi morality and how the SS reconciled those incongruities to create a moralistic unity. Bialas considers, in particular, how the average SS man was able to separate his self-image as a brave soldier carrying out his duty to the Aryan *Volk* from both the reality of the mass murder of innocent people as a member of an Einsatzgruppe and his personal conscience.

The Nazi racial program provoked further "internal" conflicts, most notably within the German diaspora. The German minority supposedly belonged to a wider Germanic community of blood and therefore was presented in popular discourse, if not scholarly accounts, as nationally and ideologically monolithic. In reality, however, the Nazi racial science program tended to highlight the tensions, antagonisms, and rivalries within ethnic German communities on the one hand and the German diaspora and the *völkisch* "motherland" on the other.

Steffen Werther looks specifically at Nazi occupation policies in North Schleswig and the complex relationship between Nazi Germany and the ethnic German minority in Denmark. He analyzes the fortunes of two National Socialist parties in Denmark between 1933 and 1945 to trace the shifting political and national allegiances of the German community. He argues that the pursuit of a "Greater Germanic Reich" by Himmler's SS often collided with the aims and aspirations of the German ethnic community in Denmark and the two main parties that represented them. Contesting traditional historiography that presents German ethnic communities in 1930s and 1940s Europe as ideologically homogeneous and receptive toward Nazi *völkisch* ideology, Werther emphasizes the multiple, and often ambiguous, identities of members of the German minority in Denmark.

Populations identified by Nazi racial theorists and officials as sharing common Germanic racial and cultural traits had an uneasy relationship with the German authorities. In her case study of the Netherlands, Geraldien von Frijtag Drabbe Künzel deals with the attempts to enlist Dutch peasants and artisans in the German colonization project in the occupied East. She demonstrates how the notion of German-Dutch racial kinship and the idealization of the Dutch farmer as the uncorrupted essence of the Dutch *Volk* moved from the margins to the mainstream of scholarly and political discussions in the 1930s, actively promoted by ethnologists and anthropologists as well as prominent personalities within the Dutch National Socialist Party. This notion of kinship culminated in the campaign to recruit Dutch willing to resettle in the occupied Soviet territories following the occupation of the Netherlands in 1940. Von Frijtag Drabbe Künzel argues that since both the collaborationist Dutch government and the German authorities believed that the Netherlands was overpopulated, for Dutch farmers to survive, the former were confident that many farmers would enthusiastically embrace the opportunity to start a new life as pioneers in the East. In addition to proving their loyalty to the Reich, the Dutch administration hoped that the colonial enterprise would enable the Netherlands to regain prestige as a colonial power as well as generate economic benefits. The reality of colonization, however, proved far more problematic. The negative experiences of settlers in the East, in particular the tensions and considerable cultural differences between them and both German officials and ethnic German fellow settlers, demonstrated that irrespective of the idealized vision of German-Dutch racial kinship, "Germanic" friendship and solidarity were actually in short supply.

Those identified as the racially purest of Nordic peoples had sometimes the most challenging relationship with the Nazi regime. Terje Emberland examines the evolution of German-Norwegian relations through the prism of Himmler's attempts to create a Norwegian SS unit. He argues that SS policies toward Norwegians were directly influenced by their racial perceptions of Norwegians as primeval Nordic farmers and fierce warriors. As a result, they placed great significance on the success of programs to enlist Norwegian support. In the wake of the occupation of Norway in 1940, Himmler initiated a formal campaign to recruit Norwegians for the SS. Emberland argues that the recruitment campaign had different aims than just providing the military manpower for Nazi Germany. Himmler, along with SS racial ideologists, believed that through recruitment in the SS Norwegian men would reawaken their racial, martial spirit. He envisaged that, after proving their mettle in battle, the SS men would return to Norway to form an elite cadre of politicians, policemen, and bureaucrats who would spearhead the transformation of Norway as an SS state; the remainder—peasant warriors making up a new Teutonic Order—would guard the borders of the Nazi empire from the "Asiatic hordes" in the East. As Emberland's chapter demonstrates, despite the effective failure of the recruitment drive, especially in comparison to similar campaigns elsewhere in Europe, the fact that Himmler never abandoned his vision demonstrates the importance of racial conceptions in SS policy toward the Norwegians.

Fascist Italy proves one of the most complex examples of concurrent resistance and acquiescence to the Nazi racial agenda. In her chapter Elisabetta Cassina Wolff looks at the phenomena of racial science and antisemitism in Fascist Italy, which until recently were viewed as contingent, alien, and unpopular innovations borrowed from Nazi Germany for political and diplomatic reasons. Wolff analyzes various racial theories—primarily those of Telesio Interlandi and the young racial theorists who published in his journal, *La difesa della razza*. She argues that, while support for and a consciousness of racial thinking remained marginal among the general population, from the late 1930s onward racial ideology became increasingly prominent in Italian intellectual fascist thought. Despite its failure to gain mass popularity, the growing intellectual disputation about race in Fascist Italy nonetheless mirrored wider debates about national identity and the future of fascist ideology. These debates in turn signaled the emergence of a more virulent form of racism as an intellectual reaction to Fascist Italy's colonial wars and the building of a fascist empire in Africa in the 1930s.

Nazi Empire and National Regeneration: Racial Science in Hitler's New Europe

Several essays in this collection address the impact of the Nazi occupation or of an alliance with Germany on the development of racial science and eugenics in Central, Eastern, and Southeastern Europe. While the rise of Nazism and the political status accorded to individual nation-states within the Nazi empire played an important role in the advance of racial thinking, many of the traditions and ideas that came to the fore in the 1940s existed long before Hitler came to power in Germany. The striking similarity between the issues of concern to racial scientists and eugenicists in East Central Europe should be contrasted with the very different social pressures, cultural traditions, and political agendas that drove them. In some cases, such as Fascist Croatia, a basic understanding of Nazi racial science was combined with radical nationalist concepts of race and nation. This created an extreme form of racial politics that even the German occupation authorities were unwilling to support. By contrast, in countries such as Romania, local cultural and scientific traditions contradicted the basic biological tenets of Nazi racial science while using a similar, pseudoscientific language. In Hungary racial science appears to have merely provided "context" for well-established local theories of national and biological origin, culminating in the mass deportation of the Hungarian Jews by the Arrow Cross regime in the summer of 1944.

The case study of Latvia by Björn Felder illustrates the complex relationship between the Nazi concept of race on the one hand and local nationalist agendas and social pressures on the other. In his chapter Felder examines the evolution of racial science in interwar Latvia and the subsequent period of German occupation. Felder analyzes the genesis of the idea that Latvians overwhelmingly belonged to the "Nordic" race. He does so by tracing the views and scientific projects of leading racial anthropologists at the University of Riga as well as the influence of radical ideologies and movements. He pays particular attention to the campaign of racial purification, negative eugenics, and persecution championed by certain Latvian scientists. A detailed examination of the Institute for the Restoration of National Vitality and its leadership brings Felder to the conclusion that prior to 1941 the Nazi concept of race had little currency within the Latvian scientific community, which emphasized "positive" rather than "negative" eugenics. According to Felder, the shift to negative eugenics occurred only under Nazi occupation. Nonetheless, Felder challenges earlier scholarship, which argued that negative eugenics had been imposed on Latvia as a result of Nazi rule, by demonstrating the utility of negative eugenics to Latvian politicians, scientists, and scholars. The concept of negative eugenics, which was part of the mainstream scientific debate in interwar Latvia, enabled these leaders to differentiate Latvians from citizens of neighboring nations like the Russians and thus escape the "East Baltic" race category.

The eugenic and racial discourse in Estonia was equally shaped by political factors, cultural traditions, and social pressures. Thus Anton Weiss-Wendt demonstrates that eugenic ideas had taken hold in Estonia long before the Nazi invasion in the summer of 1941. Ideas about living space, the improvement of the race, "healthy breeding," and the pernicious influence of the Russian minority were part of public discourse as early as the 1920s. By the late 1930s negative eugenics, including enforced abortion and sterilization of selected group of offenders, were recast as public health policy. Paying particular attention to the career of the anthropologist Johann Aul, Weiss-Wendt describes how Estonian scientists and politicians took advantage of the Nazi occupation to promote their own agenda of expelling the Russian minority and countering the negative stereotype of an "East Baltic" race with the counterimage of a "blond," racially healthy Nordic race. He also details how the research of Aul and other scientists indirectly contributed to Nazi plans for the demographic restructuring of Europe.

A distinction needs to be made between East European countries or specific provinces that fell briefly under Soviet control in 1939–40 (Estonia, Latvia, Lithuania, western Poland, and Bessarabia) and the rest of Nazi-dominated Europe. In the former territories existential fear prompted by Soviet terror became the decisive factor not only in determining political allegiances but also in the willingness to accept, for example, ethnic cleansing as a logical solution to the perceived threat of ethno-national extinction.[44] The context was obviously different in the case of larger countries, particularly those formally allied with Nazi Germany. Marius Turda examines the role that racial scientists and eugenicists played in the development of racial policy in Hungary between 1940 and 1944. Defining the nation in biological terms, according to Turda, was not merely an oversimplification of racism or a distortion of eugenics. Rather, it should be viewed within the framework of an alternative nationalist project—a new form of cultural and political modernity conditioned by the fusion of mass politics and eugenic utopias of national belonging. Although throughout the 1920s and for much of the 1930s ideas about racial purity were contested, Turda argues that by the time the Hungarian Institute of National Biology was founded in 1940 racial and antisemitic concepts of Hungarian national identity had moved into the mainstream. This was particularly true after the territorial expansion between 1938 and 1941 at the expense of southern Slovakia, northern Transylvania, and the Vojvodina that saw Hungary absorb large non-Hungarian populations. Turda describes how institutions such as the Hungarian Institute of National Biology were used to foster a sense of national belonging in an enlarged state through the concept of "biologism," a holistic form of eugenics that fused the concept of biological worth with physical and moral education of the population. It was at this point, Turda contends, that antisemitism, often an aspect of Hungarian racial and eugenic thinking, became a central preoccupation of eugenicists and racial scientists. The renewed centrality of antisemitism in Hungarian racial science culminated in mid-1944 with the official inauguration of a Hungarian Institute for Research into the Jewish Question, which heralded the launch of the Final Solution in that particular country.

While the political alliance with Nazi Germany undoubtedly contributed to the evolution of racial and biological concepts in 1940s Hungary, local biological and racial discourses played a far greater role in this respect. The chapter by Vladimir Solonari examines the similarly entangled history of racial science in Romania. An overview of Romanian eugenics in the 1920s indicates that, despite the introduction of anthropological studies departments at a number of universities, scientific racism was confined to a marginal cohort of Romanian physicians and biologists. According to Solonari, this was partly due to the fact that popular racial theories promoting racial purity and denigrating "mongrel peoples" could not easily be applied in Romania since Romanians believed they were the descendants of Roman settlers and pre-Roman Dacian populations. Given their supposedly mixed-race origins, scientific racism had the potential to undermine the biological self-worth of the Romanians. Solonari shows how national agendas frequently collided with the long-term goals of the Nazis, who experienced difficulties planting racist ideas in Romanian soil despite the emergence of more explicitly biological concepts of ethnicity and race in the 1930s and the seizure of power in Romania by pro-Nazi forces in 1940.

In Croatia, as Rory Yeomans demonstrates in his chapter, the ultranationalist Ustasha regime capitalized on both Nazi racial science and the substantial German presence to pursue its own radical agenda. Yeomans examines racial science and eugenics in the Independent State of Croatia between 1941 and 1945 as it was applied in the program of racial purification and mass murder pursued by the Ustasha regime against the native Serb and Jewish populations. Yeomans examines how the Ustasha regime used Nazi biological conceptions of race to legitimize its campaigns of national and racial regeneration. These conceptions and programs evolved over time, partly under political pressure from the Nazi and Italian occupation authorities and partly as a result of internal political pressures from within the regime itself.

Yeomans argues that the regime's abandonment of its initial policy of mass murder, deportation, and forced assimilation of the Serb population did not reflect a fundamental change in thinking. Rather, it represented a contingent tactical maneuver. This policy change was frequently contested from within, both from factions in the regime and from the grassroots Ustasha movement itself. At the same time, the pseudoracial and pseudobiological theories that animated the initial program of mass murder and deportation were intermittently revived long after they had been officially abandoned. Meanwhile, with few exceptions, the campaign of extermination against Jews continued unabated. The regime's officials, propagandists, and intellectuals explicitly linked the campaigns against the Serbs and the Jews to wider campaigns of national regeneration, moral purification, social justice, and cultural rebirth. As a result, thousands of idealistic students, young sociologists, anthropologists, and scientists rallied behind the slogan of national regeneration. The very fact that many of the racial and anthropological theories utilized by the Ustasha regime were either drawn from or ascribed to the writing or ideas of nineteenth- and early twentieth-century Croatian (rather than German) scholars and scientists enabled the regime to present its violent revolution as truly national in form and scope.

When analyzing racial science in Hitler's New Europe, as this volume has attempted, one should keep in mind the centrality of the Nazi regime. Nazi ideas of race and science are essential to this discussion for the impact they had not only on the development of society in Germany but also in Europe at large. The essays in this collection reveal as much about identity, race, and nationalism in East Central Europe as they do about Nazism and Nazi Germany—the ideas and utopian visions that preoccupied its leaders, intellectuals, and ideologists. Above all, they expose the diverse thinking that lay behind the grand designs for the new Nazi empire and the role that racial science was to play in this campaign of radical restructuring. Extensive research over the past few decades has opened up new avenues of enquiry into the ultimate social, political, demographic, and racial goals of the Nazi regime. Paradoxically, the more we know, the more ambiguous many of the early interpretations of Nazi Germany and Hitler's new European order appear. The more historical data becomes available, the more questions and problems they pose. Although this collection does not provide a definitive answer, it sheds new light on some of the most important aspects of Nazi racial science, raising a series of ethical, moral, and intellectual questions that are more relevant today than they have ever been.

Notes

1. Stein Ugelvik Larsen, Bernt Hagtvet, and Jan Petter Myklebust, eds., *Who Were the Fascists? Social Roots of European Fascism* (Oslo: Universitetsforlaget, 1980), iv–viii.
2. For an overview of recent scholarship on the connection between Nazi Germany and modernization, see Paul Betts, "The New Fascination with Fascism: The Case of Nazi Modernism," *Journal of Contemporary History* 37, no. 4 (2009): 541–58.
3. Zygmunt Bauman, *Modernity and the Holocaust* (London: Polity Press, 1989), 91–92.
4. Cf. Rainer Zitelmann and Michael Prinz, eds., *Nationalsozialismus und Modernisierung* (Darmstadt, Germany: Wissenschaftliche Buchgesellschaft, 1991); Winfried Nerdinger, ed., *Bauhaus-Moderne und Nationalsozialismus: Zwischen Anbiederung und Verfolgung* (Munich: Prestel, 1993); Shelley Baranowski, *Strength through Joy: Consumerism and Mass Leisure in the Third Reich* (Cambridge: Cambridge University Press, 2004); Irene Guenther, *Nazi Chic? Fashioning Women in the Third Reich* (New York: Berg, 2004). For a critique of the modernization theory, see Jens Alber, "Nationalsozialismus und Modernisierung," *Kölner Zeitschrift für Soziologie und Sozialpsychologie* 41, no. 2 (June 1989): 346–65; Norbert Frei, "Wie modern war der Nationalsozialismus?" *Geschichte und Gesellschaft* 19, no. 3 (1993): 367–87.
5. Michael Burleigh and Wolfgang Wippermann, *The Racial State: Nazi Germany, 1933–1945* (Cambridge: Cambridge University Press, 1991), 8–22.
6. The debate between "intentionalists" and "functionalists" is discussed more fully in Christopher Browning, "Nazi Resettlement Policy and the Search for a Solution to the Jewish Question, 1939–1941," and "Beyond 'Intent and Functionalism': The Decision for the Final Solution Reconsidered," in his *The Path to Genocide: Essays on Launching the Final Solution* (Cambridge: Cambridge University Press, 1992), 3–27, 85–122. In these essays Browning comes down cautiously on the side of functionalism before 1940 while arguing that after 1940 intentionalism was the driving force behind the Holocaust.
7. Tim Mason, "Intention and Explanation: A Current Controversy about the Interpretation of National Socialism," in *The Führerstate, Myth and Reality: Studies on the Structural Politics of the Third Reich*, ed. Gerhard Hirschfield and Lothar Kettenacker (London: German Historical Institute, 1981), 23–40.
8. See Götz Aly and Susanne Heim, *Vordenker der Vernichtung: Auschwitz und die deutschen Pläne für eine neue europaïsche Ordnung* (Hamburg: Hoffmann Campe, 1991), 9–18; Götz Aly and Susanne Heim, "Die Ökonomie der Endlösung:

Menschenvernichtung und wirtschaftliche Neuordnung," *Beiträge zur Nationalsozialistichen Gesundheits und Sozialpolitik* 5 (1982): 7–90; Götz Aly and Susanne Heim, "Sozialplanning und Völkermord: Thesen zur Herrschaftsrationalität der nationalsozialistischen Vernichtungspolitik," in *Vernichtungspolitik: Eine Debatte über den Zusammenhang von Sozialpolitik und Genozid in nationalsozialistichen Deutschland,* ed. Wolfgang Schneider (Hamburg: Konkret, 1991), 11–24.

9. Aly and Heim, *Vordenker der Vernichtung,* 9–18.
10. Browning, *Path to Genocide,* 59–76.
11. Browning, *Path to Genocide,* 86–124.
12. Burleigh and Wippermann, *Racial State;* Michael Burleigh, *Death and Deliverance: Euthanasia in Germany, 1900–1945* (Cambridge: Cambridge University Press, 1994).
13. Robert Proctor, *Racial Hygiene: Medicine under the Nazis* (Cambridge MA: Harvard University Press, 1988), 11–14; Mark Levene, *The Rise of the West and the Coming of Genocide* (London: I. B. Tauris, 2005), 186–88, 191–92, 197–99.
14. Eric Weitz, *A Century of Genocide: Utopias of Race and Nation* (Princeton NJ: Princeton University Press, 2003), 36–39, 45, 50.
15. Jürgen Simon, "Kriminalbiologie und Strafrecht von 1920 bis 1945," in *Wissenschaftlicher Rassismus: Analysen einer Kontinuität in den Human-und Naturwissenschaften,* ed. Heidrun Kaupen-Haas and Christian Saller (Frankfurt: Campus Verlag, 1999), 228–31, 242.
16. Benoît Massin, "Anthropologie und Humangenetik im Nationalsozialismus oder: Wie schreiben deutsche Wissenschaftler ihre eigene Wissenschaftsgeschichte?" in Kaupen-Haas and Saller, *Wissenschaftlicher Rassismus,* 37.
17. Sheila F. Weiss, "The Race Hygiene Movement in Germany," *Osiris* 3, no. 2 (1987): 194–95.
18. Massin, "Anthropologie und Humangenetik," 12–64.
19. Michael Wildt, "The Political Order of the *Volksgemeinschaft*: Ernst Fraenkel's Dual State Revisited," in *On Germans and Jews under the Nazi Regime: Essays by Three Generations of Historians; A Festschrift in Honor of Otto Dov Kulka,* ed. Moshe Zimmermann (Jerusalem: Hebrew University Magnes Press, 2006), 155–59.
20. Michael Wildt, *Volksgemeinschaft als Selbstermächtigung: Gewalt gegen Juden in der deutschen Provinz 1919 bis 1939* (Hamburg: Hamburger Edition, 2007), 361.
21. Michael Zimmermann, *Rassenutopie und Genozid: Die nationalsozialistische "Lösung der Zigeunerfrage"* (Hamburg: Christians, 1996), 373, 377.
22. Richard Weikart, *Hitler's Ethics: The Nazi Pursuit of Evolutionary Progress* (New York: Macmillan, 2009).
23. Robert Proctor, *The Nazi War on Cancer* (Princeton NJ: Princeton University Press, 1999).

24. Götz Aly, *Hitler's Beneficiaries: Plunder, Racial War, and the Nazi Welfare State* (New York: Picador, 2008).
25. Marius Turda and Paul Weindling, "Eugenics, Race and Nation in Central and Southeastern Europe, 1900–1940: A Historiographic Overview," in *"Blood and Homeland": Eugenics and Racial Nationalism in Central and Southeast Europe, 1900–1940*, ed. Marius Turda and Paul Weindling (Budapest: Central European University Press, 2007), 2–8, 10, 12.
26. Turda and Weindling, "Eugenics, Race and Nation," 8–9.
27. Mark B. Adams, Garland E. Allen, and Sheila F. Weiss, "Human Heredity and Politics: A Comparative Institutional Study of the Eugenics Record Office at Cold Spring Harbor (United States), the Kaiser Wilhelm Institute for Anthropology, Human Heredity, and Eugenics (Germany), and the Maxim Gorky Medical Genetics Institute (USSR)," *Osiris* 20, no.1 (2005): 253–54.
28. Weitz, *Century of Genocide*, 30–33.
29. The development of eugenics rationale into practice from the First to the Second World War can be illustrated by the extreme ideas of some American and German scholars from 1917 and 1918 to dispatch "idiots" to the front line, where they could surely perish, and the "wild euthanasia" of 1941 and 1942, when the SS Einsatzgruppen executed en masse mental patients in the occupied Soviet territories.
30. Among other things, Himmler suggested,

> In handling the foreign ethnic groups in the East we must pay heed to recognize and to show attention to as many separate peoples as possible. . . . I want to state thereby that we must have great concern not to unite the people of the East, but to dissect them into as many parts and splinters as possible. . . . We must dissolve them into innumerable small fragments and atoms. . . . It must also be possible within a somewhat longer stretch of time, to bring about the disappearance of the ethnical concepts of the Ukrainians, Gorelians, and Lemkians."

Quoted in Helmut Krausnick, "Denkschrift Himmlers über die Behandlung der Fremdvölkischen im Osten (Mai 1940)," *Vierteljahrshefte für Zeitgeschichte* 5, no. 2 (1957): 194–98.

31. Daniel J. Kevles, *In the Name of Eugenics: Genetics and the Uses of Human Heredity* (Berkeley: University of California Press, 1986); Stefan Kühl, *The Nazi Connection: Eugenics, American Racism, and German National Socialism* (Oxford: Oxford University Press, 1994); Edwin Black, *War against the Weak: Eugenics and America's Campaign to Create a Master Race* (Washington DC: Dialog Press, 2003).

32. Specifically, Hitler wrote, "The American Union, by principally refusing immigration to elements with poor health, and even simply excluding certain races from naturalization, acknowledges by slow beginnings an attitude which is peculiar to the national State conception." Adolf Hitler, *Mein Kampf* (Boston: Houghton Mifflin, 1939), 658.
33. For a comprehensive overview, see Kühl, *Nazi Connection*, and Stefan Kühl, *Die Internationale der Rassisten: Aufstieg und Niedergang der internationalen Bewegung für Eugenik und Rassenhygiene im 20. Jahrhundert* (Frankfurt: Campus, 1997).
34. Richard Weikart, *From Darwin to Hitler: Evolutionary Ethics, Eugenics, and Racism in Germany* (New York: Palgrave-Macmillan, 2004), 10.
35. André Pichot, *The Pure Society: From Darwin to Hitler* (London: Verso, 2009), 179.
36. Reproduced in Proctor, *Racial Hygiene*, 96. The poster incorrectly identified Estonia as Lithuania, whereas England, contrary to the poster's statement, never enacted sterilization law.
37. Kühl, *Nazi Connection*, 31, 36, 50, 59–60, 63; Kuhl, *Die Internationale der Rassisten*, 12, 15, 34–35, 126–27, 137–38. Ironically, the director of the American Eugenics Society, Albert E. Wiggam, used the term *biological Holocaust* in his popular book to argue against "racial mixing." Wiggam, *The New Decalogue of Science* (Indianapolis: Bobbs-Merrill, 1924), 15.
38. Somewhat ironically, the comprehensive *Oxford Handbook of the History of Eugenics*, published in 2010 by Oxford University Press, contains virtually nothing on the Second World War period, effectively treating the years 1939 to 1945 as an aberration.
39. Michael Wildt, *An Uncompromising Generation: The Nazi Leadership of the Reich Security Main Office* (Madison: University of Wisconsin Press, 2010).
40. Kühl, *Nazi Connection*, 28, 65–68.
41. Kühl, *Nazi Connection*, 53, 91–94.
42. Among the critics of this notion is Daniel J. Kevles, who has described the link between positive and negative eugenics as a "two-pronged program." Kevles, "International Eugenics," in *Deadly Medicine: Creating the Master Race* (Washington DC: U.S. Holocaust Memorial Museum, distributed by the University of North Carolina Press, 2004), 50.
43. Philip Zimbardo's now-famous Stanford prison experiment had to be prematurely discontinued for that very reason.
44. Here we are using Terry Martin's definition of ethnic cleansing: the forcible removal of an ethnically defined population from a given territory. Cf. Martin, "The Origins of Soviet Ethnic Cleansing," *Journal of Modern History* 70, no. 4 (December 1998): 817–18.

1

Defining "(Un)Wanted Population Addition"

Anthropology, Racist Ideology, and Mass Murder in the Occupied East

ISABEL HEINEMANN

Reichsführer-SS Heinrich Himmler used his wartime speeches to outline his specific idea to transform Eastern Europe into a "greater Germanic settlement space." According to Himmler, following military conquest, German authorities—the Schutzstaffel (SS) in particular—were to exploit the land to the benefit of the Reich and German citizens. The indigenous population would be segregated into two groups: a majority of slaves and helots, slowly starving and worked to death, and a minority of carefully selected people considered "fit for Germanization." Along with German and ethnic German settlers, this minority was supposed to form a new kind of elite in the East. In front of a group of the SS and policemen in Ukraine, Himmler declared in September 1942, "This Germanic East up to the Ural needs . . . to be transformed into the plantation of Germanic blood, for in four to five hundred years—if Europe's fate leaves us that much time until a war between the continents will break out—we can rely on five to six hundred million Germanic people [*Germanen*] instead of one hundred and twenty million."[1] However, only the "racially fit" should have their place in his vision of a Greater Germanic Europe, as Himmler further elaborated: "We will bring together all the good blood, all Germanic blood existing in the world. . . . Every trace of good blood—and this is the first fundamental rule you have to retain—that you encounter anywhere in the East you either have to win to the German cause or to kill it."[2]

These phantasmagorias of conquest and "Germanization" were not confined to an ambitious SS leader, naturally. As early as October 1939, at the end of the Polish campaign, Hitler had informed the German parliament of the planned reorganization of Eastern Europe's ethnographic space (*Neuordnung der ethnographischen Verhältnisse*), which he considered the primary objective of German occupation policies. What he meant was "a resettlement of nationalities for the sake of better lines of separation than those existing today."[3] Consequently, during the Second World War Himmler's SS expelled and resettled millions of Poles, Russians, ethnic Germans, and people of the Western European countries in the name of a "Greater Germanic Reich."[4] A specific group of perpetrators set up the screening procedure, performed the racial exams, and supervised the expulsions—the SS racial experts from the Race and Settlement Main Office (Rasse-und Siedlungshauptamt, or RuSHA).[5]

One victim of forced expulsion and resettlement was the Polish doctor Josef Rembacz; he had been expelled from his home in the Zamość region of the Lublin District in October 1942. In 1946 he gave the following testimony before a Polish court:

> I, together with other Poles from Skierbiszowo (where I lived and worked as doctor at the local health office), were brought to the camp of Zamość. A commission consisting of several members carried out a racial screening in this camp. . . . The population was divided into four groups. Group I was Nordic (*Nordisch*) and Group II Phalian, South Dinaric (*Fälisch, Süddinarisch*). These two groups were brought into the ethnic German settler camp in Łódź for special examination. Group III consisted of "racially mixed breeds" (*Mischrassen*). This group was turned over to a commission of the labor administration; the families were split up and sent to Germany for work. Group IV was composed of people with hereditary diseases or physical deformations as well as of so-called unwanted races (Jews, Sinti and Roma, Mongols), who were sent to an unknown destination, most likely to Auschwitz. I know that no-one from this group ever returned.[6]

Ascribed to Race Group III, Rembacz was forced to work for the German authorities as a camp doctor in the local expellee camps until the end of 1943. What is striking about this particular statement is the victim's precise description of the racial selection procedure.

Recent years have witnessed much new research on the SS operations in occupied Eastern Europe—especially regarding the implementation of the Final Solution of the Jewish Question.[7] Alongside the motivation of the perpetrators, recent publications have discussed the utilitarian and military objectives of occupation policy, the economic exploitation of the occupied countries, and awareness of the Nazi mass murder of the Jews among ordinary Germans.[8] Comparative analyses of ethnic cleansing during the twentieth century have further placed the Holocaust and Nazi occupation policy within a wider context of genocide.[9] Some scholars, meanwhile, have discussed the Nazi extermination policies as a critical example of biopolitics, drawing on a concept coined by Michel Foucault in the 1970s to describe the crucial link between racism and power in modern societies.[10] Finally, considerable advances have been made in recent years describing how German academics and experts readily contributed to the economic exploitation and ethnic reordering of occupied Europe by providing the regime with background research and detailed planning.[11] Despite scholarly progress since the mid-1990s in explaining the process of mass murder, economic exploitation, and ethnic reconfiguration, we still know relatively little about the racist foundation of occupation policies and its importance for the functioning of the Nazi state in general.[12] Recent attempts to expand the term "ethnic community" (*Volksgemeinschaft*) into an analytical tool have further demonstrated that racist tenets were more than just abstract ideological foundations to many German contemporaries.[13] Consequently, in order to fully understand the deadly effectiveness of the Nazi extermination policies; the motivation of planners, experts, and perpetrators; the attitudes of the German public in the face of mass violence; and the character of the process itself one has to carefully examine not only the concept of Himmler's "new order" of the SS but also the seductive potential of the ideas of "racial purity" and *Lebensraum*.[14]

Accordingly, this chapter argues that forced population transfers (i.e., expulsions, resettlement, and the mass murder of European Jews) not only gained momentum through economic, military, or other utilitarian objectives but mainly pursued racial homogenization of the occupied territories. Following an overview of the racial selection procedure as a product of racial anthropological discourse within German academia, the chapter describes the pattern of expulsion and resettlement in occupied Poland. Next, it examines the plans for ethnic cleansing developed by SS racial experts for other parts of Western and Southeastern Europe. Finally, it explores the causal link between the escalating racial policies and the genesis of the Final Solution of the Jewish Question.

Himmler's Visions and German Academia: Racial Anthropologists as Prophets

During the interwar period, especially in the Third Reich, anthropology and racial hygiene were considered key sciences. They provided the applied methodology and legitimization for the racist policies of the Nazi state, ranging from forced sterilization and euthanasia to the ethnic restructuring of Eastern Europe and genocide.

Nonetheless, the history of racial science goes at least as far back as the second half of the nineteenth century, gaining momentum after the end of the First World War.[15] Referring to Charles Darwin and Francis Galton but also to social Darwinists like Ernst Haeckel and Wilhelm Schallmayer, German and other Western European scholars described the biological and social disparity of human beings as a natural and thus immutable fact. By defining races as "established entities of human beings living through generations, united in their physical and intellectual qualities," as Alfred Ploetz stated, and referring to Mendel's laws of inheritance, racial anthropologists such as Hans Friedrich Karl Günther propagated the superiority of the Nordic, or Aryan, race.[16] For Günther, the outward appearance, genetic predisposition, and mental qualities of the group members demonstrated the high value of the Nordic race. Therefore, the author of the best-selling book *Rassenkunde des deutschen Volkes* advised the German people and decision makers to decisively prevent "racial mixing" and consecutive "racial degeneration."[17] Like Günther, most racial anthropologists of the 1920s and 1930s conceived of their field as an applied science. During the formative phase of applied genetics (*Erb-und Rassenforschung*) in Germany, scholars from the Universities of Jena, Leipzig, and Munich and from the Kaiser Wilhelm Institute for Human Heredity Sciences and Eugenics in Berlin (Kaiser Wilhelm Institut für Anthropologie, menschliche Erblehre und Eugenik, or KWI-A) all helped to develop racial consciousness and build up scientific networks that persisted beyond 1933. The scholars of the KWI-A (founded by Erwin Baur, one of Germany's leading geneticists in the early twentieth century)—Eugen

Fischer, Fritz Lenz, and Otmar Freiherr von Verschuer—prepared the groundwork for the acceptance of the racist paradigm in science and politics in the Third Reich.[18] The Jena "race quadriga"—Hans F. K. Günther, geneticist Karl Astel, and biologists Viktor Franz and Gerhard Herberer—for their part amalgamated current trends in racial biology, anthropology, genetics, and eugenics into an ostensibly modern blend of racial sciences, which they promoted through their contacts in political circles and the SS.[19] In Munich, psychiatrist and eugenicist Ernst Rüdin of the Kaiser Wilhelm Institute for Psychiatry gained an international reputation: in 1932 he succeeded the founder of the American eugenics movement, Charles Davenport, as president of the International Federation of Eugenic Organizations.[20] Another Munich anthropologist, Theodor Mollison, taught racial anthropology, simultaneously acting as one of the editors of the influential journal *Archiv für Rassen-und Gesellschaftsbiologie*. Founded in 1904 by Alfred Ploetz, the journal served as a platform for an elaborate exchange of ideas in the field of racial anthropology. Many of Mollison's students at Munich's Ludwig Maximilian University went on to play roles in shaping SS racial policies. Another key figure in German academic race discourse during the early years of National Socialism was Otto Reche, anthropologist and ethnologist at Leipzig University and director of the Institute for Race Research and Ethnology (Institut für Rassen-und Völkerkunde).[21] Already during the 1930s these scholars enthusiastically welcomed the new research possibilities and funding options offered by the new regime, which made racial sciences one of its top funding priorities.[22] Furthermore, most of the abovementioned scientists lent their expertise to the new regime and eagerly contributed to the fundamental race laws passed during the formative phase of the Nazi state. Thus Rüdin helped to draft the Eugenic Law of July 14, 1933 (Gesetz zur Verhütung erbkranken Nachwuchses), while Fischer and his colleagues at the KWI-A collaborated more indirectly by promoting the benefits of forced sterilization and euthanasia through their scientific writings.[23] Although the notorious Nuremberg Laws were drafted by state officials from the Reich Ministries of Justice, Health, and the Interior, the scholars from the KWI-A as well as individuals such as Reche publicly approved the discriminatory legislation or else assisted by issuing the requested racial-biological hereditary certificates.[24]

However, it was not only anthropologists, geneticists, and biologists who readily accepted and further advanced the new ideological paradigms of Nazi life sciences. Attracted by the hope for national revival, the quest for funding, and a broadened research agenda, ethnographers, population experts, agronomists, and spatial planners embraced the cause of living space and racial purity.[25] Most of them believed that social differences could be explained and ultimately "cured" through the application of biological principles of selection and breeding. These social experts not only explained away the defeat of Germany in the First World War and the economic crisis of the 1930s as a result of the racial degeneration of the German *Volk* but also offered guidelines for political decision making. Beginning in the late 1920s, and especially after 1933, their research in the fields of racial anthropology, agrarian science, and biology received generous funding from the German Science Foundation (Deutsche Forschungsgemeinschaft), thus laying the academic groundwork for the subsequent ethnic cleansing of occupied Eastern Europe.[26]

Once the war began, Himmler and his experts could instantly turn to a circle of scholars willing to collaborate in the implementation of their plans for a large-scale racial purification program. Of special significance in this context is the example of anthropologist Reche, coincidentally a member of the SS.[27] On September 29, 1939, four weeks after the German invasion of Poland, Reche approached the SS, offering guidelines for the ethnopolitical reconstruction of the East ("Leitsätze zur bevölkerungspolitischen Sicherung des deutschen Ostens"). The opening paragraph of the guidelines read as follows:

> Most of the inhabitants of the newly transferred territory are racially (as well as according to their character, intellect, and physical capacity) completely unsuitable for assimilation into the German *Volk* and body politic [*Volkskörper*]. Above all, the Jews and Jewish mixed breeds [*Judenmischlinge*] living in the respective regions have to be removed as soon as possible. . . . The Polish population must, for the most part, be considered quite an inauspicious, unfavorable mixture of elements of pre-Slavic, East Baltic, and East European races [*prä-slavischen, ostbaltischen und ostischen Rasse*], including particularly strong Mongoloid influences [*mongolische Einschläge*]. . . . Only the racial anthropologist can decide whether, occasionally, parts of the Polish population might be racially suitable [*rassisch brauchbar*].[28]

As we now know from the expulsions and racial screening of non-German civilians during the Second World War and individual cases like that of Josef Rembacz, Reche's proposals did not remain unheard. One of his students, the anthropologist Bruno Kurt Schultz, developed the SS racial screening procedure that served as a model for the later selections. From 1940 the chief of the Race Office of RuSHA, SS-Standartenführer Schultz, assumed a comparatively high position in the SS hierarchy, posing as a leading expert on issues of race. Consequently, in 1942 he was appointed chair of racial anthropology at the newly founded Reich University in Prague.[29] Combining his tasks as a scholar and head of the RuSHA Race Office, he took care of the anthropological training of both university students and SS racial experts. While in Prague, Schultz also coordinated the racial selections of the Protectorate's population.[30] In addition, he drafted plans for the Germanization of occupied Europe. At the beginning of 1942 Schultz, at this time a university professor in Berlin, joined a meeting of race specialists at the Reich Ministry for the Occupied Soviet Territories (Reichsministerium für die besetzten Ostgebiete). Among other participants at the meeting was Eugen Fischer, then head of the KWI-A. During the discussion Schultz declared "that the racially unwanted are to be evacuated to the East, whereas the racially fit shall be accepted into the re-Germanization procedure either in the Old Reich or in the East." Schultz pledged an "accurate, exact racial screening of the population of the Baltic States . . . disguised as sanitary inspection so as to prevent concern among the population."[31]

Besides anthropologists and geneticists, leading agriculturists helped to shape the program of ethnic homogenization introduced by Himmler and his experts. In October 1939 Berlin agronomist Konrad Meyer accepted Himmler's invitation to lead the Planning Office of Himmler's newly created resettlement institution, the SS Main Office of the Reich commissioner for the strengthening of Germandom (Reichskommissar für die Festigung deutschen Volkstums, or RKF). By this time Meyer had already pursued a successful career as chair of agrarian science at Friedrich-Wilhelm University of Berlin, building up a powerful position as the leading German expert, political adviser, and science manager in the field.[32] An SS member since 1933, Meyer voluntarily chose to become involved with the RKF, ultimately advancing to the rank of SS-Oberführer. Until the end of the war, he and his handpicked fellow scholars in the RKF Main Office worked out detailed plans for the agrarian, economic, ethnographic, and demographic reconstruction of Eastern Europe, financed by the German Science Foundation. Their proposals appeared between 1940 and 1943 in different versions, the most well known of which was Master Plan East (Generalplan Ost) of June 1942.[33] Meyer, like Reche, readily accepted that any "reordering of the ethnographic landscape" of Eastern Europe would include mass expulsions. Thus in 1941 he wrote in a Nazi student newspaper, "We have to bear in mind that we will not succeed in preserving the East for all times as 'German territory' unless we have removed every single drop of 'alien blood' [*fremdes Blut*] from the German settlement space, which would otherwise endanger the inner coherence of the German Volk [*einheitliche Geschlossenheit des grenzdeutschen Volkstums*]."[34] In the same year Meyer elaborated how the ethnic homogenization of the newly annexed territories should take place with the help of ethnic Germans, the German peasantry, and especially German youth. In the preface to *Landvolk im Werden* (Peasants in the making), one of his most influential publications, he thus wrote, "This book mainly addresses the young generation that carries the National Socialist idea of blood and soil in their burning hearts, pushing it forward with all the revolutionary power of the young."[35]

The cases of Reche, Schultz, and Meyer demonstrate that it was scholars from mainstream German academia—and not just those on the radical fringes, as many scholars contended after 1945—who readily set up plans to "purify" the newly colonized territories by means of racial selection and resettlement of ethnic Germans.[36]

Western Poland as "Training Ground": Ethnic Reconstruction and Racial Purification

Following the swift victory over Poland in fall 1939 and the annexation of its western part, Nazi population planners instantly engaged with the question of "Germanization" of a region with a mixed population of Poles, Jews, and Germans. In October 1939 Hitler ordered Himmler to organize the return of ethnic Germans from abroad and the Germanization of the occupied Polish territories. One of the RKF tasks was "the elimination of the harmful influence of alien segments of the population that pose a threat to the Reich and the German ethnic community."[37] Based on his notion of a "Germanic East," Himmler intended a large-scale ethnic reconstruction of Poland "on the basis of the singular and decisive racial worth," to be carried out by the SS.[38] Therefore he ordered the SS racial experts to create a procedure for racial examination that would be performed on the incoming ethnic Germans as well as the local Poles. Sufficiently trained in racial examination techniques (in the form of obligatory racial exams for SS candidates and their spouses that were supposed to guarantee the purity of the new order), RuSHA race specialists enthusiastically accepted this new and challenging task.[39] In response to the considerable boost in competencies, in October 1939 the RuSHA enlarged its staff and created a branch office in Łódź as well as several district offices in the Warthegau and other annexed territories.[40]

The first group of people subjected to racial screening was the ethnic Germans who had followed Hitler's call to return to the Reich ("Heim ins Reich!") since October 1939.[41] Altogether, over 1 million ethnic German settlers were processed by the race experts of the SS during the Second World War. The results of the racial screening determined whether families would receive a "farm in the East," as most of them had hoped to. This privilege would be granted only to people ascribed to Race Group I or II. Persons ascribed to Race Group III were considered to have too much "Polish blood" and therefore were sent to the Old Reich (Altreich) for reeducation. They had to spend their days in settler camps and were obliged to work in factories or as farmhands. Ethnic Germans ascribed to Race Group IV were considered to be "of alien ethnic origin" and "racially unsuitable." During wartime most of them were locked in camps, since logistic difficulties prevented their transportation to their country of origin, as had been originally intended. Sometimes they received permission to move into the Reich, serving as manpower without being recognized as full-blood Germans.

Racial examination was performed by the RuSHA in the transit camps at Poznan or Łódź in the Warthegau as well as by mobile commissions operating in ports of entry for the ethnic Germans.[42] Whereas many ethnic Germans were disappointed when their hopes for a farm in the annexed territories were dashed due to their deficient racial value, in the case of many Poles, like Josef Rembacz, the result of the racial examination proved a matter of life or death. For the latter group, racial selection followed forced expulsion and loss of property as a part of Himmler's settlement plans.

Between five hundred thousand and seven hundred thousand non-Jewish Poles were expelled from the annexed Polish territories in order to provide housing for the incoming German "settlers." The Polish families were routinely rounded up in the early morning hours by security police and SS troops. They had to leave their farms, their cattle, and most of their belongings behind, since only sixty pounds of luggage—excluding valuables and furniture—were allowed per person. In many cases they were not given enough time to pack at all. The expellees were then brought to collection camps where most of them underwent racial screening, as Rembacz had described it.[43] Having lost literally everything, the "racial worth" of the expelled Poles determined whether some of them would be labeled an "wanted population addition" and subsequently dispatched to the Old Reich for "re-Germanization."

The selection procedure was similar in the case of Poles and ethnic Germans. Individuals had to appear naked in front of the screening commission, consisting of at least one racial examiner and one or more assistants. The commission took no less than twenty-one anthropological measurements, which were entered on a special form, the so-called race card (*Rassenkarte*).[44] These race cards used early computer technology (*Hollerith-Verfahren*) to process and store the data thus collected.[45] Josef Rembacz painstakingly described the details of the procedure:

> Examiners considered height, weight, shape of the body, form of the skull, form of the face, color of eyes and hair, thickness of hair, etc. The examination served as a basis for calculating the racial formula. The screening was implemented by racial specialists, the so-called *Eignungsprüfer*. . . . After this examination one had to appear before the main commission. . . . The entire family that had been ascribed high racial value was summoned for a brief examination by the head of the commission. He compared the results of the screening with his own evaluation, and the commission issued its final opinion.[46]

This kind of racial selection was first performed on the Polish population; examination of other ethnic groups followed suit. The "racially fit" were to become Germans—since Nazi racial theory assumed those individuals had German roots—regardless of their ethnicity. A little more than 4 percent of the Polish population were considered a "wanted population addition" (*erwünschter Bevölkerungszuwachs*) and fit for "re-Germanization" (*wiedereindeutschungsfähig*).[47] A majority of those people were dispatched to Germany, where they were expected to undergo transformation into "Germans." In practice, they were exploited as slave labor, even though a formal distinction was made between them and Polish slave laborers. Individuals deemed "racially unfit," and thus an "unwanted population addition," were either brought to special *Rentendörfer* (villages for the children and elderly designed to cause death by starvation) or directly deported to concentration camps—Rembacz had mentioned Auschwitz—where they had to work as slave laborers, slated for imminent destruction.[48] The majority of the population, however, was ascribed to Race Group III. Those people were to perform slave labor in the Old Reich; if they were allowed to stay, then it was only as farmhands on farms that had been given to ethnic Germans. Needless to say, the expellees received no compensation for their confiscated property.[49]

One of the persons deemed an "unwanted population addition" was Ryszard W. The three-year-old boy and his family—his father was a farmer and his mother a schoolteacher—were deported in December 1939 from Kowalew near Jarotschin in the Warthegau, while German settlers took possession of the family's estate. In the General Government (Generalgouvernement), where they arrived by train, they found refuge in Zarnow near Tomaszow. They did not receive any compensation for their property but depended exclusively on the help of a local teacher who let them stay with him.[50]

Significantly, Jews were never considered for racial screening. In contrast to their Polish neighbors, they were deported without undergoing racial selection. For Himmler and his experts, the planned Germanization required the prior annihilation of the Jewish population. Thus the project of Germanization in occupied Eastern Europe generated momentum for the Final Solution of the Jewish Question.[51]

Toward a Greater Germanic Europe: Attempts at Ethnic Homogenization in Western and Southeastern Europe

In the course of the Second World War SS experts applied the techniques of racial selection and expulsion throughout occupied Europe, creating RuSHA branch offices along the way. Apart from Poland, Himmler and his race experts focused their attention on the population of Southeastern Europe, opening a RuSHA branch office in Prague in 1941.[52] In addition, during spring and summer 1941 SS experts screened no less than half a million Slovenes in search of "good blood" in occupied Slovenia, specifically in the regions of Lower Styria and Upper Carniola.[53] Although on a smaller scale, the SS also exercised influence on Nazi population policy in Western Europe. In 1942 and 1943 race and settlement experts expelled between twenty thousand and thirty thousand "racially unwanted" French from Alsace and Lorraine to occupied France.[54] People deemed "fit for re-Germanization" were brought to so-called reeducation camps in the Reich, such as the Schelklingen camp in Baden-Württemberg. Marie Louise Zimmermann, a sixteen-year-old girl from the Alsatian village of Bischwiller, was deported along with her parents in winter 1942. The father, a schoolteacher, had been accused of pro-French sentiments. Following racial examination, the family was dispatched to Schelklingen. As Marie Louise recalled years later, the SS commander of the camp told them upon their arrival that they "were to be reintegrated into the German ethnic community [*Volksgemeinschaft*], depending on our quality as persons deemed fit for re-Germanization [*Wiedereindeutschungsfähige*]. The camp would be a reeducation camp—effectively a school—where we would have to work in groups on farms, in factories, or within the camp to learn order and discipline."[55]

Further expulsions occurred between November 1942 and spring 1943 during the forced Germanization of the Zamość region in the General Government and in the course of the relocation of ethnic Germans in the Commissariat General Zhitomir in Ukraine.[56] Around the same time, in the area around Himmler's field headquarters, Hegewald, near Zhitomir, settlement experts established so-called settlement pearls (*Siedlungsperlen*). Their dream of a permanent German settlement in Ukraine was short-lived, however. By November 1943 about thirty thousand ethnic Germans had fled the region, heading for the settler camps farther west.

The objectives of Himmler and his academic planners and the sheer scope of the racial experts' activities strongly suggest that SS population policies were conceived as a European-wide project. The offices of the SS leaders on issues of race and settlement (SS-Führer im Rasse-und Siedlungswesen) attached to the staff of the local higher SS and police leaders sprouted all over Europe, from The Hague and Oslo to Mogilev and Kiev. Whereas these leaders in the Netherlands and Norway mainly screened local women who wanted to marry German SS men, searched for Jews and *Judenmischlinge*, and screened Germanic candidates for the SS, their counterparts in Ukraine and Belorussia had different tasks. They screened prospective members for the local police battalions (*Schutzmannschaften*), engaged in antipartisan warfare, and searched for pure-blood children fit for adoption by German families and for pure-blood maids to be employed in German households.

In Search of "Germanic Blood": The SS Racial Expert as a Type of Perpetrator

When Josef Rembacz appeared in front of the screening commission in Zamość in 1942, SS-Obersturmführer Hans Rihl presided over the session. A member of the RuSHA since 1936 and a member of the Nazi German Student Association since 1935, Rihl had had substantial experience in racial screening. From 1939 to 1941 he had screened Poles in Łódź, in 1941 he directed a screening commission in occupied Slovenia, and after a short intermezzo in the Waffen-SS, he oversaw the racial screening of Poles in the Zamość region.[57] Otherwise, Rihl was a bookseller by profession and had studied anthropology, philosophy, and religion at Munich University.[58] While in Zamość, Rihl wrote monthly reports to the RuSHA Office in Berlin. Among other things, he suggested splitting Polish families of allegedly low racial value—in other words, forcibly separating spouses and taking children away from their parents—in order to secure every drop of "valuable blood" for the Germans. Thus he wrote,

> It should be possible to split families composed of RuS II and RuS IV cases. . . . Quite often, a man considered suitable for re-Germanization arrives without his family in the camp, or a woman of Race Group II whose husband has fled or disappeared. These people cannot become part of the re-Germanization process, as the whole family will be excluded from the procedure once the missing spouse has turned up. Yet if it were possible to split the families, including a formal divorce, it should be possible to win many individuals of high racial value [*blutlich wertvoll*] to the German cause.[59]

The example of Hans Rihl raises the following question: Who were the men who conceived and implemented racial policy on the ground? During the war some five hundred SS officers occupied leading positions in the Berlin RuSHA Office and its branches throughout occupied Europe.[60] These officers can be considered a specific type of perpetrator who—as highly professional, rational social experts—clearly distinguished themselves from other groups of perpetrators in the Nazi state. According to a definition provided by the German historian Lutz Raphael, an "expert" is a person who possesses "scientific and specialized knowledge" considered crucial for the construction of social systems.[61] This definition fits perfectly with the SS racial specialists, who often referred to themselves as "racial experts" (*Rasseexperten*).

Most of these experts were born between 1900 and 1909, Rihl being one of the youngest. Several qualities distinguished them. First, in contrast to many other perpetrator groups they had a relatively high level of education. A sample of one hundred of these experts reveals that no less than forty of them held university degrees, twenty-one had earned a PhD, and seven had written their

Habilitation (professorial thesis). Most of them had studied either racial anthropology or agriculture, corresponding to their two main fields of activity—"race and settlement," sometimes rendered as "blood and soil."[62] Second, this group consisted of "rational" men, convinced of the validity of the racist paradigm. Additionally, they considered themselves an ideological elite within the SS elite, thus assuming a specific group identity. Third and most important, they had a double function within Nazi population-policy making: they were planners and practitioners of ethnic homogenization, experts in the literal sense. Simultaneously, they served as "architects of extermination" (to use the original term, *Vordenker der Vernichtung,* coined by German historians Götz Aly and Susanne Heim), while designing far-reaching Germanization plans. Some of them gained considerable influence as RKF planners, for example, Konrad Meyer, author of the different drafts of Master Plan East.[63] At the same time, they implemented racist ideology as members of selection commissions and resettlement commando units supervising racial screening and mass expulsions. Their twofold competence in devising and implementing racial policies characterizes best this specific group of perpetrators.

This observation highlights the connection between racial selection and the character of Nazi resettlement and extermination policy. The implementation of profoundly racist Germanization plans by SS experts was a catalyst for Nazi extermination policy against the Jews and other "unwanted races." The attempts to "reconstruct" entire populations were crucial factors prompting radicalization and eventually mass murder.

First, the results of the racial screening served as the main criterion for the attempted ethnic reordering of the occupied and annexed regions of Europe, evident from the drafts of the Master Plan East that distinguished between "racially valuable" and "less valuable" population groups. For instance, the first draft of the Master Settlement Plan (Generalsiedlungsplan) of December 1942—an expanded version of the Master Plan East that now also encompassed parts of Western and Southeastern Europe—calculated the "Germanization capacity" of the different populations corresponding to the findings of racial experts. The plan gave the following ratios: 5 percent of Poles; 15 of Lithuanians; 30 of Latvians; and 50 of Estonians, Czechs, French, and Slovenes.[64] SS population experts estimated the total population of the territory incorporating Poland, the Baltic States, the Protectorate, Lower Styria and Upper Carniola, and Alsace and Lorraine at approximately 43.5 million. According to the Master Settlement Plan, within a twenty-year period the population of the respective territory should dwindle to 28.4 million, including an additional 5.2 million German settlers. This meant that within the following twenty years about 23.3 million people had to disappear—more than half of the original population. The plan did not specify where these millions of people should go or under what circumstances they should perish. European Jews were not included in this number either. Although none of the plans made it explicit, it is evident that for the racial experts mass death posed a necessary condition for the entire Germanization process.

Second, the racial experts accelerated their activities in step with the radicalization of Nazi extermination policy, including the genesis of the Final Solution. The German victory over Poland constituted the initial phase of radicalization: by trying the techniques of selection and expulsion on millions of civilians, the SS experts assumed far-reaching competencies. The next phase began at the turn of 1941–42 when the Soviet Union and Western Europe became an operating theater for Himmler's race experts and new settlement options emerged. A third and last phase occurred in 1944 as the Soviet advance made the race experts focus on the Old Reich, Austria, and the annexed Polish territories. In those areas they were fighting against the alleged pollution of "German blood" by screening slave laborers who had sexual relations with Germans as well as identifying persons of German-Jewish descent (*Judenmischlinge*).[65] Thus the race experts reapplied their techniques of racial screening and atomization of social groups to the German heartland, introducing forced abortion, deportation, and mass murder.[66]

Finally, the classification of civilians into different race groups, which had the aura of scientific precision, helped to split the Polish people—as well as parts of the population of the Protectorate, Alsace and Lorraine, and Slovenia—into numerous groups that could then be treated differently. The enforced distinction between "ethnic Germans," people "fit for re-Germanization," and "unwanted population additions" made expulsion, expropriation, and exploitation easier to implement. It also advanced the persecution of the European Jews, who were singled out in the course of racial examination.

The number of people screened by the race experts during the Second World War corresponded to the European range of the Germanization plans: at least 2.7 million civilians, made up of 1.2 million ethnic Germans and some 1.5 million non-Germans, went through the hands of the racial examiners.[67] Although Himmler's "plantation of good blood" under the supervision of the SS was never realized, SS racial experts made a significant contribution to an ethnic reconstruction of occupied Europe.

Seduced by abundant research funding and the prospect of swift national revival in the aftermath of defeat in the First World War, many members of the German scientific establishment enthusiastically contributed to the Nazi cause by laying the academic groundwork for racial screening and ethnic cleansing during the Second World War. In the field of racial anthropology, the scholars of the KWI-A and the Jena "race quadriga," in particular, played an important role in formulating Nazi eugenic and race laws. Whereas some of these scholars engaged in criminal medical experiments—for example, Ernst Rüdin and Otmar Freiherr von Verschuer—in the course of the Second World War, others, like Otto Reche from Leipzig University and Konrad Meyer from Berlin University, resorted to racism when drafting the plans for the reconstruction of occupied Eastern Europe—regardless of human cost. Respected in their academic fields, they were convinced Nazis and members of the SS who shared Himmler's dreams of racial purity and colonial settlement in the East. Along with ruthless SS racial experts like Bruno Kurt Schultz or Hans Rihl, they acted as "architects" of population transfer and mass extermination, thus forming a distinct group of perpetrators. Racist professionals as they were, they both shaped and implemented Nazi population policy. Remarkably, most of them went on to successful careers in West Germany, with their responsibility for contributing to the racial and occupation policies of the Nazi state until recently neither publicly acknowledged nor questioned. For instance, Bruno Kurt Schultz served from 1961 onward as professor emeritus in the Department of Medicine at the University of Münster, while Otto Reche returned to issuing genetic paternity tests. Most prominently, in 1956 Konrad Meyer was appointed chair in landscape design at the Technical University of Hannover, becoming a leading postwar expert in this field.[68]

The SS racial experts succeeded in splitting specific population groups according to their "scientifically proven racial value," making the exploitation and eventual annihilation of individuals easier to implement. Their designed selection procedures, race cards, and plans for the re-Germanization of the occupied territories greatly facilitated this task. The racial parameters that they introduced were conceived as guidelines for a greater Germanic nation to be established after the war by means of resettlement plans such as the Master Settlement Plan.

Obviously, the mass murder of Jews constituted the most extreme element in the attempt to establish a "racially purified" Europe. Still, many other ethnic groups faced expulsion, deprivation of rights, confiscation of property, starvation, and exploitation as slave laborers, depending on their alleged "racial quality" as defined by the racial experts. Consequently, the Final Solution and the "ethnic reconstruction" of occupied Europe through resettlement and expulsion were intertwined processes, both based on the paradigm of racial purity and the quest for a Greater Germanic Empire in the East.

Naturally, the fundamental difference between Nazi racial policies directed against Jews and those against Slavs lies in the extent of implementation. Jews were to perish without exception, whereas the fate of Poles, Ukrainians, and Russians depended on their perceived "racial worth." The mass murder of the European Jews was carried out to the bitter end, while the Germanization plans were only partially put into effect, ultimately postponed until after the (successful) end of the war. Rather tellingly, during the so-called RuSHA Trial (1947–48)—which put RKF officials and SS racial experts in the dock—the prosecution declared the forced population transfers and "Germanization," as experienced by Josef Rembacz from Zamość and millions of other individuals, "techniques of genocide [that were] neither so quick nor perhaps so simple as outright mass extermination . . . [yet] far more cruel and equally effective."[69]

Notes

Abbreviations Used in the Notes

AGK Archiwum Głównej Komisji Badania Zbrodni przeciwko Narodowi polskiemu (Archives of the Main Commission for the Investigation of Crimes against the Polish People, Warsaw)

BAR Bundesarchiv Abteilung Deutsches Reich (German Federal Archives, Deutsches Reich Branch, Berlin)

STANU Stadtarchiv Nürnberg (State Archives, Nuremberg)

1. Himmler's speech, 16 September 1942, BAR, NS 19/4006.
2. Himmler's speech, 16 September 1942, BAR, NS 19/4006.
3. Hitler's speech before the Reichstag, 6 October 1939, quoted in Max Domarus, *Hitler: Reden und Proklamationen 1932–1945; Kommentiert von einem deutschen Zeitgenossen*, vol. 2, pt. 1 (Munich: Süddeutscher Verlag, 1973), 1377–93.

4. For a detailed analysis of the Nazi resettlement policies, see Isabel Heinemann, *"Rasse, Siedlung, deutsches Blut": Das Rasse-und Siedlungshauptamt der SS und die rassenpolitische Neuordnung Europas* (Göttingen: Wallstein Verlag, 2003).
5. Isabel Heinemann, "Another Type of Perpetrator: The SS Racial Experts and Forced Population Movements in the Occupied Regions," *Holocaust and Genocide Studies* 15, no. 3 (Winter 2001): 387–411.
6. Testimony of Dr. Josef Rembacz, Łódź, 25 April 1946, STANU, NO-5266.
7. Among recent publications are the following: Peter Klein, *Die "Gettoverwaltung Litzmannstadt" 1940 bis 1944: Eine Dienststelle im Spannungsfeld von Kommunalbürokratie und staatlicher Verfolgungspolitik* (Hamburg: Hamburger Edition 2009); Michael Alberti, *Die Verfolgung und Vernichtung der Juden im Reichsgau Wartheland 1939–1945: Die Anfänge und die Durchführung der Endlösung* (Wiesbaden: Harrassowitz Verlag, 2006); Markus Leniger, *Nationalsozialistische "Volkstumsarbeit" und Umsiedlungspolitik 1933–1945: Von der Minderheitenbetreuung zur Siedlerauslese* (Berlin: Frank & Timme, 2006); Hans-Christian Harten, *Rassenhygiene als Erziehungsideologie des Dritten Reiches: Biobibliographisches Handbuch* (Berlin: Akademie-Verlag, 2006).
8. For recent analyses of the perpetrators' motivation, see Gerhard Paul, ed., *Die Täter der Shoah: Fanatische Nationalsozialisten oder ganz normale Deutsche?* (Göttingen: Wallstein Verlag, 2002); Harald Welzer, *Täter: Wie aus ganz normalen Menschen Massenmörder werden* (Frankfurt: Fischer Verlag, 2005); Michael Wildt, *Generation des Unbedingten: Das Führungskorps des Reichssicherheitshauptamtes* (Hamburg: Hamburger Edition, 2002). See also Christian Gerlach, *Kalkulierte Morde: Die deutsche Wirtschafts-und Vernichtungspolitik in Weissrussland 1941 bis 1944* (Hamburg: Hamburger Edition, 1999); Götz Aly, *Hitlers Volksstaat: Raub, Rassenkrieg und nationaler Sozialismus* (Frankfurt: Fischer Verlag, 2005).
9. Norman M. Naimark, *Fires of Hatred: Ethnic Cleansing in Twentieth-Century Europe* (Cambridge MA: Harvard University Press, 2001); Eric D. Weitz, *A Century of Genocide: Utopias of Race and Nation* (Princeton NJ: Princeton University Press, 2003); Mark Mazower, *Hitler's Empire: Nazi Rule in Occupied Europe* (London: Allen Lane, 2008); Isabel Heinemann and Patrick Wagner, eds., *Wissenschaft, Planung, Vertreibung: Neuordnungskonzepte und Umsiedlungspolitik im 20; Jahrhundert* (Stuttgart: Franz Steiner Verlag, 2006); Donald Bloxham, *The Final Solution: A Genocide* (Oxford: Oxford University Press, 2009).
10. Michael Wildt has recently argued that the Nazi state defined itself as a biopolitical regime displaying a highly selective version of racism. Wildt, "Biopolitik, ethnische Säuberungen und Volkssouveränität. Eine Skizze," *Mittelweg* 36, no. 6 (2006): 87–106. The concept of biopolitics has been applied to the Nazi state by Michel Foucault, in *Verteidigung der Gesellschaft: Vorlesungen am College de*

France (1975–76) (Frankfurt: Suhrkamp Verlag, 1999), 300–303); see also Philipp Sarasin, "Zweierlei Rassismus? Die Selektion des Fremden als Problem in Michel Foucaults Verbindung von Biopolitik und Rassismus," in *Biopolitik und Rassismus*, ed. Martin Stingelin (Frankfurt: Suhrkamp Verlag, 2003), 55–79.

11. Götz Aly and Susanne Heim, *Vordenker der Vernichtung: Auschwitz und die deutschen Pläne für eine neue europäische Ordnung* (Frankfurt: Fischer Verlag, 1991); Michael Burleigh, *Germany Turns Eastwards: A Study of Ostforschung in the Third Reich* (Cambridge: Cambridge University Press, 1988); Isabel Heinemann, "Wissenschaft und Homogenisierungsplanungen für Osteuropa, Konrad Meyer, der 'Generalplan Ost' und die DFG," in Heinemann and Wagner, *Wissenschaft, Planung, Vertreibung*, 45–72; Gabriele Metzler and Dirk van Laak, "Die Konkretion der Utopie: Historische Quellen der Planungsutopien der 1920er Jahre," in Heinemann and Wagner, *Wissenschaft, Planung, Vertreibung*, 23–43; Dirk van Laak, "Planung: Geschichte und Gegenwart des Vorgriffs auf die Zukunft," *Geschichte und Gesellschaft* 34, no. 3 (2008): 305–26; Hans-Walter Schmuhl, *Grenzüberschreitungen: Das Kaiser-Wilhelm-Institut für Anthropologie, menschliche Erblehre und Eugenik 1927–1945* (Göttingen: Wallstein-Verlag, 2005); Uwe Hossfeld, *Geschichte der biologischen Anthropologie in Deutschland: Von den Anfängen bis in die Nachkriegszeit* (Stuttgart: Franz Steiner Verlag, 2005).

12. Regarding scholarly progress since the mid-1990s, cf. Dieter Pohl, *Von der "Judenpolitik" zum Judenmord: Der Distrikt Lublin des Generalgouvernements 1939–1944* (Frankfurt: Lang, 1993); Pohl, *Nationalsozialistische Judenverfolgung in Ostgalizien 1941–1944: Organisation und Durchführung eines staatliche Masssenverbrechens* (Munich: Oldenbourg, 1996); Bogdan Musial, *Deutsche Zivilverwaltung und Judenverfolgung im Generalgouvernement: Eine Fallstudie zum Distrikt Lublin 1939–1944* (Wiesbaden: Harrassowitz Verlag, 1999). Peter Longerich's biography of Himmler defies the trend of lack of knowledge about the racist foundations of occupation policies by focusing especially on the connection between Himmler's ideas of racial purity and German settlement and his role in the implementation of the Final Solution. Longerich, *Heinrich Himmler: Biographie* (Munich: Siedler, 2008). Götz Aly has discussed the mass murder of the European Jews and the ethnic cleansing of occupied Europe as intertwined processes, yet without analyzing the ideological foundation of the Nazi quest for racial purity and *Lebensraum*. Aly, *"Endlösung": Völkerverschiebung und der Mord an den europäischen Juden* (Frankfurt: Fischer Verlag, 1995). In my book I have specifically highlighted the racist component of Nazi occupation policies and its radicalizing potential toward the Final Solution. Heinemann, *Rasse*.

13. Frank Bajohr and Michael Wildt, eds., *Volksgemeinschaft: Neue Forschungen zur Gesellschaft des Nationalsozialismus* (Frankfurt: Fischer Verlag, 2009). See also the

report from the conference "German Society in the Nazi Era: 'Volksgemeinschaft' between Ideological Projection and Social Practice," London, 25–27 March 2010, available online at http://hsozkult.geschichte.hu-berlin.de/tagungsberichte/id=3121, accessed April 8, 2010.

14. Cf. Peter Longerich, *"Davon haben wir nichts gewusst!" Die Deutschen und die Judenverfolgung 1933–1945* (Munich: Siedler Verlag, 2006); Michael Wildt, *Volksgemeinschaft als Selbstermächtigung: Gewalt gegen Juden in der deutschen Provinz 1919 bis 1939* (Hamburg: Hamburger Edition, 2007).

15. For an overview of German traditions of racist thinking, see Ulrich Herbert, "Traditionen des Rassismus," in *Bürgerliche Gesellschaft in Deutschland*, ed. Lutz Niethammer (Frankfurt: Fischer Verlag, 1990), 472–88; Stefan Breuer, *Ordnungen der Ungleichheit: Die europäische Rechte im Widerstreit ihrer Ideen 1871–1945* (Darmstadt: Wissenschaftliche Buchgesellschaft, 2001).

16. Alfred Ploetz, *Die Tüchtigkeit unserer Rasse und der Schutz der Schwachen: Ein Versuch über Rassenhygiene und ihr Verhältnis zu den humanen Idealen, besonders zum Socialismus* (Berlin: Fischer Verlag, 1895); Sabine Schleiermacher, "Grenzüberschreitung der Medizin: Vererbungswissenschaft, Rassenhygiene und Geomedizin an der Charité im Nationalsozialismus," in *Die Charité im Dritten Reich: Zur Dienstbarkeit medizinischer Wissenschaft im Nationalsozialismus*, ed. Sabine Schleiermacher and Udo Schagen (Paderborn, Germany: Schöningh Verlag, 2008), 69–188; Schleiermacher, "Biologie und Gesellschaft: Eugenik und Rassenhygiene im medizinischen Diskurs," in *Biologismus, Rassismus, Rentabilität: Die Ambivalenz der Moderne*, ed. Hanns-Werner Heister (Berlin: Weidler Buchverlag, 2007), 187–201; Peter Weingart et al., *Rasse, Blut und Gene. Geschichte der Eugenik und Rassenhygiene in Deutschland* (Frankfurt: Suhrkamp, 1988); Paul Weindling, *Health, Race and German Politics between National Unification and Nazism, 1870–1945* (Cambridge: Cambridge University Press, 1989).

17. Hans F. K. Günther, *Kleine Rassenkunde des deutschen Volkes* (Munich: Lehmann, 1933); Günther, *Rassenkunde des deutschen Volkes* (Munich: Lehmann, 1933); Günther, *Rassenkunde des jüdischen Volkes* (Munich: Lehmann, 1930); Günther, *Der nordische Gedanke unter den Deutschen* (Munich: Lehmann, 1927).

18. Schmuhl, *Grenzüberschreitungen*; Hans-Walter Schmuhl, ed., *Rassenforschung an Kaiser-Wilhelm-Instituten vor und nach 1933* (Göttingen: Wallstein-Verlag, 2003).

19. Uwe Hossfeld, *Geschichte der biologischen Anthropologie*, esp. 206–66.

20. Sheila F. Weiss, "'The Sword of Our Science' as a Foreign Policy Weapon: The Political Function of German Geneticists in the International Arena during the Third Reich," in *Ergebnisse: Vorabdrucke aus dem Forschungsprogramm "Geschichte der Kaiser-Wilhelm-Gesellschaft im Nationalsozialismus,"* ed. Carola Sachse and Susanne Heim (Berlin: Forschungsprogramm, 2003); Stefan Kühl, *Die Inter-*

nationale der Rassisten: Aufstieg und Niedergang der internationalen Bewegung für Eugenik und Rassenhygiene im 20. Jahrhundert (Frankfurt: Campus, 1997).

21. Katja Geisenhainer, *Rasse ist Schicksal: Otto Reche (1879–1966), ein Leben als Anthropologe und Völkerkundler* (Leipzig: Evangelische Verlagsanstalt, 2002).
22. Anne Cottebrune, *Der planbare Mensch: Die Deutsche Forschungsgemeinschaft und die menschliche Vererbungswissenschaft, 1920–1970* (Stuttgart: Franz Steiner Verlag, 2008), esp. 62–91, 273–82.
23. Schmuhl, *Grenzüberschreitungen*, 280–91.
24. Schmuhl, *Grenzüberschreitungen*, 299–312; Cornelia Essner, *Die "Nürnberger Gesetze" oder die Verwaltung des Rassenwahns 1933–1945* (Paderborn: Schöningh Verlag, 2002).
25. Hansjörg Gutberger, *Bevölkerung, Ungleichheit, Auslese: Perspektiven sozialwissenschaftlicher Bevölkerungsforschung in Deutschland zwischen 1930 und 1960* (Wiesbaden: VS Verlag für Sozialwissenschaften, 2006); Susanne Heim, ed., *Autarkie und Ostexpansion: Pflanzenzucht und Agrarforschung im Nationalsozialismus* (Göttingen: Wallstein-Verlag, 2002); Friedemann Schmoll, *Die Vermessung der Kultur: Der "Atlas der deutschen Volkskunde" und die Deutsche Forschungsgemeinschaft 1928–1980* (Stuttgart: Franz Steiner Verlag, 2009); Willi Oberkrome, *Ordnung und Autarkie: Die Geschichte der deutschen Landbauforschung, Agrarökonomie und ländlichen Sozialwissenschaft im Spiegel von Forschungsdienst und DFG (1920–1970)* (Stuttgart: Franz Steiner Verlag, 2009).
26. Heinemann, "Wissenschaft und Homogenisierungsplanungen," 45–72; Cottebrune, *Der planbare Mensch*.
27. Otto Reche, "Herkunft und Entstehung der Negerrassen," in *Koloniale Völkerkunde, koloniale Sprachforschung, koloniale Rassenforschung: Berichte über die Arbeitstagung im Januar 1943 in Leipzig*, ed. Hermann Baumann (Berlin: Reimer, Andrews & Steiner, 1943), 152–67; Reche, *Verbreitung der Menschenrassen* (Leipzig: List & von Bressendorf, 1938); Geisenhainer, *Rasse ist Schicksal*.
28. Otto Reche, "Leitsätze zu bevölkerungspolitischen Sicherung des deutschen Ostens," in *Der "Generalplan Ost": Hauptlinien der nationalsozialistischen Planungs- und Vernichtungspolitik*, ed. Mechtild Rössler and Sabine Schleiermacher (Berlin: Akademie-Verlag, 1993), 351–55.
29. Charles University was reopened as the German Charles Ferdinand University of Prague. Besides the rather dated book by Teresa Wróblewska—*Die Reichsuniversitäten Posen, Prag und Strassburg als Modelle nationalsozialistischer Hochschulen in den von Deutschland besetzten Gebieten* (1984; repr., Toruń, Poland: Wydawnictwo Adam Marszalek, 2000)—there is no comprehensive history of the German universities in the occupied and annexed territories during the Third Reich.
30. For more information on Schultz, see Heinemann, *Rasse*.

31. Erhard Wetzel's report, 7 February 1942, STANU, NO-2585.

32. Heinemann, "Wissenschaft und Homogenisierungsplanungen"; Robert L. Koehl, RKFDV: *German Settlement and Population Policy; A History of the Reich Commission for the Strengthening of Germandom* (Cambridge MA: Harvard University Press, 1957). Alexa Stiller is currently writing a PhD thesis on the RKF.

33. Summary of the memo on Master Plan East, 28 May 1942, and Ulrich Greifelt to Himmler, 2 June 1942, in *Vom Generalplan Ost zum Generalsiedlungsplan*, ed. Cesław Madajczyk (Munich: Saur, 1994), 85–130.

34. Konrad Meyer, "Siedlungs- und Aufbauarbeit im deutschen Osten," *Die Bewegung* 8 (1941): 7.

35. Konrad Meyer, *Landvolk im Werden* (Berlin: Deutsche Landbuchhandlung, 1942), 3.

36. For a conventional interpretation of Nazi German academia as largely nonpolitical, see Notker Hammerstein, *Die Deutsche Forschungsgemeinschaft in der Weimarer Republik und im Dritten Reich: Wissenschaftspolitik in Republik und Diktatur 1920–1945* (Munich: Beck, 1999). For a more comprehensive analysis proving the deliberate collaboration and individual initiative of relevant parts of the German academic elite, see Karin Orth and Willi Oberkrome, eds., *Die Deutsche Forschungsgemeinschaft 1920–1970: Forschungsförderung im Spannungsfeld von Wissenschaft und Politik* (Stuttgart: Steiner, 2010). See also publications on the history of the Max Planck Society (former Kaiser-Wilhelm Institute) under National Socialism as well as recent studies of German universities in the Third Reich.

37. Hitler's Order on Strengthening the Germandom, 7 October 1939, printed in *Der Prozess gegen die Hauptkriegsverbrecher vor dem Internationalen Militärgerichtshof*, vol. 26 (Nuremberg: IMT, 1947), 255–57.

38. Himmler's speech in Posen, 24 October 1943, quoted in Josef Ackermann, *Heinrich Himmler als Ideologe* (Göttingen: Muster-Schmidt Verlag, 1970), 291–96.

39. Himmler's Marriage Order for the SS, 31 December 1931, BAR, NS 2/174. For the institutional history of the RuSHA, see Heinemann, *Rasse*.

40. Most files of this branch office can be found in the AGK, 167; others are stored in the Russian State Military Archives in Moscow, 1372.

41. This large-scale population transfer had been the result of an agreement between Germany and the Soviet Union, signed shortly before the former's attack on Poland. Further resettlement agreements were ratified, for instance, with the Baltic States.

42. See the files of the Einwandererzentralstelle, BAR, NS 69.

43. See the monthly reports of the RuSHA branch office Zamość to RuSHA in Berlin, AGK, 167/48; report of the Umwandererzentralstelle branch office Zamość, 13 December 1942, BAR, R 75/9.

44. See photos of race cards in Heinemann, *Rasse*, 64–65; original copies in BAR, NS 2/152.
45. Edwin Black, *IBM und der Holocaust: Die Verstrickung des Weltkonzerns in die Verbrechen der Nazis* (Munich: Propyläen Verlag, 2001); Götz Aly and Karl-Heinz Roth, *Die restlose Erfassung: Volkszählen, Identifizieren, Aussondern im Nationalsozialismus* (1984; repr., Frankfurt: Fischer Verlag, 2000).
46. Testimony of Dr. Josef Rembacz, Łódź, 25 August 1946, STANU, NO-5166.
47. In 1944 around thirty-five thousand out of eight hundred thousand Poles were considered a "wanted population addition." Reich Security Main Office to RuSHA and RKF, 19 December 1942, AGK, 167/38; monthly reports of the RuSHA branch office Litzmannstadt, 31 December 1941–30 April 1944, AGK 167/6; monthly reports of the RuSHA branch office Zamość, AGK, 167/48; final report on the Umwandererzentralstelle activities in the Warthegau and the General Government for the year 1943, 12 December 1943, BAR, R 75/3.
48. Final report on the Umwandererzentralstelle activities in the Warthegau and the General Government for the year 1943, 12 December 1943, BAR, R 75/3. See also report on the activities of the Umwandererzentralstelle branch office Zamość from 27 November to 31 December 1942, 13 December 1942, BAR, R 75/9.
49. For the annihilation of the Jewish population in the General Government and the war against the non-Jewish civilian population, see Pohl, *Nationalsozialistische Judenverfolgung*; Bogdan Musial, ed., *"Aktion Reinhard": Der Völkermord an den Juden im Generalgouvernement 1941–1944* (Osnabrück, Germany: Fibre Verlag, 2004); Nils Gutschow, *Ordnungswahn: Architekten planen im "eingedeutschten Osten" 1939–1945* (Gütersloh, Germany: Birkhäuser, 2001); Jacek A. Mlynarczczyk, *Judenmord in Zentralpolen: Der Distrikt Radom im Generalgouvernement 1939–1945* (Darmstadt: Wissenschaftliche Buchgesellschaft, 2007).
50. Attestation by the mayor of Topolice, Tomaszow Mazowiecki District, regarding the resettlement of the Ryszard W. family, 12 December 1940, provided to the author by Ryszard W., October 2006.
51. The Master Plan East and the resettlement carried out in Zamość highlighted the consensus among SS experts that the Jewish population of the territories slated for Germanization had to perish first.
52. Letter of RuSHA head Otto Hofmann to RFSS regarding the racial screening of the Czechs, 17 February 1941, BAB, NS 2/57; letter of the head of the RuSHA branch office in Prague, Ewin Künzel, 17 October 1942, State Central Archives, Prague, 114/25/9. See also Heinemann, *Rasse*, 151–57.
53. For a description of the SS experts' initiatives in Southeastern Europe, see Isabel Heinemann, "Die Rasseexperten der SS und die bevölkerungspolitische Neuordnung Südosteuropas," in *Südostforschung im Schatten des Dritten Reiches. In-*

stitutionen-Inhalte-Personen, ed. Mathias Beer and Gerhard Seewann (Munich: Oldenbourg, 2004), 135–57.

54. Current activities of the Einwandererzentralstelle branch office Litzmannstadt, 30 November 1944, State Archives in Łódź, 204,2/11; Friedrich Brehm to RKF, 25 February 1943, BAR, R 49/79. See also Lothar Kettenacker, *Nationalsozialistische Volkstumspolitik im Elsass* (Stuttgart: Deutsche Verlags-Anstalt, 1973), 267; Dieter Wolfanger, "Die nationalsozialistische Politik in Lothringen (1940–1945)" (PhD diss., University of Saarbrücken, 1977), 173–77.
55. Marie-Louise Roth-Zimmermann, *Denk' ich an Schelklingen . . . Erinnerungen einer Elsässerin an die Zeit im SS-Umsiedlungslager (1942–1945)* (Sankt Ingbert, Germany: Röhrig Universitätsverlag, 2001).
56. Bruno Wasser, "Die 'Germanisierung' im Distrikt Lublin als Generalprobe und erste Realisierungsphase des 'Generalplan Ost,'" in Rössler and Schleiermacher, *Der "Generalplan Ost,"* 271–93; Bruno Wasser, *Himmlers Raumplanung im Osten: Der Generalplan Ost in Polen 1940–1944* (Basel: Birkhäuser, 1993); Wendy Lower, "A New Ordering of Space and Race: Nazi Colonial Dreams in Zhytomyr, Ukraine, 1941–1944," *German Studies Review* 25, no. 2 (May 2002): 227–54; Lower, *Nazi Empire-Building and the Holocaust in Ukraine* (Chapel Hill: University of North Carolina Press, 2005).
57. SS officer file Hans Rihl, Berlin Document Center.
58. SS officer file Hans Rihl, personal questionnaire, 7 August 1935, Berlin Document Center.
59. Monthly report of the SS-Führer im Rasse-und Siedlungswesen head at Zamość, Hans Rihl, to the RuSHA branch office Litzmannstadt, 1 January 1943, AGK, 167/48.
60. This number comprises only the high-ranking and midlevel functionaries in RuSHA. Otherwise, the total RuSHA membership during the war fluctuated between five and ten thousand.
61. Lutz Raphael, "Experten im Sozialstaat," in *Drei Wege deutscher Sozialstaatlichkeit: NS-Diktatur, Bundesrepublik und DDR im Vergleich*, ed. Hans-Günther Hockerts (Munich: Oldenbourg, 1998), 231–58.
62. For a more comprehensive data analysis, see Heinemann, *Rasse*, 590–91, 601–41.
63. On Konrad Meyer, see Heinemann and Wagner, *Wissenschaft, Planung, Vertreibung*.
64. For the most comprehensive collection of documents on the Master Plan East and Master Settlement Plan, see Madajczyk, *Vom Generalplan Ost*, esp. 86–130, 234–55.
65. The entire file, BAR NS 19/1047.
66. See Isabel Heinemann, "'Until the Last Drop of Good Blood': The Kidnapping of 'Racially Valuable' Children as Another Aspect of Nazi Racial Policy in

the Occupied East," in *Genocide and Settler Society: Frontier Violence and Stolen Aboriginal Children in Australian History*, ed. Dirk Moses (Oxford: Berghahn Books, 2004), 244–66.

67. For a discussion of this data, see Heinemann, *Rasse*, 598–603.
68. Schultz records, Department of Medicine, University of Münster Archives, 52/278; Schultz records, Department of Mathematics and Natural Science, University of Münster Archives, 92/16.
69. "Opening Statement of the Prosecution," in *Trials of War Criminals before the Nuernberg Military Tribunals under Control Council Law No. 10*, vol. 4, pt. 2 (Washington DC: U.S. Government Printing Office, 1950), 627.

2

Preserving the "Master Race"

SS Reproductive and Family Policies during the Second World War

AMY CARNEY

On October 28, 1939, Reichsführer-SS Heinrich Himmler issued an order to the entire Schutzstaffel (the SS) and police in which he proclaimed, "Every war is a bloodletting of the best blood."[1] What he referred to was not just the loss of the men who had and would perish on the battlefield but also the absence of their unborn children. Having spent the past decade molding the SS into a racial elite dedicated to Adolf Hitler, Himmler now wanted to make sure that his efforts would not be undone by the Second World War. He sought to ensure that his SS men understood the gravity of the situation. The Thousand-Year Reich for which they now fought could not prevail without a strong biological legacy; a victory in the cradle had to be obtained simultaneously with a victory of arms.

By emphasizing in his late October order the necessity of children for the future, Himmler simply reinforced a concept he had long been advocating. Creating racially healthy families represented an ideal that had united all SS men since the early 1930s. The basis for this ideal primarily came from Himmler's selective application of the then-prominent discipline of eugenics.[2] For the previous half century, eugenicists in Germany and elsewhere in the West had called upon governments and individuals to heed their advice about regulating marriage. They also advocated measures designed to limit the reproduction of "less" fit members of society while encouraging procreation among the "more" fit. Of the scientists who made specific suggestions as to how a state could manage these processes, one of the most important was the founder of eugenics, Francis Galton. In 1865 he wrote an article outlining the prerequisites for founding a eugenic utopia.[3] He suggested creating an examination that took into account every important mental and physical quality that a person could possess. Once young men and women had passed the examination, they would be encouraged to select a marriage partner from among their peers. As a reward, these young couples would receive a monetary wedding present from their government, which would pay for the maintenance and education of their children. In return, these children would grow up to serve the state. Galton believed that such a system would improve the hereditary endowment of the population as well as best serve the needs of the state.

Several decades later, in 1913, Charles Benedict Davenport wrote a pamphlet describing the current status of biological research and how the information gathered from these investigations could be used to create appropriate laws to regulate marriages.[4] He suggested that a state body should be founded that could oversee implementation of the laws, once passed. To ensure the scientific and legal credentials of its work, a biologist, a physician, and a lawyer would dedicate their services full-time to working for this body. Outside of this state office, individual physicians would examine couples who wished to wed as well as scrutinize their family histories; if satisfied with the assessment, the physician would provide the couple with a medical certificate, allowing them to proceed with their union.

Beyond the work of Galton in Britain and Davenport in the United States, German scientists also promoted eugenic measures to regulate marriages and families. Among the works produced, none was as influential as the 1921 book *Grundriss der menschlichen Erblichkeitslehre und Rassenhygiene* (Human heredity and racial hygiene).[5] Written by Erwin Baur, Eugen Fischer, and Fritz Lenz, this book became the standard text on heredity and eugenics. Among the topics explored in the book, the authors put forward proposals designed to fortify the health of the population. Among them were maintaining marriages, allocating monetary allowances for children, and convincing people that more than two children per marriage were needed to sustain the population.[6]

These men represented just a few of the scientists who contributed to the diffusion of eugenics and put forth eugenics-based measures designed to regulate marriages and families. By the 1920s their ideas had not only permeated scientific literature but also popular culture. Himmler saw value in many of their ideas, and he implemented them in the SS to promote the creation of a self-sustaining biological elite. Throughout the 1930s he issued numerous guidelines to shape the sexual and reproductive choices of his men as they married and established families. However, Himmler was not the first person to insist on regulating the marital and reproductive decisions of a segment of the population. The German Society for Race Hygiene (Deutschegesellschaft für Rassenhygiene), founded in 1905 by Alfred Ploetz, had implemented a marriage pledge. Its members, who came from the well-educated upper and middle classes, agreed to submit to a medical examination before marrying; anyone who failed the examination promised to refrain from marrying and having children.[7] This agreement among the members of the society was notable but extremely limited due to the small size of the organization. In contrast, with Ploetz's guidelines, Himmler was the first person to apply eugenic measures to a large and selective group with the intention of fostering the growth of the population; his attempt was far more ambitious than the efforts of the German Society for Race Hygiene. With extensive oversight, he sought to have each SS man, along with his wife and their children, become members of the SS family community (*Sippengemeinschaft*), an exclusive racial community within the larger national community (*Volksgemeinschaft*) that the Nazi Party wished to construct.[8] Representing the biological elite of the nation, these men and women were posited to serve as the vanguard of the Third Reich.

After the outbreak of war in 1939, most of the family-related measures introduced to the SS in the 1930s remained in place, even though many of them had to be adjusted to fit the demands of the new situation.[9] These changes included modifying the engagement and marriage process, adopting new rhetoric, creating opportunities for SS men to have children, and establishing financial and educational provisions to assist their families. As in the prewar period, the majority of these changes were initiated through orders that SS men were expected to carry out without question. However, when it came to the implementation of family-related orders, Himmler permitted some leeway due to the exigencies of the war.[10] He periodically bent his own rules in order to balance his priorities. In particular, he had to balance his goal of establishing an armed branch of the

SS (Waffen-SS) that could secure a military victory with that of cultivating a family community which could uphold that victory in the future.

Binding the "Best Blood": Marriage as the Continuing Foundation of the Family Community

Based on his selective utilization of eugenics, Himmler viewed marriage as the basis of his population policies. The union of a racially healthy SS man and his equally fit bride formed the foundation not only of their individual family but the SS family community as well. These families spearheaded the cumulative expansion of Nordic blood; they had the potential to sustain the SS and the Third Reich. Therefore, the SS, through the Race and Settlement Main Office (RuSHA), regulated the marriages of its members. The official document that set down the initial guidelines for that process—the December 31, 1931, engagement and marriage order—was the first, if not the most significant, directive in the spectrum of eugenic-based mandates relating to the family and its well-being.[11] Collectively, these orders outlined a specific procedure that required an SS man and his future bride to complete several questionnaires designed to summarize their personal backgrounds and family histories. In addition, each couple had to submit to a medical examination, most often conducted by an SS doctor well versed in race hygiene; this doctor evaluated their physical and racial health and determined their suitability to become parents. Officials in the Office for Family Affairs (Sippenamt) in RuSHA reviewed the questionnaires and the results of the medical examinations before granting or denying permission to marry. However, Himmler personally made the decision for commissioned officers or in cases where the bride was pregnant. Once approval had been given by either RuSHA or Himmler, RuSHA officials completed an SS man's application and sent him a letter indicating that he could get married. The entire process was time consuming and highly bureaucratic, leading to a backlog of applications.

With the outbreak of the war, the necessity of ensuring the longevity of the SS through the marriage and families of its members grew more urgent because of the increased potential of losing the "best blood." This imminent threat led to some relaxation of the marriage regulations. The first exception was granted the very day that Germany invaded Poland.[12] Any SS member who wished to expedite the process due to mobilization had only to submit two questionnaire forms to RuSHA; the medical examination forms were essential yet not absolutely necessary. RuSHA officials made a decision within a few hours, most of the time approving the proposal as long as the questionnaires revealed no serious hereditary misgivings about either the SS man or his fiancée. If further verification exposed medical or racial problems with either applicant, the SS man had to accept the consequences.

When Himmler issued this particular order, he did not explain why he had introduced the changes. Nonetheless, he did briefly address the matter in a March 1943 letter that he wrote to SS-Obergruppenführer Udo von Woyrsch but also forwarded to the heads of RuSHA and the SS Main Office (Hauptamt). Himmler declared that he had personally wrestled with the question at the beginning of the war as to whether he should promote marriage among his SS men without first educating them about the proper racial regulations.[13] He subsequently told von Woyrsch that he wanted to promote marriage among his men and therefore was willing to allow for mistakes that might hypothetically follow from this decision. As Himmler further wrote in the letter, what was important was that each SS man had at least one child before he fell in battle. In order to facilitate the birth of these children, he was willing to risk the possibility that a minority of them would not meet SS racial standards. Even if a few of the children were not racially impeccable, Himmler argued that poor blood was better than no blood, especially when it came to having as many boys as possible, who could later serve in the armed forces.

Five months into the war Himmler issued a supplementary order that expanded the necessary paperwork back to prewar levels.[14] Under these regulations from January 1940, an SS man and his bride had to submit a number of documents: a RuSHA questionnaire with a photograph, a hereditary health form, a medical examination form completed by an SS doctor or one associated with the military, a family tree that proved Aryan descent back to 1800 for enlisted men and noncommissioned officers and their brides and to 1750 for commissioned officers and their brides, and a report clarifying the couple's assets and debts. The SS man also had to request that his immediate superior submit a statement on his behalf to RuSHA. Based on his interaction with the SS man (and, whenever possible, his bride), the supervisor had to indicate whether he supported the marriage or not.[15] Once these requisite forms were received by RuSHA, its officials evaluated the application and made an appropriate decision. As with the previous order, if further investigation revealed hereditary problems, the SS man might have to leave the organization.

These engagement and marriage regulations applied to everyone in and associated with the SS. The original 1931 order had solely applied to members of the SS and their brides, even though in July 1938 it had been extended to the police as well. In January 1940 Himmler expanded his orders to include the Waffen-SS, and in 1941 and 1942 he explicitly ruled that even those SS members who did not have German citizenship still needed to request permission to marry.[16] Furthermore, in 1943 he required female employees of the SS and the police to seek his approval before getting married. These women had to submit a genealogical tree and the medical examination forms, but their future husbands, if they did not belong to the SS or the police, merely had to submit a genealogical tree.[17]

Despite the increasing inclusiveness of the engagement and marriage order, some people were exempt from this edict. In October 1940 the RuSHA Office for Family Affairs issued a letter that clarified its position as to who did not need to request permission. According to the office, anyone who had joined the Waffen-SS, including the Death's Head (Totenkopf) units, during the war but who did not intend to remain in the SS after the war was not obligated to obtain authorization. The Office for Family Affairs also recommended that the Waffen-SS command record the names of these men. The Waffen-SS made this exception public over a month later in its ordinance gazette.[18] Although neither the RuSHA letter nor the ordinance was directly signed by Himmler, the fact of publication suggested that he tacitly approved it, at least for the time being.

Two years later the Waffen-SS rescinded this exemption policy and ordered that all members of the organization needed to have their marriage approved.[19] This order, which appeared in a February 1944 ordinance gazette, stipulated that even those men who had joined the Waffen-SS for the duration of the war were still required to obtain permission. This February 1944 order also stated that the men undergoing training needed to be informed about the basic laws of the SS, particularly its marriage regulations.[20] There was only one exception to these rules by the end of the war, namely in May 1944, when Himmler excluded men who had joined the SS guard units (Wachverbände) from the Wehrmacht.[21] Nonetheless, as the available wartime statistics show a high number of SS men requesting and receiving permission to get married, it is highly unlikely that these exemptions, or lack thereof, affected many people.

Most of those men who needed permission to marry during the war seemed to have little difficulty receiving it, even when their circumstances were less than ideal. As in the prewar period, Himmler personally approved the marriages when the fiancée was already pregnant.[22] However, unlike before, when he lambasted his men and their superiors for submitting the paperwork because of the pregnancy, he complained of this condition less frequently during the war.[23] Nonetheless, on one occasion Himmler did write to an officer who had requested permission to marry an already-pregnant fiancée; he indicated that because the officer had known for months that his bride-to-be was pregnant, it was "cavalier and unchivalrous . . . [and] unworthy of an SS man" for him to have waited so long before asking for authorization. Himmler did approve the marriage, but he also requested that the officer make up for his lapse in judgment by committing himself to his marriage.[24] Although just one letter, it does reinforce a now well-known discrepancy in Himmler's opinions on marriage and fatherhood. He accepted illegitimate children of "good" blood, such as those born in the Lebensborn homes, many of whom were fathered by his SS men.[25] Yet at the same time Himmler took offense whenever these same SS men sought to get married after their girlfriends became pregnant. He could bend traditional morality in the name of producing children of "good" blood in one instance but not in the other. In Himmler's view, violating the engagement and marriage order represented a far worse transgression than having an illegitimate child.

While Himmler's response to the pregnancy issue demonstrated a certain degree of flexibility, there were other issues that he did not overlook as easily. Approval was often denied if a bride were deemed unacceptable due to the presence of "foreign" or "Jewish" blood or because her family history revealed hereditary defects or diseases.[26] Furthermore, Himmler objected to marriages in which the bride was older than the groom, despite having married an older woman himself. His opposition in this case was based on his opinion that an older woman could have few, if any, children—a view substantiated by his own one-child marriage. He made this perspective quite clear when rejecting such requests.[27] In many of these cases, after Himmler denied a request he either directly wrote to or ordered that RuSHA contact the superior officer of the man whose application he had rejected. The Reichsführer requested that this officer explain to his subordinate why his application was refused, clarify the racial laws of the SS, and remind him of the necessity of each SS man having children.

There were a limited number of exceptions when it came to these rules, such as the approval of the respective marriages of SS-Brigadeführers Walter

Schellenberg and Peter Hansen. In the former case, Schellenberg discovered that his future mother-in-law was Polish. Knowing the strong likelihood that his application would be rejected, he elicited help from his superior, SS-Obergruppenführer Reinhard Heydrich. Heydrich intervened on Schellenberg's behalf and secured Himmler's approval, although Schellenberg noted in his memoirs that Heydrich's intercession had a high price—the head of the Reich Security Main Office (Reichssicherheitshauptamt, or RSHA) maintained a secret file on Schellenberg's wife and her family until Heydrich's assassination in 1942.[28] Conversely, Hansen directly petitioned Himmler and requested approval even though he and his fiancée had no intention of having children. He supported his argument on the basis that both he and his future bride already had children from previous marriages and wished to raise them together with a good National Socialist upbringing. Himmler approved Hansen's union. Otherwise, he rarely showed such leeway, typically noting that he had consented against his better judgment.[29]

There was one other area where Himmler reevaluated the engagement and marriage order because of the exigencies of the war. In 1935 he had issued a decree that dismissed from the SS any man who had violated the engagement and marriage order by marrying without first requesting approval.[30] Over the next three years several hundred men thus received honorable and dishonorable discharges. Shortly after the war began, however, Himmler decided to pardon these dismissed men. In short, he wanted every racially suitable, able-bodied man to serve in the Waffen-SS. After communicating with officers on his personal staff (*Persönlicher Stab*) and in the Legal Main Office (Hauptamt Gericht) and RuSHA, in November 1940 he issued an order that granted amnesty to men who had left the SS in 1935, 1936, or 1937. Each man could seek readmission if he met several conditions: first, he had to have been dismissed solely for violating the engagement and marriage order; second, an assessment of the dismissed man had to demonstrate that his character and attitude still met the standards required by the SS for admission; and third, RuSHA had to give belated approval for the man's marriage, although he did not need this authorization if he were no longer married and there had been no children from the union.[31]

This amnesty would have affected the membership of very few men, yet it demonstrates Himmler's latitude regarding his expectations of his SS men. He sought to balance his desire to establish a Nordic, blood-based elite with his need to have a strong SS military branch. The former required strict rules guiding the admission of wives, and thus children, into the SS family community, while the latter necessitated the inclusion of as many physically healthy men as possible. War forced him to readjust his priorities beyond his demands regarding SS men's marriages. Specifically, he persuaded these married men to have children, provided them with the opportunity to do so in the middle of a war, and assured them that the SS would care for their wives and children while they served and possibly died.

Ensuring the "Vitality of the *Volk*": Rhetoric to Encourage the Establishment of Families during Wartime

Rhetoric was one of the first techniques used to influence SS men to have children during wartime. This rhetoric continued to emphasize that having racially healthy children was a duty of every SS man. Speeches, orders, and booklets issued during the war indicate that this responsibility became increasingly significant. The first official document that stressed this necessity was Himmler's aforementioned order from October 1939, which urged his SS men to accept "the old wisdom" that a man can die peacefully knowing he has left behind children and that a current military victory had little meaning altogether if there were not enough children to sustain the legacy of their fathers.[32] This was also the order that ascribed to unmarried women of "good" blood the moral responsibility to carry the child of a soldier, a stipulation resisted by the military, not the SS. The military objected to Himmler's call for women to fulfill their "sacred obligation" to have children while their men—fathers, brothers, and even husbands—were fighting for Germany in the ranks of the Wehrmacht, as it appeared to be tacitly endorsing marital infidelity.

Himmler addressed this misunderstanding in late January 1940. Whereas he bluntly stated that illegitimate children existed and would continue to exist in the future, he did clarify that no decent SS soldier would ever attempt to establish a sexual relationship with a woman whose husband was at the front. Should one of the SS men (most of whom were also at the front) do so, however, there were two people involved in a seduction, Himmler explained—the seducer and the woman who allowed herself to be seduced. Thus, while "one does not approach the wife of a comrade," a woman was the best guardian of her honor. Any man who rejected this notion that a woman could sufficiently protect her honor insulted German women.[33]

These differences of opinion were further worked out when Himmler met with army commander-in-chief General Walther von Brauchitsch. The general subsequently released a statement acknowledging that he agreed with the Reichsführer that "the vitality of the *Volk* rests in the assured growth of its children."[34] He asserted that "marriage is the basis of the family" and encouraged all racially valuable people to marry and have as many children as possible because they were the "highest good" and represented the best means to preserve the *Volk*. Although extramarital relations and the children that resulted from them could not replace the family, Brauchitsch did acknowledge that the state should protect illegitimate children from any disadvantages.

With that misunderstanding resolved, Himmler continued to press his SS men to procreate. He expressed these aspirations in both private letters and speeches, stressing that he could be content to let his men wait to father children if he were certain that they would survive the war. As this certainty did not exist, Himmler wanted each man to have at least one child. Such a child would not only carry on the Nordic blood and the SS man's personal family lineage but could also, if he were a boy, take his rightful place in twenty years' time fighting for Germany.[35] To emphasize the significance of establishing a family, in November 1942 Himmler drafted new guidelines that made an SS officer's decision to start a family part of the prerequisite for promotion.[36] Among the requirements, an officer must have married by age twenty-six and have had a suitable number of children, a number determined by the age of the couple and the length of their marriage; there is nothing to suggest that the prewar requirements to have four children by age forty had been changed.

If the officer were not married or did not have children, he had to justify his single and/or childless status. In accordance with a February 1944 amendment to the directive, the officer had to explain why he had failed to have multiple children if his wife was under forty years old and/or the youngest child had been born more than two years earlier.[37] Despite these provisions, it is reasonable to assume that not having a wife or enough children did not prevent most officers from being promoted.[38] According to SS engagement and marriage statistics produced in the early 1940s, more than 50 percent of the officers who requested permission to marry were over thirty years of age, and an examination of the personnel files of officers holding one of the four highest SS ranks shows that by early 1945 less than 30 percent of the married officers had four or more children.

In his quest to produce children, Himmler was backed by other prominent

SS and party officials. He had the support of Hitler, who, on numerous occasions throughout the war, privately spoke about the need for families to have many children. The Führer wanted people to comprehend the necessity of large families: "Everyone should be persuaded that a family's life is assured only when it has upwards of four children—I should even say, four sons."[39] Hitler argued that many great men in the past had come from big families. He also proclaimed that in the future the German elite would descend from the SS because "only the SS practices racial selection."[40] He wanted this practice of producing many children within each SS family to continue and rejoiced in the knowledge that during the war, when blood was being lost in vast quantities, healthy SS men were fulfilling their duty by convincing young women to have children.[41]

Himmler also had the backing of SS-Oberstgruppenführer Kurt Daluege, the chief of the uniformed police (Ordnungspolizei). In late 1942 Daluege submitted a draft order presenting his agency as a model within the SS as far as large families were concerned.[42] Unlike Himmler, however, Daluege spoke from personal experience on family matters, as he and his wife had four children, the last of whom had been born just a few months prior to submission of the aforementioned order.[43] He proclaimed that a lack of children (*kinderarmut*), which he defined as two or fewer children per marriage, would lead to the annihilation of the German *Volk*. The excuses used by police officers in the Weimar period to justify having small families could no longer apply in the National Socialist state. In fact, the opposite must become true because, as a leading social class within Germany, the police had a duty to create families that could serve as role models for the rest of the *Volk*. Only medical reasons or advanced age should prevent a man from creating a child-rich (*kinderreich*) family. Because previous admonitions had gone unheeded, though, Daluege wanted superior officers to commit themselves to promoting the idea of having large numbers of children among their subordinates. He further sought to have officers and their wives pursue all medical avenues to ensure their ability to procreate. Although Himmler did not directly dismiss this order, he allowed Daluege permission to circulate it only as a "secret Reich matter" (*Geheime Reichssache*). As such, officers within the police could use it as an educational outline to guide them in personal conservations with their men.[44]

Despite Himmler's refusal in this particular instance, he did promote the publication and dissemination of works designed to elicit support for large families. Among those publications was "Sieg der Waffen—Sieg des Kindes" (Victory of arms—victory of the children), a pamphlet created by the Educational Office (Schulungsamt) of the SS Main Office in late 1940. The title of the pamphlet summed up its general purpose: to convince SS men through words and images that Germany's military achievement must be followed by an equivalent success in the cradle. Only an early and child-rich marriage could sustain the German *Volk* and ensure the continuation of National Socialism; otherwise, the political struggles of the past two decades and the sacrifices made in both world wars would have been for naught. In order to sustain a growing and healthy *Volk*, each SS man had to embrace the obligation to preserve and improve the quality of Nordic blood, which meant accepting the notion that his family needed to have four to six children. By becoming "the happy parents of a large and healthy flock of children," SS men and their wives would prove their commitment to the *Volk* and the nation.[45]

The SS Main Office continued publishing educational materials that underscored these themes in the following years. In 1941 it released the pamphlet "SS Mann und Blutsfrage" (The SS man and the question of blood), which emphasized the needs of the *Volk* over those of the individual.[46] Fertility alone guaranteed the survival of the genetic material of the Nordic race, but a decline in the willingness of the German people to have children in the first decades of the twentieth century combined with the demographic deterioration caused by war had threatened this survival. The vitality of the *Volk* depended on the preservation of its best racial stock, particularly through their readiness to marry early and have many children. Only if "the number of cradles [was] larger than the number of coffins" would an individual clan and the entire *Volk* be saved and the military victory in the Second World War be assured by the continuing sacrifices of those offspring.[47]

A third publication from 1943, "Rassenpolitk" (Race politics), stressed the idea of the *Volk* as a living community and the SS as the bearers of the best Nordic blood.[48] As an elite within the *Volk*, the SS had the obligation to eschew the traditional two-child system and to embrace a plentiful family of at least four children. By assuming this responsibility, each SS man had to acknowledge that the needs of the SS and the *Volk* superseded his personal decisions regarding procreation. As guardians of the Nordic race, he and his wife had to accept that passing on their blood to their offspring was their highest duty as well as honor.

Ensuring a "Victory of Children": Facilitating the Creation of Families in the Midst of War

These three SS publications represent a sample of the literature designed to elicit SS men to embrace the principle that only a "victory of arms" followed by a "victory of children" would safeguard the future of the Third Reich. However, using rhetoric to persuade an SS man and his wife to have children meant very little if the husband and wife were not in the same location and could not conceive a child. To overcome this problem, Himmler issued new policies to facilitate conjugal relations. Just under two years into the war, when the military situation was still highly favorable to Germany, Himmler ordered the withdrawal from the front of any SS man who was the last surviving adult male in his family and who did not already have at least two male descendants. He required these men to report to him to receive an assignment away from the front, such as training new troops.[49]

Less than a year later Himmler drafted another order designed to withdraw all childless SS men from the front. Having received the approval of Hitler, in mid-August 1942 Himmler issued the "SS Order to the Last Sons."[50] In three simple points, he announced the removal of every last-surviving son on the "Führer's instructions."[51] Each SS man had one year to ensure his family's lineage, as the Reich had a vested interest in ensuring that the families of its best men did not expire. Producing "children of good blood" and thus guaranteeing that each man was no longer the last son in his family represented an obligatory duty because "it has never been the way of SS men to accept fate and not contribute anything to change."[52] The future of the Third Reich depended as much on the reproductive contributions of the SS men as it did on their prowess as soldiers, once again intertwining military and familial responsibilities. After an SS man had fulfilled his paternal obligation, he had to return to his military post.

RuSHA and the Waffen-SS coordinated information to determine which men were last sons. This process of managing the last sons required filing reports with Himmler, who—as with just about every matter in the SS—wanted to remain apprised of any changes.[53] The ability to implement this order and to allow the men time to father children became increasingly difficult as the tide of war turned against the Nazi regime and every German soldier was needed in the attempt to stave off defeat. The closer the war physically came to Germany, the more difficult it became to transfer last sons to less dangerous assignments, so that by August 1944 only those last sons with "special cases of hardship" could be withdrawn from the fighting front.[54]

By removing SS men from the war for a year, the "SS Order to the Last Sons" exemplified a rather drastic measure in the attempt to ensure the continuation of the SS family community. It was not ostensibly designed for married SS men, although they might have benefited from it more than their unmarried counterparts, who had to use the same timeframe to get married and have a child.[55] However, the SS also implemented policies designed solely to aid married SS men and their wives. In particular, in allotting time for a yearly vacation, married men, especially those with children, were favored over single men.[56] In October 1943 Himmler took this concept of a vacation a step further. He wanted married SS men to have an additional opportunity to meet with their wives. The point of spending this extra time together had nothing to do with rewarding these men for their services to the SS; on the contrary, it had everything to do with Himmler's goal of increasing the birthrate in the SS.[57]

The Reichsführer's original idea was to have the men take a short leave of absence from their official responsibilities and rendezvous with their wives for five to six days because, as he noted in a letter to SS-Gruppenführer Benno Martin, "we cannot otherwise expect that the so-wished-for and necessary children sprout from these marriages."[58] To facilitate these trysts, the responsible SS division had to arrange for quarters where these men could live with their wives throughout the duration of the special visits. These SS units had to cover the men's railway and accommodation costs since, depending on the whereabouts of a husband and wife, both might have to travel in order to take advantage of this additional leave. The divisions could settle these expenses with the SS personnel staff, and with these expenditures covered, each couple was merely responsible for the cost of meals during their reunion.[59] According to the paperwork filed by the SS divisions with the personnel staff, these trips ranged in cost from 30 to 170 Reichsmarks per couple, meaning that the personnel staff routinely reimbursed each division thousands of Reichsmarks. For example, between November 1943 and June 1944, over 111,000 Reichsmarks were spent on this program.[60]

As the primary goal of this extra vacation was to produce children, each furlough had to be planned accordingly. For instance, upon requesting a vacation for her husband, Paula Wagner was informed by SS-Obersturmbannführer Rudolf Brandt, Himmler's private secretary, that he had forwarded her appeal to the appropriate SS officials. While they waited for a response, Brandt recommended that she make an appointment with a gynecologist. Such an examination would allow the doctor to determine the most favorable time for conception. With this knowledge, a vacation could be obtained for her husband when there was the highest possibility of their conceiving.[61]

The husbands also needed to understand the magnitude of the extra vacation, a point that Himmler doubted most men had grasped. Therefore, he wanted all SS commanders and troops to receive a copy of "Reminder and Obligation," an early 1943 article by Hans Sievers, a naval doctor working at the women's clinic at the university in Greifswald.[62] Sievers proclaimed that the loss of life throughout the war had already begun to reverse the rising birthrates that Germany had experienced between 1934 and 1939. According to him, every lost soldier meant a lost child, and the unborn child of each deceased soldier represented a decline in the political ascent of National Socialism. The physician further noted that a high-quality marriage remained unfulfilled without children. He defined a sterile marriage as one in which, given regular possibilities for conception, a child had not been produced after two years and suggested that young women in infertile marriages should consult with their doctors about this problem. Sievers moreover asserted that married soldiers needed the opportunity for scheduled vacations to meet with their wives at a time favorable for conception.[63] Although not directly associated with the SS, Sievers reinforced the message already introduced by SS propaganda, namely that the victory of arms had to be followed by one in the cradle.

The "Sacred Task" of Caring for the Family Community during the War

Taken together, the "last sons" order and the additional scheduled vacation demonstrate that Himmler did not want the war to impede the reproductive goals that he had set for the SS. Although neither program significantly aided in the growth of the population, the vast majority of married SS men did have at least one child. Yet even with the birth of these children, the war obstructed men's ability to influence their upbringing directly, especially if they were killed in active service. Himmler certainly recognized that many of his SS men might be reluctant to increase the size of their family if they believed that they might not be able to care for them in the future. With this in mind, he established plans designed to aid widows and orphans.

The notion of the SS having an obligation to care for the wives and children of its members predated the war. On November 9, 1936, Himmler created an oath that all SS-Obergruppenführer and SS-Gruppenführer had to swear to him personally; sealed with a handshake, this pledge committed these highest-ranking SS officers to serve as "the guardians of the blood and life laws [*Lebensgesetz*] of the SS."[64] The oath stipulated the responsibility to maintain the quality of the bloodlines of the members of the SS and their families. Himmler had created this oath because he believed that one of the "greatest dangers for the future of the SS" would be if the sons and daughters of SS men were automatically admitted into the organization as members and wives without further examination. Future generations, he argued, could not be admitted solely based on the merits of their ancestors. Just because their fathers had belonged to the SS was not reason enough to allow their admission, as stricter conditions for entry had to be imposed on each generation. By taking this oath, each officer pledged that he would ensure that every person who sought to join the SS family community had the right racial credentials, even if that meant rejecting his colleagues' children—or even his own. On a more practical level, in late 1937 the Reichsführer also committed the SS, "as a sworn community of families of SS-men," to the "sacred task" of assisting the wives and children of deceased comrades.[65] The commanders at all levels of the SS had the responsibility to help these families not just with financial aid but also with personal support, especially when it came to educating fatherless children.

The obligation of SS men to care for the well-being of the families of fellow servicemen took on greater significance immediately following the outbreak of the war. In mid-September 1939 Himmler requested that the Personnel Main Office (Personalhauptamt) compile a report comprising the names of all fallen SS soldiers. Specifically, he wanted to receive the names of married SS members, as this information would provide a starting point from which SS commanders could accept responsibility for the welfare of their widows and orphans.[66] Just over a month later the Reichsführer briefly commented on such responsibilities in the same October order in which he implored SS and police men to have children before departing for war.[67] Toward the end of this order Himmler addressed the issue of the families that each solider left behind. He noted that unlike previous eras, when soldiers did not want to burden their wives with wartime children, SS men should have no such reservations. If an SS man died fighting for Germany, Himmler would appoint a guardian to oversee the financial and educational needs of his family. This assistance applied to legitimate and illegitimate children alike, since the ultimate goal of the order was to facilitate the transmission of the best hereditary material by removing any possible pretexts for an SS man to abscond from his biological duty.

Beyond suggesting that caring for the families of the fallen represented a duty of surviving SS men, these two orders did not outline any specific measures. However, they did set the tone for future orders and likewise revealed Himmler's underlying concern for the woman and children of his SS men, a concern that Felix Kersten, Himmler's masseur, noted in his diary. According to Kersten, the Reichsführer was quite respectful toward women and "very fond of children." Himmler paid particular attention to war widows and orphans, going so far as to forbid his adjutants from turning them away with the excuse that he was too busy and did not have time to see them. "Compared with their sacrifice," Kersten reported him saying, "the half hour which I sacrifice to them is such a small matter that I would be ashamed if I failed to listen to them or [failed] to give them the feeling that there was somebody to whom they could turn."[68]

As the war progressed and the number of casualties rose, Himmler had to confront the duty of assigning tasks to various SS branches concerning the families of the deceased. Each SS division, along with the Personnel Main Office, was responsible for overseeing the development of the children until they reached maturity. In addition, the Personnel Main Office had to maintain records regarding the economic circumstances of the surviving dependents and determine whether a family needed financial help from the SS beyond what it had already received from the state. The SS Main Office advised mothers and legal guardians about the schooling and occupational training of children, particularly those boys who showed an interest in pursuing a career as a Waffen-SS officer. RuSHA managed the noneconomic needs of each family.[69] On special occasions this general care even consisted of distributing goods such as coats, mittens, stockings, darning cotton, and candles. Furthermore, RuSHA was supposed to intervene when the officers of another division or office could not satisfactorily resolve an issue.[70]

However, the branch of the SS given the greatest responsibility for the care of women and children was the Lebensborn program (loosely translated as "wellspring of life"). In conjunction with division officers, Lebensborn officials placed mothers and children in the institute's homes. The Lebensborn homes alone had facilities where expectant mothers could confine themselves until the birth of their child as well as remain with their children until they were able to return home or, in rare cases, give their children up for adoption. Himmler assigned the Lebensborn office the right to assume the guardianship of fatherless children when it corresponded with the mother's wishes or was in the best interest of an individual child.[71] Such assistance by the Lebensborn also extended to parentless children whose deceased fathers had served in the SS and whose mothers had perished during the war. In this case, though, the initiative came from neither the Lebensborn nor Himmler, but from Hitler, who wanted the Lebensborn rather than an orphanage to provide for these children. The Lebensborn took full responsibility for these parentless children, primarily assigning them to foster families, most likely SS families.[72]

Throughout its existence, the SS had a myriad of rules that governed the organization and its members. Under the leadership of Heinrich Himmler, such jurisdiction included overseeing the personal life of every SS man. A series of regulations were initiated that sought to manage the choices that each member made as he chose a wife and they subsequently decided to have children. These regulations were designed to sustain the racial and hereditary purity of the SS, which would allow it to become the new vanguard of the Third Reich. However, extenuating circumstances often necessitated that Himmler modify his population policies; such alterations were frequently required because of the Second World War.

Nonetheless, he was unwilling to allow the war to impede his long-term goal of forming a family community in the SS. To ensure the continuation of the Third Reich, he redoubled his efforts to create this community. He sought to have each SS man embrace the notion that only a child-rich family would secure the military victory for which he fought. Fathering children represented just as significant a duty to Germany as any other, and the war was not to obstruct this responsibility. Therefore, the Reichsführer repeatedly issued new guidelines and orders to guarantee the survival of the valuable genetic legacy of each SS man, although he did periodically allow some leeway in their implementation due to the exigencies of the war. Altogether, these wartime measures had the same goal: to foster racially healthy marriages that could produce numerous children of "good" blood who, as the biological elite "master race," would perpetually serve and sustain the Thousand-Year Reich.

Notes

Abbreviations Used in the Notes

BAB	Bundesarchiv Berlin (German Federal Archives, Berlin)
BDC SSO	Berlin Dokument Center, SS-Offizier Personalakte (Berlin Document Center SS Officer files)
IfZ	Institut für Zeitgeschichte (Institute for Contemporary History, Munich)
NARA	U.S. National Archives and Records Administration, Washington DC

1. SS Befehl für die gesamte SS und Polizei, 28 October 1939, BAB, NS19/3901.
2. For standard works on eugenics, see Robert N. Proctor, *Racial Hygiene: Medicine under the Nazis* (Cambridge MA: Harvard University Press, 1988); Mark B. Adams, ed., *The Wellborn Science: Eugenics in Germany, France, Brazil, and Russia* (New York: Oxford University Press, 1990); Richard Weikart, *From Darwin to Hitler: Evolutionary Ethics, Eugenics, and Racism in Germany* (New York: Palgrave Macmillan, 2004).
3. Francis Galton, "Hereditary Talent and Character," *Macmillan's Magazine* 12, no. 70 (1865): 165, 319–21. This article was not the only place where Galton laid out a plan for a eugenic utopia. He envisioned a utopia called "Kantsaywhere." However, only fragments of his text remain, primarily reproduced in Karl Pearson's multivolume biography *The Life, Letters, and Labours of Francis Galton* (London: Cambridge University Press, 1914, 1924, 1930).
4. Charles Davenport, *State Laws Limiting Marriage Selection Examined in the Light of Eugenics* (Cold Spring Harbor NY, 1913).
5. Erwin Baur, Eugen Fischer, and Fritz Lenz, *Grundriss der menschlichen Erblichkeitslehre und Rassenhygiene*, vols. 1–2 (Munich: J. F. Lehmanns Verlag, 1921).
6. Baur, Fischer, and Lenz, *Grundriss*, 2:71, 81, 133, 138, 141–42, 210.
7. Weikart, *From Darwin to Hitler*, 93, 139.
8. On the *Sippengemeinschaft*, see Peter Longerich, *Heinrich Himmler: Biographie* (Munich: Siedler Verlag, 2008); Gudrun Schwarz, *Eine Frau an seiner Seite: Ehefrauen in der "SS-Sippengemeinschaft"* (Hamburg: Hamburger Edition, 1997); James J. Weingartner, "The Race and Settlement Main Office: Toward an Order of Blood and Soil," *Historian* 34, no. 1 (1971): 62–77. The creation of the Nazi *Volksgemeinschaft* is discussed in John Connelly, "The Uses of the *Volksgemeinschaft*: Letters to the NSDAP Kreisleitung Eisenach, 1939–1940," *Journal of Modern History* 68, no. 4 (December 1996): 899–930; and David Welch, "Nazi Propaganda and the

Volksgemeinschaft: Creating a People's Community," *Journal of Contemporary History* 39, no. 2 (April 2004): 213–38.

9. Throughout its entire existence, the SS and its constituent offices frequently adjusted their priorities in order to meet the demands of the party and the state. This nonstatic nature of the SS was first articulated by Robert Lewis Koehl. See Koehl, "The Character of the Nazi SS," *Journal of Modern History* 34, no. 3 (September 1962): 275–83; Koehl, *The Black Corps: The Structure and Power Struggles of the Nazi* SS (Madison: University of Wisconsin Press, 1983).
10. There were other instances of Himmler allowing leniency or tolerating deviance. As noted by Geoffrey Giles, despite his hatred of homosexuality, Himmler never systematically persecuted homosexuals. Furthermore, as shown by Christopher Browning, the men assigned to Einsatzgruppen units were not punished for declining to participate in executions. Giles, "The Denial of Homosexuality: Same-Sex Incidents in Himmler's SS and Police," *Journal of the History of Sexuality* 11, nos. 1–2 (January/April 2002): 256–90; Giles, "The Institutionalization of Homosexual Panic in the Third Reich," in *Social Outsiders in Nazi Germany*, ed. Robert Gellately and Nathan Stoltzfus (Princeton NJ: Princeton University Press, 2001), 223–55; Christopher Browning, *Ordinary Men: Reserve Police Battalion 101 and the Final Solution in Poland* (New York: Harper Perennial, 1998).
11. SS Order A, no. 65, 31 December 1931, BAB, NS19/1934.
12. SS Order, 1 September 1939, BAB, NS19/577.
13. Himmler to Udo von Woyrsch, 22 March 1943, BAB, NS2/240.
14. Engagement and marriage authorization, 26 January 1940, BAB, NS2/21.
15. The purpose of a statement from a superior officer indicating his support for the marriage of his subordinate is also discussed in "Heiratsgenehmigung für Angehörige der Waffen-SS," 30 September 1942, BAB, NS2/231; "Heiratsgenehmigung für Angehörige der Waffen-SS," 14 December 1942, NARA, T175/152; Helmut Heiber, *Reichsführer! Briefe an und von Himmler* (Stuttgart: Deutsche Verlags-Anstalt, 1968), 152–54.
16. Engagement and marriage authorization, 26 January 1940, BAB, NS2/21; "Heiratsgenehmigung für Angehörige der Waffen-SS," 1 October 1942, BAB, NSD41/40; "Heiratsgenehmigung für Angehörige der Waffen-SS," 14 December 1942, NARA, T175/152; Edgar Erwin Knoebel, "Racial Illusion and Military Necessity: A Study of SS Political and Manpower Objectives in Occupied Belgium" (PhD diss., University of Colorado, 1965), 42–43.
17. "Heiratsgenehmigung für Angehörige der Waffen-SS," 10 September 1943, BAB, NS2/231.
18. Marriage authorization, 4 November 1940, NARA, T580/326; *Trial of the Major War Criminals before the International Military Tribunal, Nuremberg, 14 November*

1945–1 October 1946, vol. 42 (Nuremberg: Allied Control Authority for Germany, 1947), 490.

19. "Heiratsgenehmigung," *Verordnungsblatt der Waffen SS*, 1 October 1942, para. 359, NARA, T611/5.
20. "Heiratsgenehmigung," *Verordnungsblatt der Waffen SS*, 1 February 1944, para. 57, NARA, T611/6.
21. Himmler to the chief of RuSHA, 11 May 1944, BAB, NS2/231.
22. Engagement and marriage request of SS-Untersturmführer Heinrich Willmann, 8 January 1940, BAB, NS19/3481; Decision regarding the marriage request of SS-Hauptsturmführer Dr. Otto Viertbauer, 12 September 1942, NARA, T175/116; Himmler to RuSHA, 27 August 1944, BAB, NS19/3978; Engagement and marriage request of SS-Oberführer Willi Köhn, 3 October 1940, NARA, Berlin Document Center Race and Settlement Main Office file, D60.
23. Himmler's order on pregnancy and marriage, 19 March 1936, BAB, NS19/577.
24. Heiber, *Reichsführer!*, 190.
25. For the definitive account of the Lebensborn program, see Georg Lilienthal, *Der "Lebensborn e.V.": Ein Instrument nationalsozialistischer Rassenpolitik* (Stuttgart: Gustav Fischer, 1985); Lilienthal, "Ärtze und Rassenpolitik: Der 'Lebensborn e.V.,'" in *Ärtze im Nationalsozialismus*, ed. Fridhof Kudlien (Cologne: Kiepenheuer and Witsch, 1985), 153–66.
26. Himmler to Hanns Rauter, 17 May 1943, BAB, NS19/3483; Himmler to the chief of RuSHA, 3 July 1944, BAB, NS2/231; Himmler to [?] Eicke, 30 November 1942, BAB, NS19/3483; Himmler to the chief of RuSHA, 5 October 1944, BAB, NS19/3978.
27. Memo regarding the marriage request of SS-Hauptscharführer Hermann Hofmann, 24 September 1942, BAB, NS19/3482; Himmler to Gottlob Berger, 9 October 1942, BAB, NS19/3483; Himmler to Eicke, 30 November 1942, BAB, NS19/3483; Rudolf Brandt to Oswald Pohl, 22 January 1944, BAB, NS19/3978; Brandt to Bruno Streckenbach, 28 November 1942, NARA, T175/116.
28. Walter Schellenberg, *Hitler's Secret Service* (New York: Pyramid Books, 1971), 37.
29. Peter Hansen to Himmler, 9 November 1942, NARA, Berlin Document Center Race and Settlement Main Office file, C26; personnel file of Peter Hansen, BAB, BDC SSO, 062A; Himmler to RuSHA, 23 February 1943, BAB, NS19/3483; Himmler to Günther Tichatschke, 22 January 1943, BAB, NS19/3483; Himmler to RuSHA, 27 August 1944, BAB, NS19/3978.
30. Authorization for engagement and marriage, 6 June 1935, BAB, NS19/577.
31. "Verstöße gegen den Verlobungs- und Heiratsbefehl," 27 October 1939, BAB, NS19/752; "Amnestie wegen Vergehen gegen den Verlobungs- und Heiratsbefehl," 16 August 1940, NARA, T580/333; "Wiederaufnahme in die SS bei Vergehen gegen den Verlobungs- und Heiratsbefehl," 1 November 1940, BAB, NS7/216.

32. "SS Befehl für die gesamte SS und Polizei," 28 October 1939, BAB, NS19/3901.
33. "An alle Männer der SS und Polizei," 30 January 1940, BAB, NS19/3901.
34. Statement of Walther von Brauchitsch, 6 February 1940, BAB, NS18/712.
35. Himmler to Hans Jüttner, 29 May 1942, BAB, NS19/3441; Speech of the Reichsführer-SS from 19 June 1942 before the officer corps of the division, IfZ, MA312; Himmler to Udo von Woyrsch, 22 March 1943, BAB, NS2/240.
36. "Richlinien zur Einreichung von Beförderungsvorschlägen für Führer der Allgemeine-SS und der Waffen-SS für die Dauer des Krieges," 15 November 1942, BAB, NS34/21.
37. "Einreichung von Beförderungsvorschlägen," 15 February 1944, BAB, NS34/21.
38. For a few exceptions, see Herbert F. Zielger, "Fight against the Empty Cradle: Nazi Pronatal Policies and the SS-Führerkorps," *Historical Social Review* 11, no. 2 (1986): 30.
39. Hugh Trevor-Roper, ed., *Hitler's Table Talk, 1941–1944: His Private Conversations* (New York: Enigma, 2008), 58–59.
40. Trevor-Roper, *Hitler's Table Talk*, 83.
41. Trevor-Roper, *Hitler's Table Talk*, 327.
42. Family offspring of the officer corps of the Uniformed Police, 1942, NARA, T175/18.
43. Personnel file of Kurt Daluege, BAB, BDC SSO, 134.
44. Himmler to Daluege, 5 October 1942, NARA, T175/18.
45. "Sieg der Waffen—Sieg des Kindes," BAB, NSD41/130. Richard Weikart discusses the significance of this SS pamphlet in the creation and dissemination of Nazi rhetoric on evolutionary ethics in "The Role of Evolutionary Ethics in Nazi Propaganda and Worldview Training," paper presented at the conference Ideologie und Moral im Nationalsozialismus, Dresden, Germany, 20 November 2010.
46. "SS Mann und Blutsfrage: Die biologischen Grundlagen und ihre singemäße Anwendung für die Erhaltung und Mehrung des nordischen Blutes," BAB, NSD41/128.
47. "SS Mann und Blutsfrage: Die biologischen Grundlagen und ihre singemäße Anwendung für die Erhaltung und Mehrung des nordischen Blutes," BAB, NSD41/128.
48. "Rassenpolitik," BAB, NSD41/122.
49. "Zurückziehung aus der fechtenden Truppe," 4 August 1941, BAB, NS7/381.
50. Knoebel, "Racial Illusion," 75.
51. SS order, "Letzte Söhne," 15 August 1942, BAB, NS2/23.
52. SS order, "Letzte Söhne," 15 August 1942, BAB, NS2/23.
53. SS order, "Letzte Söhne," 9 April [1943], NARA, T354/519; file registration of the "Letzten und einzigen Söhne," 29 March 1944, BAB, NS2/17.
54. "Zurückziehung 'einziger und letzter Söhne' aus der kämpfenden Truppe in besonderen Härtefällen," 8 August 1944, BAB, R187/680.
55. Heiber, *Reichsführer!*, 215.

56. "Urlaub zu Weihnachten 1942 und Neujahr 1943," 5 December 1942, BAB, NS3/1080; "Erholungsurlaub für Waffen-SS/Allgemeine-SS Angehörige, einschl. Zivilian-gestellte," 5 March 1943, BAB, NS2/22.
57. "Planmäßiger Urlaub," 2 October 1943, NARA, T175/71; Lilienthal, *Der "Lebensborn e.V.,"* 137.
58. Heiber, *Reichsführer!*, 219.
59. "Planmäßiger Urlaub," 5 August 1944, NARA, T175/71.
60. "Planmäßiger Urlaub," 19 September 1944, NARA, T175/71; "Zusammenstellung der vom November 1943 bis Juni 1944 gezahlten Beträge für planmässigen Urlaub," 30 June 1944, NARA, T175/71.
61. Heiber, *Reichsführer!*, 211.
62. Himmler to Hans Jüttner, 9 September 1943, NARA, T175/71.
63. Hans Sievers, "Mahnung und Verpflichtung," NARA, T175/71.
64. "Grundgesetz über die Vereidigung der SS-Obergruppen- und Gruppenführer als Hüter des Blutes- und Lebensgesetzes der Schutzstaffel," 9 November 1936, BAB, NS19/3902; "Protokol über die Vereidigung," 9 November 1936, BAB, BDC SSO, 030A.
65. "SS Befehl für den Wintersonnenwend-Wettkampf," 1937, BAB, NS2/2.
66. Himmler's order, 16 September 1939, BAB, NS34/21.
67. "SS Befehl für die gesamte SS und Polizei," 28 October 1939, BAB, NS19/3901.
68. Felix Kersten, *The Kersten Memoirs, 1940–1945* (New York: Macmillan, 1957), 304.
69. "Ausführungsbestimmungen zum Befehl des Reichsführers-SS," 19 June 1940, BAB, NS19/3901.
70. "Ausbildungs- und Berufsförderung für Kinder gefallener SS-Männer," 19 June 1943, NARA, T175/152; "Richtlinien zur Durchführung der Ausbildungs- und Berufsförderung für Kinder gefallener SS-Männer," 30 July 1943, NARA, T175/190; Heiber, *Reichsführer!*, 147–48.
71. "Bestellung des 'Lebensborn' e.V. als Beistand für die ehelichen Kinder und Übernahme der Vormundschaft für die unehelichen Kinder gefallener SS-Kameraden," 27 October 1939, BAB, NS19/3903; Guntram Pflaum to Himmler, 11 November 1939, NARA, T175/76; "Ausführungsbestimmungen zum Befehl des Reichsführers-SS," 19 June 1940, BAB, NS19/3901.
72. Max Sollmann to Karl Wolff, 19 February 1943, IfZ, MA304; Wolff to Sollmann, 31 January 1943, IfZ, MA304; Himmler's order, 13 January 1943, IfZ, MA304; "Erziehung elternloser Kinder," 5 January 1943, IfZ, MA304; Himmler to Hans-Heinrich Lammers, 5 January 1943, IfZ, MA304; "Elternlose Kinder," 19 March 1943, NARA, T175/73.

3

Germanic Brothers

The Dutch and the Germanization of the Occupied East

GERALDIEN VON FRIJTAG DRABBE KÜNZEL

In the summer of 1942, less than three years after the outbreak of the Second World War, German rule was established in a substantial part of Europe. Nazi Germany had considerably enlarged its territory by incorporating Austria and large border areas. From the Atlantic Ocean to the Caucasus, Germans governed occupied Europe. Leading German planners regarded the occupied regions in the East (Poland, the Baltic States, Ukraine, Belorussia, and Russia) as areas fit for Germanization. Essentially, Germanization meant creating an empire ruled by Germanic people over all others—an empire based on "race" and "blood"—in which there was, at the very most, a subordinate place for only a fraction of the indigenous Slavic population and certainly none for the millions of Jews. It also meant the transfer of people from "Germanic" offspring to this area.

Since the collapse of the Communist bloc and the subsequent opening of archives, there has been substantial academic interest in the policy of Germanization. The general trend in recent historiography has been to place Germanization in a broader historical perspective. Some studies associate Germanization with the rise of state engineering, planning, and population policy, while others compare it either with past imperialist projects or specific instances of ethnic cleansing.[1] Other scholars, by contrast, concentrate on the longer history of German colonial aspirations toward this part of Europe.[2]

Remarkably, many recent studies omit the involvement of those whom the Nazis considered as fellow Germanic peoples in the colonization project. In the occupied Netherlands, however, German authorities made serious efforts to shift a substantial segment of the Dutch population to the East and to establish Dutch colonies along the way. In order to attract volunteers and to coordinate settlers' employment in the German Reich Commissariats Ostland and Ukraine (Reichskommissariat Ostland/Ukraine), the occupation authorities launched a series of propaganda campaigns and established several state agencies. Between November 1941 and July 1944 over five thousand Dutch settled in Belorussia, the Baltic States, and Ukraine. They were offered a year's contract, but the prospect of permanent settlement was held out to them too. In the press they were often referred to as "pioneers."

The Dutch involvement in the Germanization project raises several important issues. Specifically, this chapter engages with the question of whether the Dutch—leaders, fellow travelers, and pioneers alike—had motivations similar to those of their German comrades and whether they considered themselves to be "of kindred blood" and members of the same Germanic "master race" and therefore the rightful rulers of the occupied East. German ideas about the Dutch are equally relevant to this discussion: where on the Nazi racial curve did the Dutch stand, and why did German authorities seek the Dutch contribution in the first place?

In order to comprehensively answer these questions, this chapter begins the story in the years leading up to the outbreak of war, examining, in particular, developments in the fields of ethnology and folklore in prewar Dutch society. This chapter asserts that the advancement of racial thought and, more specifically, the rediscovery of a supposedly Germanic descent not only laid the groundwork for Dutch academic collaboration in Nazi racial policy but also paved the way for Dutch participation in the Germanization project.

Racial Science in the Netherlands Before the Second World War

At first glance, the history of racial science in the Netherlands seems to have been a history of deviation. Whereas racial science made its appearance in European academia in the second half of the nineteenth century, it was hardly noticed in the Netherlands. The reception of Charles Darwin's *On the Origin of Species* (1859) and *The Descent of Man* (1872) was lukewarm, and it took more than six years before the former book was translated into Dutch. It seems plausible to explain this Dutch reaction by the profoundly religious character of Dutch society. Pious Christians, numerous and influential in nineteenth-century Holland, disagreed with Darwin's ideas on evolution by natural selection.[3]

In several European countries and in the United States eugenics gained influence in academic circles before the turn of the century. Scientists and physicians, convinced of the direct link between social behavior and physical condition, were determined to increase the numbers of those deemed to be of "valuable hereditary material" and reduce the proportion of those of less "valuable" genetic stock. The question was, in other words, whether the process of natural selection could, and should, be encouraged by state intervention. In a number of countries eugenics movements gained a political foothold. In some Scandinavian countries, for example, the central government decided to sterilize citizens who were considered "of lesser value."[4] In Nazi Germany eugenics eventually led to the killing of hundreds of thousands of "unwanted" people.[5]

Neither of these developments occurred in the Netherlands. Debates on improving the human stock began in the Netherlands decades after they had proliferated elsewhere in Europe and the United States. The number of proponents of eugenics was, and would remain, relatively small. Moreover, the Dutch state was not inclined to intervene in what was understood to be the private domain of its citizens. Suggested measures such as the introduction of an obligatory medical inspection of people who intended to get married never became actual policy.[6]

To a degree, Dutch advocates of eugenics envied their German counterparts who were profiting from 1933 onward from the Nazi regime's institutional support. However, the majority of Dutch condemned the actual Nazi policy of so-called racial hygiene. The sterilization law (Gesetz zur Verhütung erbkranken Nachwuchses), which introduced compulsory sterilization of citizens suffering from hereditary diseases, was heavily criticized in the Netherlands. Prominent members of the eugenics movement also disagreed with the unambiguously racist dimension of German eugenics. From 1935 onward, relations between Dutch and German proponents of eugenics grew increasingly tense and, at several conferences, the tensions culminated in public rows.[7]

The perception of the Dutch academic community as immune to racial science and disparaging of Nazi policy has been questioned in some recent studies, however. For example, in the case of alcoholism, generally considered a hereditary disease, Dutch and German physicians were more in agreement. As far as the treatment of alcoholics was concerned, Dutch politicians across the political spectrum supported eugenic measures. One of the foremost Dutch geneticists, Petrus Johannes Waardenburg, recommended social isolation; a famous Dutch professor of psychiatry Gerbrand Jelgersma favored a policy aimed at preventing the biological reproduction of alcoholics.[8]

However small the eugenics movement was, its main advocates were influential men of science. One of them was Sebald Rudolf Steinmetz, the founding father of Dutch sociology. Although a lawyer by training, Steinmetz's main area of interest was ethnology. In 1907 he was appointed professor of political geography and *Volkenkunde* (the study of foreign peoples) at the University of Amsterdam.[9] According to Steinmetz, humanity was defined by biological, hereditary differences; hence his public fulminations against feminists and their struggle for women's emancipation. He believed that men and women differed by nature and that the latter were simply not equipped to perform the work that was traditionally done by the former: women's sole occupation was reproduction, he insisted.[10] As a concept, *race* was not discussed by Steinmetz, despite the fact that in 1938 he edited a Dutch volume on the "races of mankind."[11] Regarding the connection between racial and behavioral qualities Steinmetz was more ambiguous, even though his work exhibited numerous examples of racial prejudices.[12] Women and African children, according to Steinmetz, were predisposed to submission and, intellectually and morally, stood far behind European men.[13] Steinmetz was fascinated by war, which he regarded as a natural, indeed necessary, method of selection for entire nations. In war, he argued, the strongest nations and men with the strongest hereditary qualities would triumph.[14]

The notion of selection featured prominently in most of Steinmetz's works. Whereas war could, hypothetically, be perceived as natural, he also favored "man-made" selection. The strength of a nation depended on the strength of each of its citizens. Steinmetz put it rather poetically: a nation was a building made of individual bricks, whereas the state was the architect responsible for selecting the right bricks.[15] In fact, as early as 1917 Steinmetz was pleading for a state-sponsored policy that would encourage the reproduction of "valuable" people, in particular healthy farmers, and would financially support large families.[16]

Even though not all of Steinmetz's colleagues shared his eugenic ideas, biologism enjoyed wide support within the Dutch academic community.[17] According to the principles of biologism, the world's population was classified by race, with racial differences not only manifest in skin color and physical appearance but also in character traits, social qualities, and mental capacities. Needless to say, the white man was deemed to be at the summit of the racial pyramid.

Biologist thinking was also part of the mainstream *Volkskunde* (the study of peoples and their folk culture). Dutch ethnologists and folklorists were manifestly fond of old customs and traditions, which they believed expressed the true, unchanging nature of the Dutch people. Where the Dutch came from and what practices could be called typically "Dutch" were questions that had been debated for centuries, yet after the First World War such enquiries acquired a sense of urgency. Although the Netherlands had successfully maintained a policy of neutrality during the hostilities, the massive scale of destruction in neighboring countries and the subsequent redrawing of international borders caused a certain disorientation in the Netherlands. The national identity appeared to be under threat—a sentiment that was in part reinforced by Belgian claims to Dutch territory in the aftermath of the war. In this atmosphere of uncertainty, folk culture seemed to offer a path to stability.[18]

One of the most famous folklorists in the interwar period was Dirk Jan van der Ven. Van der Ven, who lacked academic training and was unaffiliated with a university, published several pamphlets and organized numerous seminars on folk culture. The folk dance performances that he and his wife organized were well known and well attended, including by foreign experts in the field. Through his work van der Ven showed his admiration for the peasant lifestyle. Van der Ven considered Dutch peasants, untouched by modernity, the true bearers of Dutch (folk) culture. His growing interest in Germanic history was reflected in the 1930s in his invitation policy: French dancers, for example, were no longer welcome. The participants now came exclusively from Scandinavia, Germany, and Belgium—the countries considered to have a shared Germanic past.[19]

Van der Ven's work was noticed and promoted by Jan de Vries, who had been a professor of ancient Germanic language and literature since 1926. Four years later de Vries published a book, *De Germaansche oudheid* (The Germanic antiquity), in which he argued against the typical representation of ancient, "barbaric" Germanic tribes, pointing to the emergence of Dutch civilization during the years of Germanic rule. With scarcely concealed admiration he described the old Germanic communities and the way they were held together by "blood ties": loyalty to the tribe, the family in the broadest sense, and the leader.[20] De Vries regarded the peoples of the Netherlands to be the descendants of these Germanic tribes. He contended that Dutch customs, culture, and "character" (*volksaard*) were to a large extent cultivated within these tribes, which shared them with other Germanic descendants such as the Flemish and the Germans.[21]

In the 1930s de Vries established his reputation as the Netherlands's most influential ethnologist. In 1935 he was appointed chairman of the International Association for European Ethnology and Folklore, and three years later he became president of the ethnological committee of the Royal Netherlands Academy of Arts and Sciences (Koninklijke Nederlandse Akademie van Wetenschappen)—the highest academic position in the field of ethnological research. Consequently, Dutch ethnological research of the late 1930s bore his stamp.[22] It was during this period that racist notions subtly entered his scholarship. For example, as early as 1930 he had described the archetypal Germanic man as being "of large and brawny build, with blue, twinkling eyes and blond, wavy hair."[23] He also stated that the European battles of the nineteenth and twentieth centuries (referring to the Franco-Prussian War of 1871 and the First World War) exemplified the clash between the Germanic and Romanic races.[24]

The early contributions of some members of the Dutch Nazi Party (Nationaal-Socialistische Beweging, or NSB), too, betray racism. From the time it was established in 1931, the NSB was split into two factions: one oriented toward Italian fascism and another promoting a "moderate" version of German Nazism (i.e., shunning its racist, antisemitic message). The latter faction was averse to modernity and glorified agrarian society.[25] One of their main spokesmen was Evert Jan Roskam, a farmer's son who had joined the NSB in 1933 and would acquire substantial authority as the Dutch advocate of "blood and soil" (*Blut und Boden*) dogma. A protégé of Anton Mussert, the Dutch leader of the NSB, Roskam quickly found a platform for his ideology—*Volk en Vaderland* and *Het Nationaal Dagblad*—two newspapers with direct links to the NSB.

In the following years Roskam wrote numerous pamphlets and popular articles on Dutch folk culture, demonstrating his affinity for the traditional peasant way of life. In 1934 he produced a pamphlet in which he compared the task of the Dutch government with that of a gardener, who "had to clean up the weeds in order to encourage natural growth."[26] Who he meant by the "weeds" became clearer in the summer of 1938. In a series of articles in *Het Nationale Dagblad* Roskam openly attacked Jews. According to Roskam, Jews were "Oriental strangers" who dominated Dutch economic, political, and cultural life. Of particular concern to him was their control over the trade in dairy and meat products. Dutch farmers had become Jewish captives, Roskam stated.[27] By then Roskam had consolidated his power base within the NSB, and the *völkisch* trend he represented and promoted was generally accepted.[28]

German-Dutch Alliances

These new developments in a neighboring country were followed attentively by German experts in racial science and ethnology. *Westforschung*, the study of Germany's western borderlands—an area where Germany had lost large chunks of territory to France and Belgium after the First World War—grew in popularity, and in the early 1930s some German universities received financial support to study the customs, language, and history of the people in these areas.[29]

Even in the period before Hitler came to power there had been a lively exchange of opinions between Dutch and German ethnologists. De Vries and van der Ven were highly esteemed by their German colleagues, whom they met regularly at various international scientific conferences. From January 1933 onward German interest in Dutch ethnology grew further. Some of the main tenets of Nazi ideology were quite close to those shared by de Vries and van der Ven. Thus they spoke of *Volk* and community instead of nation; they asserted that the Dutch people were united not by means of national borders but by their Germanic roots, which they shared with people across the border. The Nazi regime, which began conceiving of a Greater Germanic Reich at the end of the 1930s, fully approved of these ideas. Although it was too urbanized and densely populated to their taste, the German authorities appreciated the Netherlands as the homeland of a people that could be seen as an offshoot of the same Germanic tree that the Germans had sprouted from. As a result, various German ethnological institutes sought contact with Dutch scientists on the matter of launching joint research projects.[30]

The SS agency Ahnenerbe came to play a central role in these efforts. Established in 1935 to promote the study of German ancestral heritage, this agency rapidly expanded the scope of its operations. Ahnenerbe attached great importance to the new orientation in the Netherlands toward its Germanic roots and ancient Germanic customs and culture. Hans Ernst Schneider, who had been conducting ethnological research in East Prussia, became the Ahnenerbe's eyes and ears in the Netherlands. Schneider's mission was to increase the awareness among the Dutch of their Germanic history and culture—a history and a culture they allegedly shared with Germans. Eventually this awareness was expected to translate into a closer political affiliation. To make it happen, the ties between Dutch and German scholars in the fields of ethnology and folklore were strengthened. Several times Schneider traveled to the Netherlands on behalf of Ahnenerbe. On these occasions he visited, among others, van der Ven, trying to convince him of the necessity of creating a Greater Germanic Reich.[31]

Schneider's involvement did not end after the German invasion of the Netherlands in May 1940. In August, just a few months after the Dutch army had been defeated and a German Reich Commissariat had been installed to administer the occupied Netherlands, Schneider arrived in The Hague. He did not change his tactics now that the Netherlands was under military occupation. He cautiously tried to get renowned Dutch ethnologists and folklorists on his side. Steinmetz had died in December 1940, but van der Ven and de Vries received Schneider's full attention. Whereas van der Ven resisted the idea of complete incorporation of the Dutch into the Greater Germanic Reich and soon disappeared from Schneider's orbit, de Vries did not. Initially Schneider was worried about his political orientation, but de Vries grew ever more ideologically close to Schneider, warming to the ideas promoted by Ahnenerbe. One of the foremost Dutch ethnologists and professors, de Vries became even more involved in Schneider's plans and, ultimately, by 1943 was openly associating himself with the SS.[32] Tellingly, when Allied troops entered Dutch territory in September 1944, de Vries and his wife fled by train to Germany. On their journey they were accompanied by Schneider.[33]

In May 1940, just after the Netherlands had become German-occupied territory, de Vries wrote about the feeling of relief and liberation he had experienced. The new political configuration in Europe and in the Netherlands created possibilities for a better future in which *Volk* and community would be the central notions.[34] In the years that followed he continuously published on the Germanic roots of the Dutch and Germans. In 1942 he was commissioned by the head of the Ministry of Education to revise the official schoolbook history of the Dutch people. Needless to say, the version he produced emphasized the

Germanic heritage of the Dutch nation. This and similar publications that de Vries produced between 1940 and 1945 indicate that he had come to understand the concept of *Germanic* almost exclusively in biological terms. The inner self of the Germanic man, de Vries proclaimed in one of his writings, found its expression in a strong drive for action, loyalty, and a well-developed sense of belonging to his tribe.[35]

De Vries published in various pro-Nazi newspapers, of which *Hamer* had the closest connection to Ahnenerbe. *Hamer*'s editorial board consisted of members of the Volkse Werkgemeenschap, a community of Dutch ethnologists who supported National Socialism, many of whom shared de Vries's views. For instance, his assistant Gerda Schaap wrote in 1943 that the "genuine character of the people" could be found among peasants who had from time immemorial belonged to the Nordic race (*Noordras-mens*).[36] Chief editor of *Hamer* Nico de Haas, meanwhile, suggested an official cultural policy that would be determined by race (*een rasgebonden kultuurpolitiek*). Ethnologists, he argued, should dedicate themselves to destroying "weeds" that were foreign to the Dutch people (*volksvreemd*), purifying folk customs of all features alien to the "racial soul" of the Dutch (*rasziel*), eliminating all degenerate aspects, and finally, "pruning all neglected branches."[37]

Antisemitism was equally present within the circles of admirers of Germanic folk culture. In the first issue of *Hamer*, printed in October 1940, one of the leading figures within the Volkse Werkgemeenschap, Jan Nachenius, wrote a piece titled "The Peasant and the Nomad." Although not stated explicitly, it was clear that the "peasant" referred to the Dutch and the "nomad" to the Jews. The author was making the point that the Jews differed fundamentally from the Dutch and that the two did not mix.[38] Likewise, de Haas welcomed the yellow badge enforced upon Jews in 1942. A few weeks later, after the deportations had begun, de Haas warned the ethnic Dutch who might be tempted to assist their fellow Jewish citizens by means of baptizing or marrying them that they were "dishonoring the race," "betraying the Volk," and thus deserved severe punishment.[39]

In other words, under the direct influence of German Nazism, mainstream Dutch ethnology was developing in a more explicitly racist direction. Some of the leading Dutch ethnologists, specifically Jan de Vries, publicly embraced Nazism and sided with German officials such as Schneider. The latter had come to ideologically prepare the Dutch in occupied Netherlands for effective incorporation into the Greater Germanic Reich, based primarily on racial principles. However, essential first steps had already been taken prior to the German occupation, in the interwar period, when biologic reasoning entered the Dutch academic world, an appeal for the revalidation of the nation's Germanic roots grew louder, and collaboration with German fellow scientists became ever more profound.

For pragmatic Dutch scientists, too, collaboration appeared advantageous from a career perspective. It was generally known that racial science, ethnology, and folklore enjoyed the substantial financial support of the Nazi regime. In a way, the German occupation brought the money and recognition that had previously eluded Dutch scholars. It is noteworthy that the same desire for social and professional mobility also applied to a handful of Dutch physicians who had favored eugenics and now could work with German occupation authorities, the SS in particular, to see their ideas implemented.[40]

The Germanization of the Occupied East

The conquest of Czechoslovakia and Poland in 1939 realized a long-cherished dream of the Nazis: German expansion into the eastern border regions. In this dream, military victory was only a condition, albeit crucial: the real objective was the permanent extension of the German frontiers into the East. In the occupied and partially annexed Polish territories a ruthless policy of Germanization soon ensued. According to a regime that attached much value to the racial purity of its population, permanent rule required an aggressive reshuffling of populations in favor of members of the Germanic race. This reshuffling included reducing the percentage of the so-called Polish element and the resettlement of full-blooded Germans in their stead. In his capacity as Reich commissioner for the strengthening of Germandom, Heinrich Himmler coordinated this population policy. The dramatic consequences of this policy became apparent in the new German province of the Warthegau, where thousands of Jews and Poles were expelled, crammed in ghettos, or murdered to make way for German settlers.[41]

It did not take the German authorities long to realize that enthusiasm among the Germans to move to annexed Polish territories was far lower than they had anticipated. Most of the new settlers were ethnic Germans from outside the Altreich (Germany proper) who had been more or less forced to be part of a population transfer.[42] A solution seemed at hand when the Wehrmacht invaded countries of Northern and Western Europe. Himmler praised the racial qualities allegedly exhibited by most of the inhabitants of countries such as Denmark, Norway, Luxemburg, the Netherlands, and Belgium. In a letter dated 7 January 1941 he described the Dutch as "pure-blooded Germanics" who for centuries had been alienated from their biological roots: with "a steady, yet soft hand" they now would be brought back into the fold.[43] In another letter, addressed to one of his close associates, Himmler suggested using these "by blood exceptionally highly qualified people" (*blutsmässig unerhört wertvolle Kräfte*) in the East.[44]

The reason the Nazis thought that recruitment in the Netherlands promised to be a success was not just because the Reichsführer-SS valued the Dutch more highly than, for example, the Norwegians. In equal measure it reflected his belief that the pressure to emigrate was extraordinary high in the Netherlands, supposedly because of the scarcity of land. In this overpopulated and urbanized country the Dutch peasantry had no future. Perhaps this is why Himmler assumed that Dutch farmers would be eager to emigrate.[45]

Himmler's idea of including the Dutch in the project of Germanization of the occupied eastern territories was revived in the summer of 1941, shortly after the launch of Operation Barbarossa. In the initial months of German military advance the scope of the Germanization project was expanded substantially. Various secret plans that had been circulating among the leaders of the Nazi regime from 1940 onward now seemed possible: Poland would just be the beginning. The notorious Master Plan East, for instance, encompassed the Baltic States, Ukraine, and large parts of Russia and Belorussia. The Reich Security Main Office, in charge of most of the utopian plans, prescribed the deportation of millions of Slavs to Siberia and the settlement of millions of German farmers in their stead. According to estimates, most of these plans would take at least twenty years to realize. On various occasions it was stressed that implementation should wait until the war's end.[46] Nevertheless, attempts were made to implement this grand experiment. In Ukraine, to cite just one example, Himmler devoted himself to the creation of the German agrarian colony Hegewald.[47]

By the summer of 1941 Hitler was envisaging the mass immigration to the occupied eastern territories of anywhere between 2 and 3 million people, not only from Germany but also from Scandinavia, Western Europe, and even America.[48] The Dutch would also be welcome. A few months later, in his Ukrainian headquarters Himmler received Reich Commissioner Arthur Seyss-Inquart, head of the German administration in the occupied Netherlands. The two men paid a visit to Hans Frank, the German governor general of the rump Polish state, to discuss a potential Dutch contribution to the exploitation of Polish soil.[49]

Apparently, as a result of these talks, an assistant to Seyss-Inquart, Hermann Roloff, received an assignment in November 1941 to examine the possibility of Dutch emigration to Poland and occupied territories farther east. As Seyss-Inquart's expert on population planning, Roloff supervised the prestigious "Zuiderzee project" that aimed to reclaim land in the Zuiderzee, the largest Dutch inland sea. Like Himmler, Roloff perceived the Netherlands as overpopulated and Dutch peasants as an endangered species. The reclaimed land in the Zuiderzee, 170,000 hectares in total, would offer some relief so that "a new, Germanic peasantry" (*ein neues, germanisches Bauerntum*) could be installed there. However, it would apparently not be enough to alleviate the demographic crisis. Eastward emigration would thus be necessary to solve the problem of Dutch overpopulation.[50]

According to Roloff's estimates, some one hundred thousand Dutch farmers could (read: should) leave the Netherlands and become employed in the occupied East. Less than a year later, an alternative estimate projected the figure of 3 million.[51] When exactly this massive emigration drive should start, how long it would take, or where the emigrants would go, however, remained unspecified. In January 1942 Reich Minister for the Occupied Eastern Territories Alfred Rosenberg proclaimed that "in the next five years no one was going to settle" in the territories under his rule.[52] Himmler stated a few months later that it was completely unclear when nationals other than Germans would emigrate to this part of the continent.[53] In March 1942 Himmler stipulated that even the number of temporary foreign workers should not exceed two or three thousand.[54]

Confusion reigned within Dutch circles, too. As early as July 1941, the Dutch Department of Agriculture and German Trusteeship Ostland, Ltd. (later renamed Ostdeutsche Landbewirtschaftungsgesellschaft) had reached agreement on the employment of Dutch farmers in the occupied Soviet territories. In return, the Dutch state would receive a share of the harvest, including wheat crops and fodder. A special Commission for the Employment of Farmers in Eastern Europe (Commissie tot Uitzending van Landbouwers naar Oost-Europa, or CULANO) was established to coordinate the enterprise. The commission stressed that emigration was not the aim, yet it considered permanent settlement "at a later stage highly probable." Not coincidentally, one of the three members of the commission's board was the director of the Stichting Landverhuizing, a Dutch state institution that was established in 1913 to organize Dutch emigration.[55]

The Dutch Department of Agriculture and CULANO viewed the enterprise as a strictly economic affair, even if the Dutch colonization efforts clearly had political implications. Even though the German and Dutch authorities were not, at least not yet, thinking of permanent settlement, the economic exploitation of the occupied East was intrinsically political. Furthermore, as traditional Dutch agricultural institutions refused to carry out recruitment, CULANO became largely dependent on Agrarisch Front (Agrarian Front), the Dutch Nazi agricultural organization that was led by Roskam. Inevitably, most of the recruits came from within Nazi circles.[56] Contrary to the intention of the Dutch authorities, Roskam and other Dutch Nazis made no attempt to keep the recruitment secret. Before the war the colonization of foreign territory had already been on the social agenda of the NSB. New land had been sought, not so much for trade purposes as to absorb the Dutch surplus population. For some years, New Guinea was identified as a region for potential Dutch settlement; "the white, Dutch tribe," Mussert predicted, would find its living space there.[57] The war prompted a change of plans. Now that the path to settlement in New Guinea was blocked and German troops were rapidly advancing through formerly Polish and Soviet territories, the focus shifted to Eastern Europe.

Roskam was one of the early advocates of Germanic colonization of the occupied East. He saw the salvation of the Dutch peasantry in the conquest of Eastern Europe. The vast plains of Ukraine, he believed, could offer the land that Dutch farmers so badly needed. Less than two weeks after the German attack on the Soviet Union, Roskam announced the start of a campaign to recruit farmers for emigration. In the newspaper *Nieuw Rotterdamse Courant* he emphasized that Germany was building a new "Central Europe" stretching from the Baltic to the Black Sea. Dutch farmers could easily find employment in Ukraine, famous for its "black soil" ready for cultivation. A Dutch peasants' colony, like that of the Boers in South Africa, would come into existence.[58] When the first trainload of recruits left the Netherlands in November 1941, Roskam was present at the station to wave them off.[59]

Although CULANO disapproved of Roskam's interference, it was unable to prevent the politicization of the whole enterprise.[60] By March 1942 CULANO effectively ceased its activities and was soon dissolved. Officials under Alfred Rosenberg took charge of the recruitment of Dutch workers, with the local

Nazi Meinoud Rost van Tonningen becoming their most influential Dutch ally.[61] Rost van Tonningen, who had been appointed president of the Dutch National Bank and head of the Dutch Department of Finance a year before, was fully aware of the difficult situation under which international Dutch trade had labored since the beginning of the war. The Japanese occupation of the Dutch East Indies in March 1942 had made things even more difficult. According to Rost van Tonningen, Dutch colonial splendor and economic prosperity could be restored in the occupied East. His enthusiasm for Dutch projects in the occupied East in part stemmed from his political affinity with greater Germanic thought (*De Groot Germaanse Gedachte*). An ardent racist, he shared the views of his personal friend Himmler and envisaged a larger empire under Germanic rule covering most of northern continental Europe. Germanization of the East enjoyed his full support.[62]

On Rosenberg's order, in June 1942 Rost van Tonningen established a semi-governmental agency called the Dutch East Company (Nederlandse Oost Compagnie, or NOC).[63] The NOC, often at the invitation of German local authorities, organized several fact-finding missions and drew extensive plans to develop economic projects of all sorts. Although the Dutch press kept stressing the economic importance of Dutch settlement in the occupied East, the whole enterprise now had a distinct political and racist agenda. Thus the NOC board publicly discussed and largely approved of the eradication of Jews: in July 1942 Rost van Tonningen spoke enthusiastically about the "extermination" of the Jews in Eastern Europe.[64] According to another member of the board, the Ukranian population was pleased to be rid of "Jewish influences." He added that at least 10 percent of the indigenous Slavic population had also been "removed."[65]

These and similar statements implied that there was enough room for Germanic people, including the Dutch, to settle in Eastern Europe. In an attempt to legitimize the Dutch presence in the East, Dutch Nazi propaganda paid ample attention to historical Germanic settlements in Eastern Europe. Jan de Vries, who had been promoting the idea of the Germanic roots of the Dutch nation since the 1930s, published a pamphlet in 1942 on this very matter. De Vries stated that between the fourth and the sixth century AD the Germanic tribes, which had ruled the European continent from the North Sea all the way to the Ukraine, had started to migrate to the west. As soon as they left, the land had been taken over by the Slavs. The time had come, de Vries insisted, for the Germanic people to reclaim the land that historically had been rightfully theirs:

"This is a task that concerns all Germanic people, so that a superior Germanic force will come into existence in Europe, a force that will protect our continent against the threatening Asian chaos."[66]

Experiencing Germanic Rule

Between November 1941 and July 1944 several thousand Dutch workers went to the East voluntarily. Most of them, 5,216 in total, traveled under the banner of CULANO or NOC.[67] Although the majority were employed in the agricultural sector, all kinds of other professions were present among settlers: craftsmen, fishermen, dredgers, accountants, secretaries, and car mechanics. Some tradesmen and experts in the cultivation of cash crops such as tobacco, coffee, and tea also tried their luck in the conquered territories.[68]

All of the 5,216 were employed in the territories conquered in Operation Barbarossa, mainly in Ukraine. Within Ukraine, Dutch activities were localized in and around Rivne. A group of approximately eight hundred Dutch craftsmen were housed in barracks in the city and dozens of Dutch farmers lived at a nearby farmstead in Rohachiv that was leased by the NOC.[69] Almost all recruited for agricultural work first came to Rohachiv, where they received some training before heading off to other farmsteads in Ukraine. Pockets of Dutch settlement were established elsewhere, for example, in Belorussia (around Vileyka), Lithuania (around Vilnius and Kaunas), and Latvia (around Riga).

According to the top officials, only skilled men and "idealists" should be selected for the mission. Therefore, before departure, all recruits were politically and physically screened.[70] Many of them tended to agree with the ideological dimension of their mission; the majority were members of the NSB and thus, presumably, shared the party's main tenets.[71] Some had become notorious in their Dutch hometown for their hatred of Jews or their assistance in hunting down people in hiding.[72] On the way to their new destination they sang about the liberation of Germanje, the word they used for a Greater Germanic Reich.[73] Apparently the Dutch recruits also saw themselves as soldiers who would, by their mere presence in the East, free this region from its political (i.e., Bolshevik) and ethnic (i.e., Slav and Jewish) enemies. At least some were dreaming of permanent settlement. The Dutch supervisor of a peat moor near Vilnius, for example, vividly imagined the creation of a Dutch rural community, with families living in neat brick houses surrounded by large vegetable gardens.[74]

Extensive propaganda material issued for the occasion explained how to relate to other ethnic groups in the occupied East. In one brochure member of the NOC board Henri van Maasdijk warned the recruits against "blending" with those of different blood.[75] Racist stereotypes dominated NOC descriptions of the local population. Willem Goedhuys, for example, portrayed the Ukrainians as "lazy as the Scythes," "tanned as the Indians," "cruel as the Sarmatians," and thus extremely dangerous.[76] Johannes Lingmont, the NOC representative in Vilnius, characterized the Poles as slow and untrustworthy: "They talk and boast enormously, but [when it comes to] making progress—forget it!"[77] The Lithuanians, with their "broad faces and mouths of predators," looked like animals to him.[78]

Once they arrived in the occupied East, the Dutch workers became part of a racially structured society overseen by the Germans. They witnessed the unequal treatment, indeed persecution, of Slavic and Jewish people, often to their own advantage. Near Kaunas, for example, Dutch gardeners were living on farms expropriated from Jews; thirty Russian prisoners of war stood at their disposal. At a rural estate and at several peat moors near Vilnius, Dutch supervisors made use of Jews and POWs as forced laborers.[79] In Ukraine, Dutch enterprises equally profited from the POWs' slave labor. In some localities the local population was also coerced to work.[80] It is very likely that in larger cities such as Minsk and Rivne Dutch craftsmen came to replace Jewish artisans who had been murdered.[81] Many Dutch took it for granted that, due to their supposed racial superiority, they would receive a higher salary than locals in similar positions.[82]

Available evidence includes several incidents in which Dutch workers mistreated the local population. In Lithuania, when in June 1943 six Jewish forced laborers escaped from their barracks near a peat moor, Dutch men took an active part in the hunt. They also witnessed the German reprisal four days later, when sixty-seven Jews were shot.[83] In October 1943, when Ukranian villagers did not appear for work, two Dutch supervisors entered their village and shot seven inhabitants "as a warning." One of the two Dutch men was also notorious for sexually harassing Ukranian women.[84]

While the Dutch obviously profited from their supposed racial superiority, living in the occupied East turned out to be extremely difficult. The conditions of living and working were much harder than most had expected. Equipment was often insufficient, housing extremely poor, and logistics—operating horse-driven wagons on dirt roads—were in most places problematic. As the Dutch were situated in areas infested with Soviet partisans, they were constantly living under threat of an attack. Their fear was not groundless. Just like the German settlers, the Dutch were easy targets. By the summer of 1943 fifty-five Dutch, mostly farmers, had already been killed.[85]

What made things worse for the Dutch settlers was the often cool, sometimes openly hostile, attitude of Germans toward them. The relations between Germans and Dutch in this area could be accurately described as tense. The tensions rose as soon as the first groups of approximately four hundred farmers and six hundred craftsmen arrived in the early spring of 1942. The craftsmen went directly to Rivne, while the farmers made a stop in Łódź (Litzmannstadt), where they were vaccinated and instructed on how to go about their work assignments. After weeks of waiting the Dutch farmers proceeded to Riga. It took a while before they were finally employed. The majority went to a former Soviet state farm in Vileyka that was managed by the German Trusteeship.[86]

For many of these early recruits the introduction to the "colonial" lifestyle was a bitter experience. Within a year more than one-third of the farmers had returned to the Netherlands.[87] They complained that they had not been taken care off in the transfer camps and that German supervisors had tried to push them into jobs they had not applied for.[88] Craftsmen in Rivne also held the German management responsible for their poor living conditions.[89] The German authorities were equally disappointed with the Dutch newcomers. According to some German officials, discipline was something totally lacking among the Dutch. The former noticed that in the first transfer camp in Łódź the Dutch had simply ignored the curfew and spent their time in brothels. In Riga the Dutch gained a reputation as black marketeers. The complaints about the Dutch in Rivne were similar: the German authorities claimed cases of Dutch workers selling their clothes and shoes to Ukrainians on the black market for profit.[90]

Mutual accusations and grievances continued in the years that followed.[91] NOC, worried about the friction with the German authorities, sent several reporters to the area.[92] Those reporters universally acknowledged the loose morals of quite a few Dutch "pioneers" but simultaneously suggested that the German attitude toward the pioneers was a possible cause of misbehavior. Specifically, they criticized the German management in Rivne: one could not expect the Dutch to behave in a soldierly manner commensurate with their racial superiority, one reporter stated, as long as they did not receive proper uniforms and were nicknamed "white Jews!"[93] According to one other report, the Dutch behavior was directly triggered by the Germans treating the Dutch like "a slightly better sort of natives."[94] In short, Germanic brotherhood was hard to find in the occupied East.

Originally Germanization was a policy invented, led, and executed solely by the Germans. The idea of creating living space in the eastern part of Europe had originated in Germany; the Nazi regime had given this idea its distinct racialist character. Forced population transfers, ethnic cleansing, and mass murder were the direct results of this policy. Most of the protagonists of the Germanization project were Germans: settlers, soldiers, policemen, civil servants. Yet the Dutch also became involved in the creation of a Greater Germanic Reich organized along racial lines. The German authorities regarded the Dutch as racially closely related and therefore obvious collaborators in the project. They suspected that the Dutch, Dutch peasants in particular, would also be interested in working and eventually settling in the occupied East because there was not enough arable land in their home country.

This idea was shared by men like Roskam and de Vries. In a way, they personified the direct link between certain developments in Dutch ethnological research during the prewar years and the Germanization project in the Second World War. Long before the war, they had concluded that Dutch peasants—perceived as the cultural and biological backbone of Dutch society—had hardly any future in the densely populated Netherlands. German plans aiming at the (re)creation of agrarian communal living in the East therefore seemed promising for social as well as racial reasons.

The idea of Germanic kinship by blood was something the Dutch were familiar with. Since the 1930s Dutch ethnographers and folklorists had been emphasizing the supposed Germanic roots of the Dutch people, roots that they apparently shared with their German neighbors. Scholars like de Vries had called for a reassertion of their Germanic cultural heritage. Although initially defined in terms of customs, shared experiences, and language, *Germanic* became increasingly perceived as a racial qualification. This shift was prompted by the gradual emergence of racial scientific ideas in Dutch academia. Although racial science never developed into an independent academic discipline in the Netherlands, some of its elements imperceptibly entered many other fields of study. By May 1940, when the Netherlands was occupied by Nazi Germany, scientists and scholars by and large accepted the need to define the Dutch identity in terms of race. For those in favor of Dutch participation in the Germanization project, these prewar developments offered an inspiring precedent. Being "of kindred blood," the Dutch and Germans should indeed, as de Vries asserted, stand side by side in the conquest of their "lost" territories.

Contrary to these ideas of equality, the actual relationship between the Dutch and Germans in the occupied East proved to be troublesome. Some Germans considered the Dutch to be nothing better than undisciplined economic profiteers, adventurers, and thieves. The Dutch, for their part, often felt discriminated against by the Germans. It appears that in the tough daily life of the occupied East national differences were of more significance than "racial" similarities.

Notes

Abbreviations Used in the Notes

NA — Nationaal Archief (National Archives of the Netherlands, The Hague)

NIOD — Nederlands Instituut voor Oorlogsdocumentatie, Instituut voor Oorlogs-, Holocaust- en Genocidestudies (NIOD Institute for War, Holocaust, and Genocide Studies, Amsterdam)

1. Mechtild Rössler and Sabine Schleiermacher, eds., *Der "Generalplan Ost": Hauptlinien der nationalsozialistischen Planungs- und Vernichtungspolitik* (Berlin: Akademie Verlag, 1993); Christian Gerlach, *Krieg, Ernährung, Völkermord: Forschungen zur deutschen Vernichtungspolitik im Zweiten Weltkrieg* (Hamburg: Hamburger Edition, 1998); Götz Aly, *"Final Solution": Nazi Population Policy and the Murder of the European Jews* (London: Arnold, 1999); Dirk A. Moses, ed., *Empire, Colony, Genocide* (New York: Berghahn Books, 2008); Mark Mazower, *Hitler's Empire: Nazi Rule in Occupied Europe* (London: Allen Lane, 2008).
2. Michael Burleigh, *Germany Turns Eastwards: A Study of Ostforschung in the Third Reich* (Cambridge: Cambridge University Press, 1998); Timothy Snyder, *Bloodlands: Europe between Hitler and Stalin* (New York: Basic Books, 2010).
3. Janneke van der Heide, *Darwin en de strijd om de beschaving in Nederland 1859–1909* (Amsterdam: Wereldbibliotheek, 2009).
4. Gunnar Broberg and Nils Roll-Hansen, eds., *Eugenics and the Welfare State: Sterilization Policy in Denmark, Sweden, Norway, and Finland* (East Lansing: Michigan State University Press, 1996).

5. For a comprehensive study of eugenics and euthanasia in Nazi Germany, see Michael Burleigh, *Death and Deliverance: "Euthanasia" in Germany, 1900–1945* (Cambridge: Cambridge University Press, 1995).
6 Jan Noordman, *Om de kwaliteit van het nageslacht: Eugenetica in Nederland 1900–1950* (Nijmegen, Netherlands: Sun, 1989), 93–118.
7. Noordman, *Om de kwaliteit van het nageslacht*, 27; Rob van Ginkel, *Op zoek naar eigenheid: Denkbeelden en discussies over cultuur en identiteit in Nederland* (The Hague: SDU, 1999), 73, 127–34.
8. Stephen Snelders et al., "Alcoholism and Hereditary Health in Dutch Medical Discourse, 1900–45: Biology versus Psychology in Coping with Addiction," *Social History of Alcohol and Drugs* 22, no. 2 (Spring 2008): 130–43.
9. André Köbben, "Sebald Rudolf Steinmetz (1862–1940): Een hartstochelijk geleerde," in *Brandpunt van geleerdheid in de hoofdstad: De Universiteit van Amsterdam rond 1900 in vijftien portretten*, ed. Luuc Kooijmans et al. (Amsterdam: Amsterdam University Press, 1992), 313–40.
10. Quoted in Köbben, "Sebald Rudolf Steinmetz," 328.
11. Sebald Rudolf Steinmetz, *Rassen der menschheid: Wording, strijd en toekomst* (Amsterdam: Elsevier, 1938).
12. Steinmetz, *Rassen der menschheid*, 308.
13. Sebald Rudolf Steinmetz, *Soziologie des Krieges* (Leipzig: Barth, 1929), 46. See also Bart van Heerikhuizen and Nico Wilterdik, "Conservatisme en sociologie in de jaren dertig," in *Toen & thans: De sociale wetenschappen in de jaren derig en nu*, ed. Frank Bovenkerk et al. (Baarn, Netherlands: Ambo, 1978), 182–207.
14. Köbben, "Sebald Rudolf Steinmetz," 316.
15. Steinmetz, *Inleiding tot de sociologie* (Haarlem: Bohn, 1931), 159.
16. Noordman, *Om de kwaliteit van het nageslacht*, 64–69.
17. Van Heerikhuizen and Wilterdik, "Conservatisme en sociologie"; Ineke Mok, *In de ban van het ras: Aardrijkskunde tussen wetenschap en samenleving 1876–1992* (Amsterdam: Asca Press, 1999), 59–107.
18. Barbara Henkes, *Uit liefde voor het volk: Volkskundigen op zoek naar de Nederlandse identiteit 1918–1948* (Amsterdam: Athenaeum-Polak & Van Gennep, 2005), 14, 22.
19. Henkes, *Uit liefde voor het volk*, 35–38, 102.
20. Jan de Vries, *De Germaansche oudheid* (Haarlem: Tjeenk Willink, 1930), 50.
21. Henkes, *Uit liefde voor het volk*, 24–25.
22. Barbara Henkes and Björn Rzoska, "Volkskunde und 'Volkstumspolitik' der SS in den Niederlanden: Hans Ernst Schneider und seine 'grossgermanischen' Ambitionen für den niederländischen Raum," in *Griff nach dem Westen: Die "Westforschung" der völkisch-nationalen Wissenschaften zum nordwesteuropäischen*

Raum, 1919–1960, ed. Burkhard Dietz et al. (Münster, Germany: Waxmann, 2003), 291–323.

23. De Vries, *De Germaansche oudheid*, 13.
24. Henkes, *Uit liefde voor het volk*, 151.
25. Robin te Slaa and Edwin Klijn, *De NSB: Ontstaan en opkomst van de Nationaal-Socialistische Beweging, 1931–1935* (Amsterdam: Boom, 2009), 384–86, 697–98.
26. Evert Jan Roskam, *De Nationaal socialistische beweging in Nederland in verband met ons christelijk karakter en onze Germaansche volksaard* (Utrecht: NENASU, 1934), 21.
27. These ideas can be found in various articles published in *Het Nationale Dagblad* in 1938. See Evert Jan Roskam, *De Wereldbeschouwing van den Nederlandschen boer* (Utrecht: Nenasu, 1939), 61–96.
28. Te Slaa and Klijn, *De NSB*, 701-2.
29. Peter Schöttler, "Die deutsche 'Westforschung' der 1930er Jahre zwischen 'Abwehrkampf' und territorialer Offensive," *Tijdschrift voor geschiedenis* 118, no. 2 (2005): 158–59.
30. Henkes and Rzoska, "Volkskunde," 295–96; Ton Dekker, *De Nederlandse volkskunde: De verwetenschappelijking van een emotionele belangstelling* (Amsterdam: Aksant, 2002), 203.
31. Henkes and Rzoska, "Volkskunde," 305–21; Joachim Lerchenmueller, "Hans Ernst Schneider/Hans Schwertes Niederlande-Arbeit in den 1930er bis 1950er Jahren," in Dietz et al., *Griff nach dem Westen*, 1111–40.
32. Henkes, *Uit liefde voor het volk*, 171–75.
33. Dekker, *De Nederlandse volkskunde*, 255.
34. Jan de Vries, *Naar een betere toekomst* (Amsterdam: Elsevier, 1940), 5–6.
35. Jan de Vries, "De Germaanse achtergrond van onze Kultuur," in *Het herwonnen verleden: Opstellen en voordrachten* (The Hague: De Schouw, 1944), 7–23.
36. Gerda Schaap, "Illusies van het eeuwig onveranderlijke: Folkloristen, ideologie en cultuurpolitiek," in *Volkscultuur als valkuil: Over antropologie, volkskunde en cultuurpolitiek*, ed. Rob van Ginkel (Amsterdam: Het Spinhuis, 2000), 1–32.
37. Schaap, "Illusies van het eeuwig onveranderlijke," 29.
38. Henkes and Rzoska, "Volkskunde," 318.
39. Dekker, *De Nederlandse volkskunde*, 206.
40. Stephen Snelders, "Op weg naar een 'germaansche' volksgezondheid: Nationaal-socialisme, erfelijkheidsleer en eugenetica in Nederland 1940–1945," *Gewina* 30, no. 2 (2007), 62–74.
41. Mazower, *Hitler's Empire*, 1–52, 189–98. For more details on the Warthegau, see Catherine Epstein, *Model Nazi: Arthur Greiser and the Occupation of Western Poland* (Oxford: Oxford University Press, 2010).

42. Koos Bosma, "Verbindungen zwischen Ost-und Westkolonisation," in Rössler and Schleiermacher, *Der "Generalplan Ost,"* 198–214; Mazower, *Hitler's Empire,* 216–21; Epstein, *Model Nazi,* 160–92.
43. Heinrich Himmler to Arthur Seyss-Inquart, 7 January 1941, reprinted in Nanno in 't Veld, *De SS en Nederland: Documenten uit SS-archieven 1935–1945* (The Hague: Martinus Nijhoff, 1976), 532.
44. Heinrich Himmler to Richard Hildebrandt, 7 January 1941, in In 't Veld, *De SS en Nederland,* 531.
45. In his letter to Richard Hildebrandt, Himmler explicitly mentioned the possibility of solving the overpopulation issue in the Netherlands by stimulating emigration of Dutch peasants to the East.
46. See Cezław Madajczyk, ed., *Vom Generalplan Ost zum Generalsiedlungsplan* (Munich: Saur, 1994).
47. On the Hegewald colony, see Wendy Lower, *Nazi Empire-Building and the Holocaust in Ukraine* (Chapel Hill: University of North Carolina Press, 2005), 162–79.
48. Mazower, *Hitler's Empire,* 149.
49. Bosma, "Verbindungen," 201; Peter Witte et al., ed., *Der Dienstkalender Heinrich Himmlers 1941/1942* (Hamburg: Christians, 1999), 220.
50. Bosma, "Verbindungen," 198–200.
51. Bosma, "Verbindungen," 202.
52. Hanns Albin Rauter to Himmler, 6 January 1942, in In 't Veld, *De SS en Nederland,* 624.
53. Bosma, "Verbindungen," 201.
54. Niederschrift Besprechung betr. Wirtschaftsprobleme, inbesonders über Beteiligung des Auslandes am Wiederaufbau der besetzten Ostgebiete, 20 March 1942, NIOD), 265/28.
55. Note of Franz Graf Grote, 10 December 1941, NIOD, 265/31; minutes on the first meeting of CULANO, 22 July 1941, NIOD, 120a/1.
56. Note for Hans Max Hirschfeld, 25 September 1941, NA, 2.11.07.01/12.
57. Te Slaa and Klijn, *De NSB,* 747–48.
58. "Naar Oostland willen wij varen," *Het Vaderland,* 7 July 1941.
59. "Een eerste groep boeren van ruim 100 naar Oostland," *Nieuw Rotterdamse Courant,* 23 November 1941.
60. Note of Hirschfeld, 27 February 1942, NA, 2.11.07.01/12.
61. On Rost van Tonningen, see David Barnouw, *Rost van Tonningen: Fout tot het bittere einde* (Zutphen, Netherlands: Walburg, 1994); E. Fraenkel-Verkade, ed., *Correspondentie van Mr. M. M. Rost van Tonningen,* vol. 1 (The Hague: Martinus Nijhoff, 1967).

62. Report of Meinoud Rost van Tonningen, 1 September 1943, NIOD, 176/32.
63. Report of F. B. J. Gips, 29 June 1942, in David Barnouw, ed., *Correspondentie van Mr. M. M. Rost van Tonningen*, vol. 2 (Zutphen, Netherlands: Walburg, 1993), 34–44.
64. Quotation in Barnouw, *Rost van Tonningen*, 101.
65. Quotation of Frederik de Kock van Leeuwen, 6 October 1942, NA, 2.09.09, 41591.
66. Jan de Vries, "Naar Oostland willen wij rijden," in de Vries, *Het herwonnen verleden*, 35–50.
67. From May 1943 the NOC also kept track of forced laborers working at its subsidiaries, 1,572 in total. Report of activities, NOC, first quarter of 1944, NIOD, 176/15.
68. For a personal story of a tobacco tradesman, see Maarten van Bommel, *Verborgen schaduw: De tabakshandelaar van Nikolajev* (Amsterdam: Aspekt, 2011).
69. Report on activities of the NOC in 1943 and 1944, NIOD, 176/15; lists of recruits (various dates), NIOD, 176/84.
70. Heinrich Sellmer to Arthur Seyss-Inquart, 6 July 1942, and Meinoud Rost van Tonningen to Heinrich Sellmer, 10 July 1942, in Barnouw, *Correspondentie*, 2:46–47, 52.
71. A fairly good impression of the political orientation of NOC recruits can be obtained from the approximately five hundred personal cards that were filled in by the NOC administration and are collected at the NIOD, 176/435-442.
72. See, e.g., the postwar judicial files of Pieter Grimm, Jan Habing, and Jan Vos, NA, 2.09.09/205, 2.09.09/55961, and 2.09.09/75665.
73. *Polonaises*, notes by Cornelis Paape, 12 December 1946, private archives of Casper Paape.
74. Jan Habing, 8 August 1942, NIOD, 176/707.
75. Report of Henri van Maasdijk concerning Dutch possibilities in the East, undated, NIOD, 176/132.
76. Report of Willem Goedhuys on the history and future of Ukraine, undated, NIOD, 176/595.
77. Entry in the diary of Johannes Lingmont, 15 February 1943, NIOD, 1.611/2180.
78. Entry in the diary of Johannes Lingmont, 14 March 1943, NIOD, 1.611/2180.
79. Report of Willem Frederik Gerhardt concerning the Dutch mission with regard to Baltic agriculture, 15 July 1942, NIOD, 176/597; report of Henri van Maasdijk concerning Dutch possibilities in the East, undated, NIOD, 176/132.
80. Reports of Jan Barendrecht, undated, NIOD, 176/84 and 176/598.
81. Wilhelm Kube to Meinoud Rost van Tonningen, 9 September 1942, in Barnouw, *Correspondentie*, 2:80–81. Barnouw thinks that Koch had similar reasons for requesting Dutch artisans to come to Ukraine. David Barnouw, *Oostboeren, zeegerman-*

nen en turfstekers: Kolonisatie tijdens de Tweede Wereldoorlog (Amsterdam: Bert Bakker, 2004), 143.

82. Report on economic problems, in particular the participation of foreign countries in the rebuilding of the occupied Eastern territories, 20 March 1942, NIOD, 265/28.

83. Irina Guzenberg, "The 1942 General Population Census in Lithuania: The Labor Camps of Vilnius Ghetto," in *Vilniaus getas: Kalinių sąrašai*, vol. 2, ed. Žydų Muziejus (Vilnius, Lithuania: Valstybinis Vilniaus Gaono žydų, 1998), 52. On the attitudes of the Dutch toward Jews in Lithuania, see my article, "The Dutch in the Occupied East and the Holocaust," *Yad Vashem Studies* 39, no. 2 (2012): 55–80.

84. Reports of Jan Barendrecht, undated, NIOD, 176/84 and 176/598; Police file of Renske Winkler, 20 December 1945, NA, 2.09.09/109439. I am grateful to Jocelyn Krusemeijer for this reference.

85. Report of T. van der Zee, undated, NIOD, 176/331; Report of Meinoud Rost van Tonningen, 24 October 1942, and Meinoud Rost van Tonningen to Heinrich Himmler, 27 March 1943, in Barnouw, *Correspondentie*, 2:97–99, 173–74.

86. Report of Alfred Meyer, 20 January 1942, and report of Graf Franz Grote, 10 December 1941, NIOD, 265/31; Report of Willem Hendrik van Eek, 26 October 1943, NIOD, 176/362.

87. Report of T. van der Zee, undated, NIOD, 176/331.

88. Johan Scharringa to Meinoud Rost van Tonningen, 24 November 1942, and report of Meinoud Rost van Tonningen, 15 March 1943, in Barnouw, *Correspondentie*, 2:123, 184; Johan Wouter van IJzeren (the letter *ij* is unique to the Dutch alphabet) to Nederlandse Landstand, 7 August 1942, and to his wife, 18 May 1942, NA, 2.09.09/70297; Gewestelijk Arbeidsbureau Zwolle to CULANO-head office, 5 March 1942, NIOD, 120a/51.

89. Travel report concerning talks in the Netherlands with Generalkommissar Fritz Schmidt, 2 April 1942, NIOD, 265/32; Report concerning the visit of Augustus Wilhemus Johannes Borggreven to the Ostministerium and his talk with Alfred Meyer, 20 January 1942, NIOD, 265/31; Rost van Tonningen to W. H. Nimtz, 23 June 1942, in Barnouw, *Correspondentie*, 2:30–31.

90. "Bericht über die im Büro des Werkdienstes Holland in Rowno am 10: Dezember 1942 abgehaltene Besprechung," 10 December 1942, NIOD, 176/81; "Vermerk über die Betreuung Ausländer im Zwischeneinsatzlager Litzmannstadt," 8 December 1941, NIOD, 120a/58.

91. Report of Clara Palm, 23 October 1942, NIOD, 176/ 333; Report of Jan Habing, 14 August 1943, NIOD, 176/719; Liestert to Jan Barendrecht, June 20, 1943, NIOD, 176/598.

92. Travel report of Y. Vennik, undated, NIOD, 176/569; "Bericht über die Niederländer, die aus dem Werkdienst Holland stammen und z. Zt. in Rowno tätig sind," 9 December 1943, NIOD, 176/81; Report of B. van den Tempel, undated, NIOD, 176/362; Report of Willem van Eek, 26 October 1943, NIOD, 176/362.
93. Report of T. van der Zee, 21 January 1943, NIOD, 16/138.
94. Report of M. Silvergieter-Hoogstad, undated, NIOD, 176/84.

4

Pure-Blooded Vikings and Peasants

Norwegians in the Racial Ideology of the SS

TERJE EMBERLAND

This chapter examines perceptions of Norwegians within the racial ideology of the Schutzstaffel (SS), as shaped by the "Nordic idea" of Hans Friedrich Karl Günther and Richard Walther Darré. The collective image of the "Norwegian tribe" as purebred, primeval farmers and fierce, bellicose Vikings, it argues, influenced SS policies vis-à-vis Norwegians, both before and during the German occupation of Norway. During the war this racial mythology prompted the establishment of the Lebensborn institutions in Norway, the recruitment of Norwegians into the Waffen-SS, and plans for Norwegian participation in the Germanization of the occupied East.

The existing literature on the recruitment of Norwegian volunteers into the Waffen-SS rarely posed the basic question as to *why* Heinrich Himmler and the SS leadership decided to recruit them in the first place. Most historians have assumed that it was merely a matter of securing enough military manpower for the anti-Bolshevik crusade. Indeed, the need for manpower is one of the main explanations typically given for the decision to expand the Waffen-SS—the combat troops of the SS—into a pan-Germanic army. According to this line of argumentation, enlisting foreign volunteers was a way of overcoming the restrictions on recruitment imposed on the SS by the Wehrmacht. By including ethnic Germans (*Volksdeutsche*) and racially kindred Germanics in their ranks, the Waffen-SS could tap into a human resource that was out of reach of the Wehrmacht.[1] However, there are several reasons why this particular hypothesis does not provide a sufficient explanation in the case of Norway.

When launching a recruitment campaign in Norway, SS officials expected only a small contingent of volunteers.[2] What they were seeking was a select group of racially and politically superior young men who would automatically qualify for General SS (Allgemeine-SS) membership and would serve as ideological standard-bearers upon returning home. This select group was to become what the SS called the Staatsschutzkorps (corps for the protection of the state) in the new, Nazified, Norway, modeled according to the principles of the SS.[3] That is how Head of the SS and German Police Heinrich Himmler put it in a conversation with the Oslo bishop Eivind Berggrav immediately after swearing in the first Norwegian volunteers to Division Wiking in January 1941: "Take the regiment Nordland as an example. Do you believe that we need these men as soldiers? We can do without them! But we mustn't block these men from freely pursuing their desires. I can assure you that they will return as free and committed supporters of our system."[4]

Although the need for military manpower did indeed increase over the course of the war, the SS had indicated that the recruitment effort in Norway was never primarily motivated by military expedience. In fact, Norwegians were not even expected to contribute significantly to the combat operations of the Waffen-SS.

Even assuming the most modest of expectations as to the number of potential Norwegian SS volunteers, it soon became evident that the recruitment drive in Norway was a failure. After a temporary boost following the invasion of the Soviet Union in June 1941, the numbers soon went down. In spite of renewed efforts in 1942, including the reorganization and centralization of the recruitment apparatus and improved conditions and terms for volunteers, the momentum had been lost. After visiting Norway in April 1942 the leader of the SS Main Office, Gottlob Berger, complained that "the SS idea is dead" in Norway. Yet Berger did not intend to abandon the project: the SS idea "must now be reawakened," he demanded.[5] Consequently, the SS continued its recruitment drive throughout the war, obviously regarding it to be of vital importance. In any event, the end result was meager indeed. Whereas the SS managed to recruit close to thirty thousand Dutch and ten thousand Flemish volunteers, no more than five thousand Norwegians joined the ranks—the lowest number for any of the occupied countries.[6] This disappointing result was in spite of the fact that the recruitment effort in Norway was just as intensive as it was in Flanders and the Netherlands.

This observation prompts the question as to why the SS leadership did not eventually give up. This chapter contends that the answer has to do with the racial ideas of the SS, the place Norwegians occupied within the SS worldview, and the connection between those ideas and SS policies toward Norwegians, both before and during the war.

The Nordic Idea

Evidence shows that Norwegians held the position at the pinnacle of racial superiority in SS racial ideology, superseding the Germans themselves. According to manuals used in the ideological education (*Weltanschauliche Erziehung*) of the SS, the ratio of pure Nordic blood in the German population was somewhere between 50 and 60 percent.[7] In Norway and Sweden this percentage was supposedly much higher, in excess of 70 or 80 percent.[8] These estimates came from the race researcher Hans F. K. Günther, as did many of the basic racial ideas of the SS. In fact, Günther's writings served as one of the main sources of inspiration for the racial thought of the SS. In a 1935 reading list of essential ideological literature required of the SS man, Günther's *Der nordische Gedanke unter den Deutschen* (The Nordic idea among the Germans) was listed as number four, surpassed only by Adolf Hitler's *Mein Kampf*, Alfred Rosenberg's *Der Mythos des 20. Jahrhunderts*, and the Nazi Party program.[9]

Günther idealized Norwegians: he was married to a Norwegian woman and conducted his racial research in part while living in Skien in Norway.[10] According to his book *Adel und Rasse* (Nobility and race), Norwegians had managed to preserve their pure Nordic blood better than any other ethnic group belonging to the Nordic race, largely due to their relative isolation from the continent and primitive living conditions on remote farms. Even in the current time, Günther wrote, it was easy to find simple peasants and common people in Norway who displayed pure Nordic racial features otherwise only to be found in the old, pure-blooded nobility in Germany.[11]

Some of his colleagues in Germany criticized Günther for exalting the Norwegians beyond all reason. According to them, the Norwegians had long since racially degenerated due to continuous exposure to liberal democracy and insulation from war and struggle—the prerequisite experiences of warrior nations. In addition, the Norwegians had lost much of their noble Viking spirit through mass immigration to the United States.[12] Günther fiercely denied these claims. The only reason the Norwegians displayed such a docile and pacifist posture, he argued, was because they were inherently a nation of noblemen and hence could not countenance rule over each other. "But give them a people to conquer," Günther insisted, "and they soon will show their inborn capacity for domination."[13]

Günther was a major influence on the leading ideologist of the SS in the early 1930s, Richard Walther Darré, the first head of what later became the SS Race and Settlement Main Office (RuSHA).[14] In his youth Darré, along with Himmler, had belonged to the Artaman League (Artamanen-Gesellschaft), a *völkisch* youth group committed to furthering the Nordic race by literally and figuratively returning to the land.[15] With this kind of ideological baggage, Darré further developed Günther's theory concerning the racial purity of the Norwegians and Swedes by linking it to his core concept of *Blut und Boden*. According to Darré, the ancient Germanic concept of *Odal*, the heritage farm, was the source from which the *Volk* drew its racial purity and rejuvenating strength (*Lebensquell der nordischen Rasse*). This form of agricultural organization, typical of the ancient Germanic tribes, had prevailed throughout the ages in Norway and Sweden, relatively unaffected by feudalism and later agricultural developments. That is why, Darré argued, the Norwegians and Swedes had remained the purest bearers of the Nordic race.[16]

This belief also explained why, in September 1933, following his appointment as minister of agriculture and leader of the Reich Food Estate (Reichnährstand), Darré introduced the Farm Heritage Law. This law was modeled after the Norwegian *Odelsrett* (allodial rights), as one of the first important measures for securing the necessary racial "Nordicization" (*Aufnordung*) of the German people. Thus Darré claimed that "the current, officially named, 'Odelsrett' of the Norwegians, the 'Reicherbhofgesetz' of the German people, and the 'Odal' of the Germanic farmers are laws in the spirit of one common worldview."[17]

Darré further linked the ancient Germanic *Odal* principle to the legendary Viking spirit. Throughout the ages, the rigid principles of inheritance had forced younger sons out of extended families to conquer and cultivate new land and hence had given the Nordic race, though strict selection, a natural capacity for colonization.[18] In Darré's mind, one of the main tasks of the Nazi movement was to reclaim the ancient Germanic homeland in Eastern Europe through a large-scale resettlement of farmers. Obviously, then, the Scandinavians should play an important role in this grand design. In a January 1936 speech Darré reassured his staff at the Reich Farmer Office (Reichbauernkontor) that it would not take long before Nazi Germany could acquire the necessary racial material for this endeavor: "The people of Scandinavia will, then, to a certain degree, be put at our disposal. The precious genetic material of the Scandinavian countries can be mobilized for the benefit of our people, in particular through their daughters."[19] As head of RuSHA, Darré was responsible not only for the racial screening of members of the SS but also for their ideological training. As a result, the "Nordic idea," including the idealization of the Nordic *Odal* farmer and Viking, became an integral part of SS ideology.

The SS and Norway Prior to 1940

The following two examples illustrate how this "Nordic idea" influenced the SS and Himmler's image of Norwegians in the years prior to the occupation of Norway in 1940. In the summer of 1936 Dr. Wilhelm Saure was visiting Norway as German representative to the annual meeting of the international agricultural commission.[20] Saure served as legal expert to the leader of the Reich Farmer Office and had participated in drafting the Farm Heritage Law.[21] However, it was in his capacity as a RuSHA official that he took a tour through the Gudbrandsdal Valley. His choice of travel route was not accidental. The Norwegian racial researchers Halvdan Bryn and Jon Alfred Mjøen—who to a large extent had shaped Günther's racial view of the Norwegians—concluded that it was in this particular valley that one could encounter the Nordic type in its purest form.[22] Hence Saure's expedition was a form of "racial safari," whereby he not only could study the Germanic heritage preserved nearly intact in the customs and culture of the local farmers but also observe living specimens of this almost extinct racial type.[23]

Later that year Saure presented his impressions of Norway in a lofty speech at the Reich Farmer's Convention (Reichbauerntag) in Goslar: "Since the dawn of Germanic history, these Nordic farmers have developed an unsurpassed rural culture. Confronted with these cultural treasures preserved by providence, we can only stand in awe."[24] In a private report to Darré, however, he communicated a peculiar observation that he had made during his recent visit: Norwegian farmers' daughters seemed very eager to find German husbands! Saure himself could not account for this fact, but his boss felt immediately "obliged" to report to Himmler and offer an explanation:

> Personally, I think there is quite natural a reason for this. Among all big mammals, the selection process is carried out among the males. The female finds herself instinctively drawn toward the fittest and most successful male. . . . With this in mind, the problem is that the Norwegian male no longer, in posture and lifestyle, fulfills the requirements of the Germanic Man, whereas the Norwegian female today finds real racial manhood embodied in the German men. I may be wrong, but I believe I'm on the right track. Otherwise it would be impossible to explain this striking observation, which can also be made with respect to Sweden. If these women preferred foreigners in general, Englishmen and the like, then it must be a result of emotional problems. However, since these Norwegian women only direct their interest toward Germany, I think my explanation is correct.[25]

According to Darré, the problem seemed to be that the racial appearance (*Erscheinungsbild*) of the Norwegian male—due to the lack of a contemporary martial tradition and excessive exposure to democracy and other racially alien influences—no longer matched the pure Nordic hereditary qualities (*Erbbild*) of the Norwegian woman. Hence the farmers' daughters instinctively felt drawn toward German men, who fitted better into the ideal Nordic type due to their culture and education.

In his letter to Himmler, Darré suggested that one should take advantage of this factor in order to secure good, Nordic blood for Germany. One could either send eligible young German farmers to the Norwegian valleys, he speculated, or encourage Norwegian girls to travel to Germany to work as secretaries and teachers. He informed Himmler that the Reich Food Estate already had made efforts in this direction by furthering exchange programs between young farmers from the two countries. At the annual training camp in Goslar, co-organized by the Reich Farmer Office and the Nordische Gesellschaft (Nordic Society) headed by Alfred Rosenberg, youngsters from Germany, Norway, and Sweden were brought together to learn about their (allegedly) common racial heritage.

However, this was not enough, Darré concluded. He thought it would be much better if something specific were done by the SS, adding, "But at the moment, I do not know exactly how one should consider the matter."[26]

Even if the SS, through Darré's officials, kept nurturing a close contact with the Norges Bondelag, the Norwegian farmers' interest organization, throughout the 1930s, nothing more concrete came of his proposal.[27] However, the idea had been planted, and it gained new impetus during the subsequent occupation of Norway. Beginning in 1940 Darré's suggestion of racial improvement through the use of Norwegian blood materialized in one of the biggest projects initiated by the RuSHA in Norway: Lebensborn.

In the mid-1930s the idea of securing "good blood" from Scandinavia seems to have frequently occupied the mind of Darré. In the spring of 1937 this line of thinking resulted in a proposal to allow Scandinavians into the SS ranks.[28] Reichsführer-SS Himmler apparently approved of this proposal. Since the purest Nordic blood could be found among the peasant population, it was only natural to look for suitable recruits among young Scandinavian farmers sympathetic to the "Nordic idea." Hence the task was passed on to Dr. Alfred Thoss at the Reich Farmer Office in Goslar.[29] In line with the notion of the Norwegians and Swedes being ferocious Vikings, they were to be recruited into the military branch of the SS. A few weeks after receiving his order, Thoss was able to send his first candidates to the SS-Verfügungstruppe (SS combat support force). Although their nationality was not explicitly stated, it is likely that they were recruited among the Norwegians and Swedes attending the annual training camps for farming youth in Goslar.[30]

Contrary to Darré and Thoss's expectations, however, the experiment did not prove a success. When the first two candidates appeared for the medical and racial examination, it became evident that they did not meet the standards of Himmler's elite corps. Even if their inherent racial *Erbbild* might have been excellent, their outwardly racial *Erscheinungsbild* left much to be desired: one was overweight, the other flat-footed, and both were prone to drinking. The head of the SS Personnel Main Office, SS-Gruppenführer Walter Schmitt—who was probably more of a traditional Prussian officer than a Nordic romantic—later wrote that his encounter with these two Scandinavian candidates "greatly diminished" his trust in Dr. Thoss's judgment.[31] However, this setback did not dampen the enthusiasm of Thoss. The next year he suggested the establishment of a special recruitment office for Scandinavians. This idea was probably not his own. The previous summer Thoss had had a conference with Himmler at the annual convention of the Nordische Gesellschaft in Lübeck.[32] Before their meeting Himmler had talked to several Norwegians and Swedes who were eager to promote the ss worldview in their home countries.[33] It is therefore rather likely that Himmler came up with this idea himself.

Thoss's plan was met with derision in the ss Personnel Main Office. "ss-Obersturmführer Dr. Thoss's plans appear to me totally fantastic," ss-Gruppenführer Walter Schmitt wrote, pointing out that his organization possessed neither the financial means nor the manpower required for such an enterprise. His main objection, though, was of a political nature: "My objection is also that the Germanic idea presently has not been spread widely enough for people from the North to leave their homeland and sign up for the ss in Germany. I am afraid that as a result of these measures, we are therefore only likely to get mediocre, untrustworthy elements who want to exchange their nationality for a uniform."[34]

When the matter was handed over to the discretion of the Reichführer-ss, he also appreciated the financial and organizational problems involved. Yet he must have been completely dismissive of the political objections presented by Schmitt. When the same official, a year later, repeated his arguments in connection with the admission of the Swiss Dr. Franz Riedweg into the ss, Himmler replied sternly, "Don't you think that every person of Germanic descent who comes from Sweden, Norway, Denmark, Switzerland, Flanders, or any other place and who wants to join the ss will be misunderstood by the well-wishers among his countrymen and labeled a scoundrel, traitor, opportunist, etc. by the ill-wishers?"[35]

Rather than taking such objections seriously, Himmler was now more eager than ever to realize his dream of transforming the ss into a truly Germanic order. In a speech in November 1938 he proclaimed that the ss Regiment Germania should consist only of Germanic volunteers.[36] When placed within the general context of his speech, this particular statement of the Reichführer-ss apparently had little to do with military expansion but rather was as an integral part of his racial Nordicization project. This is probably why he explained his decision as follows: "I do indeed have the intention, wherever I can, to fetch, rob, and steal Germanic blood from all over the world. The Regiment Germania has not been given its name for nothing."[37]

This last statement was a clear reference to both the idea presented to

Himmler a year earlier by Darré and the experiment conducted by Thoss: the securing of good, Nordic peasant blood for the SS through the recruitment of Scandinavians. As indicative is the fact that just before making this statement Himmler spoke of the importance of recruiting young German farmers into the SS: "The best farmers must belong to the blood and life community [*Blut-und Lebensgemeinschaft*] of the SS." If recruiting racially fit young Germanic farmers into the SS was so crucial, why limit it to Germany, particularly since Günther and Darré had both so convincingly demonstrated that the best racial stock was to be found among the *Odal* farmers of Norway and Sweden?

The SS in Norway during the German Occupation

During the German occupation of Norway from April 1940 to May 1945, the SS came to exercise greater influence than in any other Western European country under Nazi rule, probably with the exception of the Netherlands. This was mainly due to the establishment of a German civil administration under Reich Commissioner Josef Terboven. Among his main responsibilities was to secure order and stability and to "win the Norwegian people" for the Führer. This goal was to be achieved through the establishment of a functioning, collaborationist Norwegian government and the subsequent handover of the powers of the German civil administration in Norway to a plenipotentiary (*Reichsbevollmächtiger*) of the German Reich.

To manage his assignment efficiently, Terboven needed to secure executive power outside the Wehrmacht; hence his cooperation with the SS. This power configuration put the Reichführer-SS in a position where he could demand large concessions, including the establishment of the office of a higher SS and police leader (Höhere SS-und Polizeiführer, or HSSPF) in Norway under his direct control. Furthermore, Himmler insisted that his influence should not be limited to security and police matters but should also include population policies (*volkstumspolitische Fragen*), in line with his ideologically motivated interest in Norwegians. Anticipating his appointment in Norway would be of short duration, Terboven had few problems accepting these demands. As a result, the SS immediately established the HSSPF Office, subsequently placing two SS regiments and six Order Police battalions in Norway.[38] With a ratio of 1 soldier to 831 civilians, this constituted the heaviest presence of German Order Police in any Nazi-occupied country. In addition, Himmler was able to place his men in central positions both within Terboven's administration and as advisers to the Norwegian fascist Nasjonal Samling party under Vidkun Quisling.

Due to the resistance offered by the Norwegian king, government, and political establishment (the king and the government escaped to England), it soon became clear that the Office of the Reich Commissioner had to become a permanent solution. Much to his dismay, Terboven was ordered by Hitler to stay put in Norway, where he was forced to rule with the SS as an ever more powerful and independent partner.

Establishing Lebensborn in Norway

Himmler efficiently used his power to implement measures deemed relevant to the improvement of the Nordic race. Among the first issues on his agenda was the fate of the children born from liaisons between German soldiers and Norwegian women. As early as May 1940, one month after German troops occupied Norway, this matter came up in a discussion between Reich Minister of Health Leonardo Conti and head of the Lebensborn, SS-Standartenführer Max Sollmann.[39] In Norway Himmler intended to use the Lebensborn—which he had established in 1935 to provide maternity homes and financial assistance to the wives of SS members—for his project of "stealing good Germanic blood." He therefore ordered the SS and Police Court Nord to sort out the legal formalities connected to the future citizenship of these children. Himmler took up this issue with Terboven during his visit to Norway in January and February 1941. During that meeting the Reichführer-SS expressed his particular interest in "the expected increase of good, Nordic blood" and emphasized that care for the mothers and their offspring was the responsibility of the Lebensborn.[40] Terboven had no objections, effectively granting the SS full control in this matter. Hence the first Lebensborn branch office outside Germany was established. Very soon, under the auspices of the RuSHA official Dr. Gustav Richert, Norway had ten different Lebensborn institutions—more than any other occupied country—including birth clinics, mother and children's homes, and schools for women who had married German soldiers.[41]

The recruitment of Norwegians into the Waffen-SS, which commenced in January 1941, was also racially motivated and vital to Himmler's plans for Norway as an important part of the Greater Germanic Reich. The overarching idea was that the volunteers would undergo thorough racial selection, receive intensive political and military training, and obtain firsthand combat experience before returning home to take up central positions in the government, the Nasjonal Samling, and the police force. This comprehensive training would secure the gradual transformation of Norwegian society in accordance with SS ideals and interests. In short, the Norwegian Waffen-SS volunteers were to form the core of a fanatical Staatsschutzkorps loyal to the SS.

The history of the SS in Norway gives ample examples of how the mythology of the racially pure Vikings greatly influenced the SS perception of the country and its people. Since SS racial ideology idealized the "Norwegian tribe," it was expected that the local population would instantly as well as wholeheartedly embrace the Greater Germanic Reich idea. As a consequence, it would mean, among other things, an influx of volunteers into the Waffen-SS. When this did not happen, the SS desperately sought explanations, albeit without challenging the basic tenets of their racial worldview. In 1942 Himmler attributed the failure of the Waffen-SS recruitment in Germanic countries not to local resistance to the SS ideology and propaganda but to the tactless treatment that the first volunteers had received at the hands of German commanding officers.[42] In an effort to address the purported cause of the problem, the head of the SS Main Office, Berger, ordered officers supervising Germanic volunteers to attend a special training seminar. At the seminar the officers learned that Norwegians and Swedes, unlike the Germans, Danes, Dutch, and Flemish, were free and stubborn "*Odal* farmers" who were not used to the collective and disciplined life of the village community. When dealing with the Norwegians, the training material instructed, one had to take into account their individualism and treat their tribal peculiarity with respect, especially since they had "preserved an extraordinary purity in their blood."[43]

Recruiting Norwegian Warrior Peasants

The mainstream literature on Norwegian volunteers mentions only in passing the promise given to them by the SS regarding the possibility of acquiring a farm in the East. Altogether, this particular aspect of the recruitment of Germanics into the Waffen-SS has never been systematically researched. It is nevertheless clear that SS recruitment of Germanic volunteers in general and Norwegians in particular was intimately connected with Himmler's vision concerning the prospective Germanization of the East—the great project he had taken upon himself as Reich commissioner for the strengthening of Germandom. Just as the Waffen-SS was to be a pan-Germanic army under German leadership, so, too, would the postwar colonization of the East be a pan-Germanic enterprise under German control.

In order to comprehend the specific plans that Himmler had for the Norwegian volunteers, one has once again to turn to the racial discourse of Darré, who linked the Germanic *Odal* principle to the mythic Viking spirit. The legend of the ancient Germanic peasant warrior shaped Himmler's idea of *Wehrbauern*: after the end of the war, the best of the Waffen-SS soldiers were to become armed farmers supervising fortified estates in the borderlands. Himmler had envisaged that these volunteers—selected on the basis of their superior racial qualities, trained by the SS, and hardened by combat experiences—would become part of a racial and military elite, a new Teutonic Order protecting the borders of the Reich from the "Asiatic hordes." To believe Günther and Darré, the Norwegians were then among the most racially suited for this task. They even had a historic claim in this colonization project, Himmler believed. In ancient times the Norse Vikings had colonized and ruled large parts of Russia; hence their participation in the imminent military campaign against the Soviet Union was just as much a "homecoming" for them as it was for the German soldiers.

Himmler wasted no time trying to enlist the Norwegians in this project. During his first visit to Norway in January 1941 he informed his old acquaintance the Norwegian businessman Olav Willy Fermann that he had handpicked him for a special mission that would render a great service to his home country. Namely, Fermann was asked to plan for and establish a Norwegian settlement in the occupied East; he responded enthusiastically.[44] However, the elitist ethos of the SS demanded that anyone who was to receive an important commission first had to prove his commitment to Germanic ideals by partaking in the business of racial warfare. Therefore, if Fermann was to play a prominent role in the future SS administration of the occupied eastern territories, he first would have to earn his stripes. Consequently, he was sent to Berlin for "special training" (*Sonderaussbildung*) and, immediately after Operation Barbarossa commenced, ordered to Himmler's headquarters in Arys.[45] Here he composed a report titled "Planung und Aufbau im Osten" (Planning and reconstruction in the East), which most likely dealt with the recruitment of Norwegians into the Waffen-SS as part of the drive to establish colonies in the occupied East.[46]

Having been promoted to the rank of SS-Untersturmführer, Fermann accompanied Himmler on a trip to Ukraine and the General Government of Poland in late September and early October 1941. Even though the Reichsführer-SS was very busy supervising the murderous work of the mobile killing units, the Einsatzgruppen, he also found time to discuss the Norwegian colonization scheme with Fermann.[47] In Lublin they met with Konrad Meyer, whose interlocutors told Fermann that his commission was to establish and run a colony of one hundred Norwegian farmers in the Lublin district.[48]

Efforts to recruit Norwegian farmers for the colonization project were also made in Norway. While Himmler was visiting Norway in January 1941, the local newspapers reported that a delegation of forty-two young Norwegian farmers was invited to Germany.[49] The purpose of the trip was to acquaint them with German farming practices and agricultural development. The trip was arranged jointly by the Reich commissioner in Norway, Terboven, and the leader of the Reich Farmer Office, Darré. During his stay in Oslo Himmler instructed the RuSHA head in Norway, Gustav Richert, to revise the travel agenda as to make it fit better with the objectives of the SS.[50] Consequently, when the young farmers arrived in Germany in early February 1941 their travel agenda included not only tours of model farms but also visits to the headquarters of Waffen-SS division Leibstandarte Adolf Hitler in Berlin and the grave of the legendary medieval king Henry the Fowler in the village of Quedlinburg, which Himmler had turned into an SS shrine.[51]

At Quedinburg the delegation received a gala treatment: to greet them came none other than SS-Brigadeführer Otto Hofmann of RuSHA. A report to Himmler assessed the visiting farmers with respect to their racial qualities and potential fighting abilities. The report also noted that they were much impressed by Hofmann's speech. While some guests inquired about the possibility of volunteering for the Waffen-SS, all of them wanted to know whether Norwegians would be permitted to take part in the colonization of the East—something that Hofmann was able to instantly confirm.[52]

Richert considered the exchange program to have been a success, and later that year he asked Hofmann's permission to send two new delegations to Germany. Hofmann replied, "I have conferred with the Reichsführer, and he regards it as imperative to send as many of these Norwegians as possible to Germany."[53] As a consequence, eight young Norwegian farmers were sent to a Waffen-SS training camp and later pursued their agricultural education at the University of Posen, paid for by the SS Main Office. During the summer break some of these students worked in Lublin, preparing the ground for a Germanic settlement in that particular region.[54]

Fermann soon withdrew from the colonization project, apparently due to his shocking experience witnessing SS atrocities on the Eastern Front, while some of the students in Posen later joined the Waffen-SS.[55] Those who had experienced life in Lublin under the leadership of the infamous SS and police leader Odilo Globocnik, however, were not very eager to accept the offer serving as colonists in the East. Regardless, Himmler personally decided to continue financing their education: after all, they were Norwegians![56]

The year 1942 marked the culmination of Himmler's effort to gain political dominance in the occupied Germanic countries. Having concluded successful negotiations with Nazi Party Chancellor Martin Bormann over a division of power and with National Treasurer Franz Xavier Schwarz over financial support, Gottlob Berger and the SS Main Office received more or less free rein to implement their project of reshaping occupied Germanic countries in accordance with the SS vision.[57] Their goal was no less than the immediate realization of the Greater Germanic Reich, as Peter Longerich has concluded in his recent biography on Himmler.[58] Himmler had reached an important milestone in this project when in August 1942 he received a formal commission from Hitler granting him exclusive executive powers over so-called Germanic work in occupied Western Europe.

Since Norway, on racial grounds, was ascribed a prominent place in Himmler's vision for the Greater Germanic Reich, the year 1942 also marked the high point in the SS struggle to take control over Norway: this was the year when the recruitment apparatus of the Waffen-SS in Norway was reorganized through the establishment of the Germanic Central Office (Germanische Leitstelle), when the Germanske SS Norge—the Norwegian branch of the Allgemeine-SS—was restructured, and when the SS Ahnenerbe branch office was established.[59] Finally, it was the same year the drive to enlist Norwegians in the Germanization campaign in the East was publicly launched though a lecture in Oslo by Konrad Meyer.

Himmler's newly established control over "Germanic work" also included power over national youth organizations. It was mainly here, among the young and malleable, that Himmler sought to recruit and indoctrinate his future Germanic peasant warriors. An important first step was to establish the Germanic Land Service (Germanische Landdienst).[60] Through this organization several thousand boys and girls between the ages of fourteen and seventeen from occupied Germanic countries had been sent to the Warthegau in Poland to be trained as prospective settlers in the East. Among these trainees were more than three hundred Norwegians—the second-largest national contingent.[61] As a token of the importance that Himmler ascribed to the Norwegians in this scheme, the Norwegian minister of sports and labor service and the youth leader Axel Stang was put in charge of the whole project, including recruiting and training youths from the Netherlands, Flanders, Denmark, and Sweden.[62]

Having concluded their basic education in the Warthegau, boys were supposed to undergo a thorough military and ideological training in the Waffen-SS. To achieve this objective, the SS set up premilitary training camps (*Wehrertüchtigungslager*) where youth could be groomed for future military service.[63] Only after volunteering for the Waffen-SS and acquiring fighting skills through service at the front would they become full-fledged *Wehrbauern*—peasant warriors—ready to be deployed (along with their wives, trained in the Germanic Land Service) in the fortified settlements in the East. Due to the development at the front, Himmler's grand Germanization project in the occupied East was put on hold. Consequently, Norwegian participation in this scheme came to naught too.

Nazi ideology was an inconsistent admixture of elitism and populism, as Christopher M. Hutton has pointed out, wherein the main fault line ran between elitist racialism on one side and more "egalitarian" glorification of the German *Volksgemeinschaft* on the other.[64] From this perspective, the SS ideology can be placed within the racial and elitist extreme. The idealization of the Nordic race made Himmler emphasize the *Blutsgemeinschaft*, that is, the community of blood, rather than the *Volksgemeinschaft*. Consequently, his ultimate goal was not just a strong and territorially enlarged Germany but a Greater Germanic Reich encompassing all the "tribes" belonging to the Nordic-Germanic race. Since this vision could not be backed by reference to history and/or deep-rooted nationalist sentiments, SS ideology acquired strong utopian and eschatological characteristics.

Until recently, historians have been loath to examine SS ideology.[65] This has only gradually changed, and there remains the tendency to focus solely on the genocidal aspects.[66] This chapter has attempted to demonstrate that looking from such a narrow perspective on SS racial thinking would render many of their policies in Norway simply incomprehensible. One can make sense of those priorities only by developing a wide-ranging perspective on SS ideology, including the cult of Nordic Man and the utopia of a pan-Germanic Reich. In short, research on the SS must focus not only on those ethnic groups that occupied the lowest place in the Nazi racial hierarchy but also on those who were placed at the top.

Notes

Abbreviations Used in the Notes

BAB	Bundesarchiv Berlin (German Federal Archives, Berlin)
RAO	Riksarkivet (Norwegian State Archives, Oslo)
RGVA	Rossiiskii Gosudarstvennyi Voennyi Arkhiv (Russian State Military Archives, Moscow)
SSO	SS-Offizier Personalakte (SS Officer file)

1. Bernd Wegner, *Hitlers politische Soldaten: Die Waffen-SS 1933–1945—Leitbild Struktur und Funktion einer nationalsozialistischen Elite* (Paderborn: Ferdinand Schöningh, 2006), 274.
2. Berger to Himmler, 7 August 1940, copy of a document presented at the Nuremberg Military Tribunal, RAO, PA Ingebriktsen. According to Berger, "with regard to Danes and Norwegians the annual quote amounts to 2 percent, or 1,234, of the total number of military-aged men."
3. Mark P. Gingerich, *Toward a Brotherhood of Arms, 1940–1945* (Madison: University of Wisconsin Press, 1991), 92.
4. Eivind Berggrav, *Da kampen kom* (Oslo: Land og kirke, 1945), 178.
5. Berger to Himmler, "Germanischer Arbeit," 29 May 1942, U.S. National Archives and Records Administration, T-175/126/2651914-5, copy at the Center for the Study of the Holocaust and Religious Minorities in Oslo, PA Skilbred.
6. The number five thousand is based on the estimate given by Gunnar Sverresson Sjåstad. See Sjåstad, "Nordmenn i tysk krigsinnsats: En kvantitativ undersøkelse av frontkjemperne under den andre verdenskrig" (master's thesis, University of Bergen, 2006).
7. Der Reichsführer-SS Hauptamt, *Rassenpolitik* (Berlin: Der Reichsführer-SS/SS Hauptamt, [1942]), 19.
8. Der Reichsführer-SS/SS Hauptamt, *SS-Mann und Blutsfrage* (Berlin: Der Reichsführer-SS/SS Hauptamt, 1941), 13; Der Reichsführer-SS/SS Hauptamt, *Rassenpolitik*, 17–19.

9. RuSHA, *Verzeichnis wertvoller Bücher für den SS-Mann* (Berlin: RuSHA, 1935).
10. Hans-Jürgen Lutzhöft, *Der nordische Gedanke in Deutschland 1920–1940* (Stuttgart: E. Klett, 1971).
11. Hans F. K. Günther, *Rasse und Stil* (Munich: J. F. Lehmanns Verlag, 1926), 18; Günther, *Adel und Rasse* (Munich: J. F. Lehmanns Verlag, 1927), 78–79.
12. Lutzhöft, *Der nordische Gedanke*, 344–47. For the idea of Nordicism in German racial anthropology, see Christopher M. Hutton, *Race and the Third Reich: Linguistics, Racial Anthropology and Genetics in the Dialectic of "Volk"* (Cambridge MA: Polity Press, 2005), 119–28.
13. Hans F. K. Günther, *Der nordische Gedanke unter den Deutschen* (Munich: J. F. Lehmanns Verlag, 1925), 89.
14. In the preface to his *Das Bauerntum als Lebensquell der nordischen Rasse*, Darré portrayed Günther as the foremost expert on racial matters. Furthermore, Darré asked Günther to comment on the manuscript. Günther is also frequently cited in his book. On the relation between Günther and Darré, see Anna Bramwell, *Blood and Soil: Richard Walther Darré and Hitler's "Green Party"* (London: Kensal Press, 1985).
15. Michael H. Kater, "Die Artamanen: -Völkische Jugend in der Weimarer Republik," *Historische Zeitschrift* 213 (1971): 576–638.
16. Richard Walther Darré, *Das Bauerntum als Lebensquell der nordischen Rasse* (1929; repr., Munich: J. F. Lehmanns Verlag, 1938), 84.
17. Richard Walther Darré, "Unsere Weg," *Odal*, 1 April 1934, reproduced in Darré, *Um Blut und Boden* (Munich: J. F. Lehmanns Verlag, 1940), 103. Darré initiated historical research to support this theory. See Christian Sigmundt aus Achen, "Rechtsgewinnung und Erbhofrecht: Eine Analyse der Methoden in Wissenschaft und Rechtsprägung des Reichserbhofrechts" (PhD diss., University of Munich, 2005), 120.
18. Darré, *Das Bauerntum*, 367.
19. Quoted in Andrea d'Onofrio, "Rassenzucht und Lebensraum: Zwei Grundlagen im Blut-und Boden-Gedanken von Richard Walther Darré," *Zeitschrift für Geschichtswissenschaft* 49, no. 1 (2001): 141–57.
20. "Det internasjonale landbruksmøte avsluttes på Lillehammer," *Nationen*, 30 July 1936.
21. In an interview with the Norwegian author Barbra Ring, this "wise young SS leader" explained how he was inspired by the Norwegian *Odelsrett* when drafting the German legislation. Ring, "Slekten og garden," *Nationen*, 5 September 1936.
22. Jon Alfred Mjøen, "Rassentypen und Rassenmischung in Norwegen," *Der Norden* 6 (September 1934), 175.
23. In January–February 1941 Himmler took exactly the same route during his visit

to Norway. "Program für Besuch des Reichsführer SS in Norwegen," Norwegian Resistance Museum, PA Skodvin. His trip included a stop at the ancient Tofte farm, "which, with its people, lifestyle, and buildings, is characteristic of the ancestral home of the Germanic race in central Norway." *Deutsche Zeitung in Norwegen*, 4 February 1941.

24. Wilhelm Saure, "Demokratie als System zur Vernichtung des Bauerntums," in *Der 4. Reichbauerntag in Goslar vom 22.–29. November 1936* (Berlin: Reichnährstand Verlag, 1936), 176.
25. Darré to Himmler, 18 October 1936, German Federal Archives in Koblenz, NL 94, Walter Darré, II/54, copy at RAO, PA Ingebriktsen.
26. Sigurd Bjoner, "Inntrykk fra ungdomsturen i Tyskland," *Nationen*, 11 December 1936; Norwegian translation of a letter from Alfred Thoss to Olga Bjoner of Norges Bondelag, the Norwegian farmers' organization, October 1936, Archives of the Norwegian Farmers' Organization in Oslo, N 44. Nordische Gesellschaft (Nordic Society) was an association founded in 1921 in Lübeck, Germany, with the objective of strengthening German-Nordic cultural and political cooperation. The association had both German and Scandinavian members. After the Nazi takeover Nordische Gesellschaft came under the control of Alfred Rosenberg and became closely associated with the "Nordic idea" of H. F. K. Günther. Rosenberg intended to use the Nordische Gesellschaft to further the Nazi cause in the Nordic countries. Walther Darré and later Himmler both became board members.
27. The Norwegian farmers' association was established in 1896 as Norges Landmandsforbund. The leadership in the organization was influenced by *völkisch* ideas and hence sympathetic to the *Blut-und-Boden* ideology of Darré and the Hitler regime during the 1930s.
28. Darré to Himmler, 18 May 1937, quoted in Bramwell, *Blood and Soil*, 72.
29. Alfred Thoss's CV, 3 January 1934, BAB), SSO, PA Alfred Thoss.
30. The recruits to the SS combat support force had to be young and unmarried. Among the participants at the youth training camp in Goslar a majority would meet these criteria.
31. SS Personnel Main Office, statement, 9 June 1938, BAB, SSO, PA Alfred Thoss.
32. The meeting took place on 20 June 1937 at 4:30 p.m. Himmler's diary, entry from 20 June 1937, RGVA, 1372k/5/218.
33. Among the people present at the meeting were Hans S. Jacobsen, the editor of the radical, neo-pagan, pro-Nazi magazine *Ragnarok*, and Olaf Willy Fermann, a Norwegian businessman based in Germany and longtime friend of Himmler.
34. SS Personnel Main Office, statement, 9 June 1938, BAB, SSO, PA Alfred Thoss.
35. Himmler to Schmitt, 30 June 1938, BAB, SSO, PA Franz Riedweg.

36. "Rede vor den SS-Gruppenführern zu einer Gruppenführerbesprechung im Führerheim der SS-Standarte 'Deutschland,'" 8 November 1938, reproduced in Bradley F. Smith and Agnes F. Peterson, eds., *Heinrich Himmler: Geheimreden 1933 bis 1945 und andere Ansprachen* (Frankfurt: Propylän Verlag, 1974), 25–49.
37. "Rede vor den SS-Gruppenführern,'" 8 November 1938, in Smith and Peterson, *Heinrich Himmler*, 38.
38. K. G. Klietmann, *Die Waffen-SS: Eine Dokumentation* (Osnabrück, Germany: Verlag "Der Freiwillige," 1965), 348–49; Head of Order Police in Norway, "Organisation, Ausrüstung und Arbeit der Ordnungspolizei seit Beginn des Krieges," 16 May 1940, BAB, R19/395/138.
39. Kåre Olsen, *Vater: Deutscher* (Frankfurt: Campus, 2002), 23.
40. "Aktenvermerk über eine Besprechung betr. die Behandlung der Frage 'unheliches' Kind und Errichtung von Lebensborn-Heimen in Norwegen," undated, BAB, R2/11470. Kåre Olsen has assumed that the meeting took place in February 1941. *Vater*, 28. Himmler's appointment calendar for this period gives no hint of such a meeting in Berlin. During his time in Norway, however, there was ample time for a conference on the Lebensborn. The presence of personnel from the Reich Commissioner's Office also indicates that the meeting took place during Himmler's visit to Norway.
41. For an account of the Lebensborn in Norway, see Olsen, *Vater*.
42. Himmler to Berger and Jüttner, 13 April 1942, BAB, NS19/2305, NS19/3522.
43. "Richtlinien für die Behandlung germanischer SS-Freiwilliger," especially the section "Richtlinien zur Behandlung von Freiwilligen aus dem Norden," undated, BAB, NS31/455.
44. Testimony of Herbert Noot, leader of Department III of the SD Office in Oslo, 27 January 1948, RAO, L-sak Fermann.
45. Fermann's statement, RAO, L-sak Fermann.
46. Himmler's Office, K. H. Bürger to Fermann, 16 October 1941, RAO, L-sak Fermann. Unfortunately the report itself is missing from Fermann's file.
47. Fermann to Vidkun Quisling, 27 September 1941, RAO, L-sak Fermann.
48. "Rettsbok," doc. no. 156, RAO, L-sak Fermann.
49. "Representanter for norske bonder til Tyskland," *Aftenposten*, 31 January 1941.
50. Hofmann to HSSPF's South, Center, and Fulda-Werra, 4 February 1941, BAB, NS19/521. Hofmann stated, "During the visit of the Norwegians, the Reichsführer-SS wants the young farmers to see some of the SS institutions and reserves for himself the right to address them." However, Himmler had to cancel the meeting due to a tight schedule. Himmler's Office to Hofmann, 2 April 1941, BAB, NS 19/521.
51. "Bericht über Besuch der norwegischen Jungbauern an der Königsgruft zu Quedlinburg am 23.2.41," 25 February 1941, BAB, NS19/123/20-22. Henry the Fowler

(Heinrich der Vogler, 875–936) is widely considered to be the founder and first king of the medieval German state. He was revered for leading the German tribes to victory against invading Magyar forces, subduing the northern Slavic tribes, and pacifying the Danes, whom he made German subjects. Himmler spoke of Henry as the epitome of Germanic valor and swore to continue his mission in the East; there are even suggestions that he saw himself as the German king's reincarnation. See Nicholas Goodrick-Clarke, *Black Sun: Aryan Cults, Esoteric Nazism, and the Politics of Identity* (New York: New York University Press, 2003), 124–25.

52. "Bericht über Besuch der norwegischen Jungbauern an der Königsgruft zu Quedlinburg am 23.2.41," 25 February 1941, BAB, NS19/123/20-22.
53. Hofmann to Richert, 8 January 1942, BAB, NS19/521/93.
54. Richert's report, 12 January 1943, BAB, OSS, PA Gustav Richert.
55. Note by Bjørn Østring on conversation with Fermann's widow, undated manuscript, Box Fermann, RAO, PA Alfa-bibliotek.
56. Correspondence between Himmler, Hofmann, Richert, and HSSPF Rediess, BAB, NS 19/521.
57. Harold Skilbred, "The SS and 'Germanic' Fascism during World War II" (PhD diss., University of California at Berkeley, 1974), 202; Hans-Dietrich Loock, "Zur 'Grossgermanischen Politik' des Drittten Reiches," *Vierteljahrshefte für Zeitgeschichte* 8, no. 1 (1960): 59.
58. Peter Longerich, *Heinrich Himmler: Biographie* (Munich: Siedler, 2008), 658.
59. Malte Gasche, "Norge og prosjektet Germanische Wissenschaftseinzats: Hans Schwalm og Ahnenerbes fiasko i Norge 1942–1944," in *Drømmen om Germania: Fra nordensvermeri til SS-arkeologi*, ed. Terje Emberland and Jorunn S. Fure (Oslo: Humanist, 2009), 202–26.
60. Micael Buddrus, *Totale Erziehung für den totalen Krieg: Hitlerjugend und nationalsozialistische Jugendpolitik* (Munich: K. G. Saur, 2003), 727–28.
61. Dr. R. Gaensechke, "Germanische Jugend in neuen Europa," *Deutsche Monatschrift in Norwegen* 8–9 (August–September 1943), 7.
62. Gaensechke, "Germanische Jugend in neuen Europa." For the recruitment of Danes into the Germanic Land Service, see Kikkel Kirkebæk, *Beredt for Danmark: Nationalsocialistisk Ungdom 1932–1945* (Copenhagen: Høst og Søn, 2004), 234n.
63. Einar Rustad, "Germansk landtjeneste" (lecture, Samlerforeningen Militaria, Oslo, 25 March 1992).
64. Hutton, *Race and the Third Reich*, 15.
65. Two pioneers who deviated from this tradition were Hans Buchheim and Hans-Dietrich Loock. See Buchheim, "Die SS in der Verfassung des Dritten Reiches," *Vierteljahrshefte für Zeitgeschichte* 2 (1955): 127–57; Loock, "Zur 'Grossgermanischen

Politik,'" 57–63. Cf. biographies of Himmler by Josef Ackermann, *Himmler als Ideologe* (Göttingen: Musterschmidt, 1973), and Longerich, *Heinrich Himmler*.

66. This is certainly the case with the recent anthology *Die SS: Himmler und die Wewelsbrug*, edited by Jan Erik Schulte (Paderborn, Germany: Ferdinand Schöningh, 2009). Ruth Bettina Birn's contribution in the volume mainly focuses on the SS treatment of so-called *Fremdvölkischen*.

5

"Nordic-Germanic" Dreams and National Realities

A Case Study of the Danish Region of Sønderjylland, 1933–1945

STEFFEN WERTHER

When on 9 April 1940, German troops marched into Denmark they were enthusiastically greeted by members of south Denmark's German minority, many of whom sympathized with the Nazi regime in Germany. Local Germans believed that the long-awaited revision of the Danish-German border was at hand. They hoped that the border, which had been shifted after the First World War, would be now returned to its original position and North Schleswig (known in Denmark as Sønderjylland) would once again become part of the German Reich. However, the German minority's hopes proved in vain. The Nazi leadership in Berlin had more far-reaching foreign policy goals. In particular, Reichsführer-SS Heinrich Himmler was obsessed with the idea of redrawing the map of Europe in order to unite the allegedly blood-related (*blutsverwandten*) Germanic peoples within a Greater Germanic Reich (*grossgermanischen Reich*).[1] The occupation of Denmark and Norway, Himmler hoped, was a stepping-stone toward making this vision a reality.

The Nazi goal was actually not an extension of German borders but Denmark's voluntary incorporation into the Third Reich. In this regard the demands of south Denmark's German minority and of its branch of the Nazi Party, the NSDAP-N (Nationalsozialistische Deutsche Arbeiter Partei-Nordschleswig), were more of a hindrance than a help because they caused strife between Danes and Germans.[2] Therefore Berlin demanded that the NSDAP-N and its leader, Jens Möller, show great restraint when it came to the border issue. Indeed, the German minority was directed to behave as a sort of mediator. Its mission was to improve understanding between Danes and Germans. This meant that the German minority had to perform a complete volte-face when it came to its goals and policies.

In its attempt to implement its "Greater Germanic" ideology, the Schutzstaffel (SS) also drew upon the support of Denmark's National Socialist Workers' Party (Danmarks Nationalsocialistiske Arbejder Parti, or DNSAP). The Nazi invasion placed the Danish Nazi Party's members in a difficult position. The DNSAP's leader, the south Danish physician Frits Clausen, admired Hitler. Yet he was also an ardent Danish nationalist who had opposed the NSDAP-N's demands for a border revision for many years. Collaboration with the German authorities offered Clausen and his small Danish party definite opportunities but also involved dangers. On the one hand, the German authorities might help Clausen gain significant political power. On the other, working with the Germans also meant risking being dismissed as a traitor by the Danish population. Consequently, Clausen's cooperation with the SS proved conflict ridden, not least because Himmler's supranational "Greater Germanic" ambitions clashed with the former's hopes to reclaim Danish sovereignty.

This chapter will open with an overview of the SS's "Germanic" policy. It will then look specifically at the Danish region of Nordschleswig/Sønderjylland.[3] A case study of this southernmost part of Denmark, which until 1920 was part of the German Reich, is significant not only because of the presence of a sizable German minority but also because this region was the organizational hub of the Danish Nazi Party.[4] The SS, which defined itself as both Nordic and Germanic, had to deal with two separate groups that were at once sympathetic to Nazism and anchored in Denmark as Germany's collaborationist partners, yet with a history of mutual antagonism dating back to the beginning of the 1930s. The leaders of the SS, and in particular the leaders of the two local Nazi parties, were faced with difficult political and ideological choices.

This chapter will examine the ensuing conflicts, ingrained in south Denmark's competing and sometimes overlapping national, racial, and in part regional concepts of identity. It will also look at how Danish Nazis and the *Volksdeutsche* responded to the "Greater Germanic idea" both before and during the German occupation. Their ambivalent reactions to SS demands suggest that the SS racial ideology could not effectively erase the nationalist passions that separated these otherwise "blood-related" peoples. Rather, the SS's supranational Greater Germanic vision entangled both collaborating parties in a political dilemma, since national identity was a fixed constituent in party members' self-conception.

The "Germanic" Politics of the SS

Although today the SS's racial ideology is primarily associated with degradation and racial exclusion leading ultimately to the genocide of the Jews, in the context of North Schleswig it also involved lesser-known, inclusive, and idealizing elements of SS racism, which were applied to the allegedly blood-related Germanic peoples of north and west Europe. The alleged existence of a Nordic race and a strictly Germanic space was an intrinsic part of Nazi ideology. In his capacity as head of the German police and the SS, Himmler sought to make his organization into a consistent exponent of "Germanic" politics, defining the SS as "an Order of Nordic-destined men" (*Orden nordisch bestimmter Männer*).

The construction of the "Nordic race"—supposedly superior to others in terms of character and physical prowess—can be roughly situated in the first decades of the twentieth century. One of the dominant ideas was the fear that the "Nordic race" faced the prospect of extinction or, at the very least, degeneration. Racial hygiene was thus perceived as a crucial weapon in the survival of the race. Considered the main exponent of Nazi "Nordic thought," Hans F. K. Günther belonged among those who believed that the "Nordic race" was endangered.[5] He advocated an "all-Nordic movement" (*allnordische Bewegung*), a sort of "blond International," to revive the threatened "Great Race."[6] Beginning in 1922 Günther published a series of treatises, which created a foundation for the SS's racial ideology.[7]

On 11 November 1938 Himmler gave a speech to SS officers in which he explicitly referred to his "Greater Germanic" ambition: "All good blood in the world, all Germanic blood that is not on the German side, might at some time be our ruin. . . . I really intend to gather Germanic blood in the whole world, to rob and to steal. . . . The Germanic Reich is the homeland of the Nordic blood, and that will be the strongest magnet with which one can draw this blood here. . . . What lies ahead for Germany is either the Greater Germanic empire, or oblivion."[8]

As far as "Germanic" work was concerned, during the Second World War Himmler attempted to centralize power in the SS, in competition with other Nazi leaders and offices such as the Foreign Office. In 1942 Hitler authorized Himmler to start building volunteer Waffen-SS legions in the countries of "kindred blood" (*artverwandten Blutes*). In the summer of that same year Himmler was given the exclusive right to enter into "negotiations with all Germanic-völkisch groups in Denmark, Norway, Belgium, and the Netherlands."[9] With this permission, the SS won what amounted to a monopoly on "Germanic work," even though the specific range of responsibilities remained vaguely defined, fueling conflict between the SS and other Nazi agencies. Himmler, for his part, considered himself commissioned by Hitler to "promote the concept of a Germanic Empire" and to amalgamate (*verschmelzen*) the "Germanic" peoples.[10] Gottlob Berger stressed that the "Germanic" countries should be drawn closely to Germany "first and foremost for racial reasons, and only then for territorial reasons."[11]

The Waffen-SS was supposed to play the central role in this process of amalgamation. It was in the Waffen-SS that the vanguard of a "Greater Germanic" movement was to be educated and it was here that allegiance to the cause would be won. Those who sought to become members of the Waffen-SS had to be more than just "Aryan"; they had to prove their racial credentials as "pure Nordic" or at least "predominantly Nordic and 'Phalian' [*fälisch*]."[12]

Just days after the occupation of Denmark and Norway, Hitler sanctioned the formation of a new Waffen-SS unit with the evocative name Nordland. This unit was to consist primarily of Danish and Norwegian volunteers. Himmler intended to "educate these volunteers from Denmark and Norway to be self-conscious *Germanen* and committed proponents of both the National Socialist worldview and the Germanic imperial ideal."[13] In September 1940 the SS Division Wiking came into existence in order to merge volunteers recruited from the various "Germanic" nations into one major unit with the purpose of promoting the resurrection of a "pan-Germanic spirit." Nonetheless, following the attack on the Soviet Union in the summer of 1941 Himmler permitted the creation of exclusive national SS legions such as Freikorps Danmark.[14] The constitution of these national units showed that Himmler had to forego parts of his original

Germanization plan. A majority of "Germanic" volunteers preferred joining SS legions that reflected their national identity—something that a supranational unit could not claim. After 1941 the Waffen-SS followed a dual policy: mobilizing the "pan-Germanic" Division Wiking with its integrating aims on the one hand and the national SS legions on the other.[15]

In Himmler's view, the national legions were compromised by their association with local nationalist movements. Indeed, the SS and the national Nazi leaders differed significantly over the political purpose of these military units. Himmler saw the use of nationalist symbols as a tactical concession, necessary to maintain support in occupied countries. In the long run, however, he preferred a pan-Germanic unit à la Division Wiking.[16] Various national Nazi leaders, by contrast, sought control over the legions, linking them to their respective parties. Himmler was determined, though, neither to allow the legions to be used as local Nazi parties' private armies nor to diminish his freedom of action in a postwar "Germanic Order." Given these contradictory ideas, it is hardly surprising that the SS leadership and the national Nazi movements distrusted each other. The local collaborators advanced national agendas that had very little in common with Himmler's ideas of a Greater Germanic Reich, a fact that informed the 1942 proposal to dissolve the national legions and return to the Germanization ideology that had framed the Wiking Division. At about the same time the so-called Germanic SS—an equivalent of the *Allgemeine-SS* in Germany proper—had been introduced in certain countries with the purpose of creating a less nationalist and more manageable alternative meant to rival the various collaborationist parties and their increasingly troublesome leaders.[17]

The German Minority and the "Greater Germanic Idea"

In 1920, in accordance with the Versailles Treaty, a referendum was held to determine where the new border between Germany and Denmark was to be drawn. Three-quarters of the inhabitants of Sønderjylland/Nordschleswig had voted to join Denmark, projecting a strong Danish identity. The remainder (about thirty thousand), who identified themselves as German and had voted to remain in Germany, found themselves in a minority in the postwar Kingdom of Denmark. The German minority's organizations set as their foremost goals the preservation of their German culture, the continual ownership of their land (*Boden*), and a revision of the border. (The third demand was also the most important one.) Hitler's seizure of power in Germany was followed by a Nazi takeover of minority organizations throughout Europe. Under the leadership of veterinarian Jens Möller, the NSDAP-N became the German minority's sole political party.[18] Predictably, political developments in the German Reich nurtured the party's hope for an early revision of the border.

The German Nazi Party had supported *Volksdeutsche* revisionist demands in Poland and Czechoslovakia. However, it was rather different in North Schleswig, where the German authorities seemed determined to silence this kind of sentiment from the 1930s onward. One possible reason for this policy might be the abovementioned vision of a Greater Germanic Reich, a vision that demanded that the German minority cooperate with, not confront, the Danes. Following the Nazi occupation, German minority leaders were expected to demonstrate absolute obedience to the interests of the Reich, which entailed introducing a distinct "Germanic" overtone into their internal propaganda, limited cooperation with the Danish Nazi Party, and the renunciation of the public demand for a border revision.[19] After all, the SS saw the ethnic German population of North Schleswig as a kind of ideological bridge to the rest of Scandinavia, the so-called Germanic North.

This long-term vision was not well received by the German minority's politicians. Having spent the past twenty-odd years vilifying the Danes as "national adversaries," they could not easily recast their fellow countrymen as a "Germanic brother people" (*germanisches Brudervolk*). Indeed, a majority of ethnic Germans in North Schleswig bluntly refused to cooperate with the Danish Nazis, not least because the latter strongly opposed a border revision. In addition, German minority leaders voiced concern about the fact that after 1932 some *Volksdeutsche* were increasingly attracted to Frits Clausen's DNSAP, partly because a minority Nazi party had come into existence only in the spring of 1933.[20] It had therefore been necessary to brand the DNSAP "national-Danish" and to forbid members of the German minority to join it.[21] German minority newspapers continuously stressed the DNSAP's "national-Danish" bias and its negative attitude toward border revision in order to dampen *Volksdeutsche* enthusiasm.[22]

Despite this, there remained a small number of ethnic Germans who were willing to accept the "Nordic idea"—as promoted south of the border—at face value and thus advocated cooperation with the Danish Nazis. These groups and individuals stood in opposition to the German minority's NSDAP-N, which they accused of failing to understand the idea behind National Socialism and of refusing to relinquish outmoded nationalist thinking. In 1935, for example, Emil Seidelmann formed a Standarte Nordschleswig, which was open only to "those with Danish passports who belong to the German cultural community [*Kulturgemeinschaft*]."[23] His organization subsequently joined the Danish DNSAP as a self-contained entity.[24]

Although political renegades like Seidelmann were few, NSDAP-N leaders nevertheless viewed them with great concern. They were attacking one of the NSDAP-N's most obvious weak spots, namely the dissonance between its nationalist politics and the "Greater Germanic idea." For example, Christian Paysen, like Seidelmann, sharply criticized the NSDAP-N's attacks on "our North Schleswig countryman Dr. Frits Clausen."[25] "North Schleswig," he maintained, "is a community of blood [*Blutsgemeinschaft*]. It is also a community of fate [*Schicksalsgemeinschaft*], even if at present the population, blinded by democratic hatred and the experience of surviving national imperialism, is resisting this community of fate."[26]

Originally the concept of *Schicksalsgemeinschaft* appeared as a popular Nazi propaganda slogan. Hitler, for example, used it as early as 15 July 1932, in a speech before the German federal election.[27] Moreover, since the early 1930s Nazi chief ideologist Alfred Rosenberg had applied the term to describe German-Scandinavian relations and their supposedly shared racial foundation. After the occupation of Denmark and Norway he continued to speak about the "Nordic" and "Germanic" *Schicksalsgemeinschaft*.[28] The SS adopted Rosenberg's concept of a community of fate as a natural and indissoluble bond binding all "Germanic" peoples.

Both Clausen and Möller read their own regional agendas into the idea of a community of blood, interpreting it as legitimizing their nationalist demands. Paysen, by contrast, viewed North Schleswig as a "model of the Greater Germanic community of fate."[29] Even though his rhetoric brought Paysen into line with official SS ideology, he was eventually reined in at Möller's instigation. In this particular case SS leaders acted pragmatically. Indeed, they had little use for those who sowed unrest within the German minority while a recruitment drive for the German Waffen-SS was in progress.[30] The SS tried hard to attract volunteers to the Waffen-SS across all of Denmark, including the areas settled by ethnic Germans. However, in contrast to the German minorities of Eastern and Southern Europe, in North Schleswig the Nazis appealed to

Volksdeutsche racial, not national, sentiments. In a confidential note dated 16 May 1939 the German Foreign Office related that the SS recruitment efforts in North Schleswig "have nothing to do with national issues [*Volkstumsfragen*]. Rather, the Reichsführer-SS seeks to win exceptionally pure Nordic types, in terms of character and race, for the Waffen-SS."[31]

Accordingly, for a long time the SS authorities refrained from differentiating between "real" Danes and those belonging to the German minority. This led to conflict, for example, when members of the German minority (with Danish citizenship) were assigned to the SS Freikorps Danmark. To serve under the command of Danish officers, under the Danish flag, was nothing short of humiliation. Furthermore, whenever *Volksdeutsche* parents received an official letter from the SS, written in Danish, notifying them that their son had fallen in the battle "for Denmark's future and honor," protests came flooding in.[32] Ordinary *Volksdeutsche* refused to be "made" into Danes, or Germanics, arguing that they were Germans and, therefore, refused to fight for Germany as anybody but Germans. Designated as Volksgruppenführer of the German minority, Jens Möller reproached the SS for training ethnic German SS volunteers alongside Danish volunteers as well as for conducting training in the Danish language and "in the spirit of Frits Clausen's thinking." Disagreeing on principle, he made a forthright claim: "The mission . . . of the Greater Germanic idea . . . can only be achieved as long as the ethnic group [*Volksgruppe*] remains a close unit."[33]

The antagonism between the German minority leadership and the SS was based on more than a mere disagreement concerning a border revision. The truth was that during the recruitment campaign in North Schleswig the SS had shown little psychological finesse. In May 1940 SS-Standartenführer Johann Hinrich Möller, who handled the recruitment of Danish volunteers on behalf of Himmler, went so far as to ask Volksgruppenführer Jens Möller to introduce him to the latter's archrival, Frits Clausen.[34] Furthermore, the SS required that the German minority find "accommodation with Frits Clausen's Danish National Socialist Workers' Party and abandon hate-generated ideas, as far as these ideas are instantly directed against the Danish people."[35]

This hard-line attitude on the part of the SS gave Jens Möller the impression that that organization was not only working against him but in fact was actively seeking to "realize the Greater Germanic ideal in such a way in Denmark that belonging to Germandom is secondary to a commitment, exacted equally from Germans and Danes, to a National Socialist Germanic community of fate."

According to Möller, the SS was demanding of the German minority that it "increasingly abandon its closed *völkisch* unity and fuse together . . . with those Danish forces whose goal, based on the principle of shared blood, is . . . to spiritually unite with the German people and merge into the Greater Germanic space."[36]

In spite of the intensity with which Möller articulated his fears, his words amounted to nothing more than desperate attempts to mollify his own people, whose expectations and hopes had been thwarted. The enormous difference in the power exercised by the SS and by the NSDAP-N meant that German minority leaders were often obliged to include the SS-popularized "Greater Germanic idea" in their own propaganda. They sometimes resorted to ingenious arguments in an attempt to link "Germanic ideology" to their own demand for a border revision. For example, an NSDAP-N educational dispatch, which was supposed to introduce members to and bring them into line with the new "Germanic" ideological course, vehemently assaulted the obvious incompatibility between the fight for a "German homeland" (i.e., border revision) and the "construction of the Greater Germanic Reich."[37] Rather, the dispatch contended, "the former task is a precondition for accomplishing the latter." The necessity of a border revision was justified by the fact that the "Schleswig issue" had been poisoning relations between Germans and Danes. Until this issue was resolved—it necessarily had to be a "German solution"—it would be impossible to make Scandinavians believe in the promise of the "Germanic community." Otherwise, North Schleswig had always been a "bridge between the Germanic South and the Germanic North," the dispatch continued. It could reclaim its role, but only after the border had been revised. The German minority's *völkisch* work thus was "to a great extent a service to Adolf Hitler's Greater Germanic Reich."[38]

The next NSDAP-N educational dispatch went so far as to exploit the "Greater Germanic idea" to advance the German minority's claim to the entire North Schleswig borderland. This logical leap was accomplished by drawing a distinction between *Volk* and *Rasse*. Thus, the dispatch declared, "The theory of German and Danish blood is a fable. There is no such thing as German and Danish blood. We are racially alike and therefore equal. *This equality thus applies to the race, but not to the* Volk. As a *Volk* we are superior to the Danes [emphasis in the original]." This German superiority was also supposed to define the NSDAP-N's relationship to the DNSAP. Since the German nation was "the leading force in the new Greater Germanic space," it was absurd gathering behind Frits Clausen. Leadership did not belong to the majority but to those who were the fittest in the *völkisch* sense. Finally, despite earlier claims to the existence of *völkisch* differentiation, the dispatch aimed to win the Danes to Germandom: "Basically, every person native to Schleswig can be won over to our side."[39] This kind of reasoning was expected to deflect anticipated protests by NSDAP-N members against the new course imposed by the SS. At the same time, it also meant to placate the SS; after all, the propaganda had mentioned the "Germanic idea," albeit in a modified form. Nonetheless, the explicit claim that not all of the "Germanic peoples" were equal and the designation of the Germans as the leading people (*Führungsvolk*) constituted a potential threat to Himmler's ambitions in Denmark, not least because the SS was trying to convince Danish Nazis to endorse a common future as a "Germanic" society based on "racial equality." The depiction of the Danes as a kind of highly ranked vassal nation scarcely tallied with this vision. Yet it was tolerated, it seems, in spite of some rather weak criticism, some of which was on display in subsequent educational dispatches.[40]

Frits Clausen, the DNSAP, and the "Greater Germanic Idea"

The DNSAP was founded in 1930 by Cay Lembcke. Under the leadership of Frits Clausen, who took over as leader in 1933 and decisively influenced the party's political orientation, the DNSAP developed from "a sect to a mini-party."[41] In 1939 the party acquired three seats in the Danish parliament. Clausen's life story may help to explain his critical attitude toward the "Greater Germanic idea." Clausen came from North Schleswig, which at the time of his birth belonged to the German Reich. Until 1920, when the border was revised, he was formally a German citizen. Yet he was brought up as a "Dane"; he had openly professed his Danish loyalties while still a schoolboy. His forced recruitment into the German army in 1914 (his refusal to enter the military academy earned him the label of "fanatic Dane," according to his personal file) reinforced his rejection of all things German.[42] After returning from Russian captivity in 1918, he was active in the border referendum, campaigning for a pro-Denmark vote and for Schleswig's return to the Kingdom of Denmark. The 1920 reunification of North Schleswig with Denmark did not entirely satisfy Clausen, who had hoped that reunification would encompass all of Schleswig, not just its northern part.

Clausen acquired a physician's degree and subsequently set up a medical practice in the south Danish town of Bovrup. In 1931 he gave up his membership in the Danish Conservative Party in order to join the DNSAP. Once a member, he assumed leadership—with great success—of the party's Sønderjylland division and in 1933 replaced Lembcke as national party leader. Under Clausen's leadership the DNSAP took on a more nationalist outlook, including loyalty to king and fatherland, the idealization of Danish history, and the extensive use of national symbols. Clausen's earlier involvement in the so-called referendum battle of 1920 helps to explain his subsequent distrust of the SS's policy. He possibly still remembered Prussia's Germanization campaign of 1864–1919 and feared that the actual goal of the much-publicized "Greater Germanic Reich" was to make Danes into Germans.[43]

Sønderjylland was the DNSAP's organizational hub. It was from Sønderjylland that a majority of party members and sympathizers hailed. Since Clausen lived and worked there, he was well versed in and willing to battle with the realities of local German-Danish relations. He was aware of the paradox inherent in the DNSAP defining itself as a Danish national party yet espousing an ideology invented by Denmark's reviled and powerful southern neighbor; indeed, he knew full well that this contradiction in terms offered his political adversaries an excellent vantage point for attack. In order to justify the DNSAP's oft-criticized use of German symbols, for instance, the party organ *National Socialisten* maintained that the swastika was actually an ancient Danish symbol.[44] In a similar vein, he maintained that the Nazi salute was an ancient Nordic greeting.[45]

Clausen tried to simultaneously blunt and make use of Sønderjylland's German-Danish polarization. For instance, he published several pamphlets in which he referred to Günther's race theory. Danish and German were, according to Clausen, not opposites: they were merely different expressions of the same blood.[46] The border between the two nations ran "through the individual person," he wrote.[47] In the early 1930s he promoted an "absorption theory," according to which all those who remained loyal to Germany should also be recruited to the Danish Nazi Party.[48] Once the success of his recruitment efforts was, at least in a few cases, made public, protests streamed in from all political camps. The German minority fiercely attacked him, and so did those Danish parties that wanted to exploit a golden opportunity to cast doubts on his national loyalty—an accusation that hit hard, particularly in Sønderjylland. Although Clausen was able to fend off such accusations by referring to his patriotic past, the charge of national betrayal could still be invoked against him, especially after the German occupation in 1940.[49] He never managed to rid himself of this stigma, in spite of his efforts to argue for a nationalist line while collaborating with Nazi Germany.

By 1935, after German minority leaders had founded their own political party, it was patently useless to make further attempts at "absorption." Clausen opposed the NSDAP-N's demands for border revision and accused the party of making illicit, "imperialist" use of that Nazi idea.[50] After the war he credited himself with having prevented any revision of the border, relating with a certain pride that the German minority leader Jens Möller had called the Danish Nazi Party the most dangerous enemy of North Schleswig's Germandom.[51] In 1940, after Germany had unexpectedly occupied Denmark, the DNSAP reiterated accusations that the government had followed a defeatist defense policy. Clausen was nonetheless willing to collaborate with the German occupation authorities in the hope of gaining the latter's backing for his bid to take over the Danish government. Despite this, Clausen had initially refused the SS request to start recruiting volunteers to the regiment Nordland from the ranks of the DNSAP. Not until Clausen was shown an order signed by Hitler himself did he acquiesce to the recruiting campaign, albeit with certain reservations. He stipulated, among other things, that those recruited should remain under his direct control. After the war, Clausen wrote that he had feared that if the SS had recruited only from the German minority, a border revision might have become a reality.[52]

It may have been intimated to Clausen that the future status of both the DNSAP and Denmark would be decisively influenced by the number of volunteers provided to the SS.[53] His cooperation in the subsequent recruitment campaign seems to indicate that he gave at least some credence to these German hints. The Nazi invasion of the Soviet Union in the summer of 1941 provided Clausen with the opportunity to form a purely Danish volunteer legion and by so doing gain Germany's goodwill and simultaneously lay the foundation for a postwar Nazi Danish army. The eventual permission from both the Danish government and the king to form the SS Freikorps Danmark was hailed as a triumph of political will.[54] Participation in "the anti-Bolshevik crusade" was also seen as service to the Danish fatherland and a guarantee of national and military sovereignty. The underlying hope—reinforced by a promise from the SS leadership—was for Denmark to secure its place in Hitler's New Europe by fighting alongside other Germanic peoples on the Eastern Front.[55] In this respect the DNSAP—generally ostracized as traitorous by the Danish population—perceived itself, ironically, as a defender of Danish national interests.[56]

Himmler and Gottlob Berger, by contrast, saw the national legions as nothing more than a temporary compromise. Indeed, the SS increasingly viewed the DNSAP's nationalist agenda as a problem. At the same time, most ordinary Danes regarded members of the DNSAP as "traitors of the fatherland," further illuminating the political quandary in which Clausen found himself. Clausen's nationalist rhetoric brought him into conflict with the SS, while his political concessions to Germany made the Danes reject him. This predicament led, in turn, to additional criticism from SS leaders, who blamed Clausen for the poor results in recruiting volunteers beyond the DNSAP's ranks. By the end of 1942 German authorities began viewing the DNSAP primarily as a means of exerting leverage on other Danish parties; the DNSAP no longer enjoyed the status of a political alternative.[57] Consequently, the SS moved toward the creation of a "Greater Germanic" organization, the so-called Germanic Corps (also known as the Germanic SS), with recruitment efforts directed specifically at current and past members of the DNSAP.

The Germanic Corps (later renamed the Schalburg Corps) was inaugurated on 1 April 1943. It was at this moment that Clausen first publicly defied the SS. He told his party members that the DNSAP was "fighting for the Greater Germanic community [but] on a national [Danish] foundation" and this was why the Germanic Corps had to be established outside the party. Using the opportunity, he went ahead and renounced economic support from the German Reich, ostensibly in order to preserve the DNSAP as a "purely Danish" party.[58] However, Clausen's opposition proved short-lived. The DNSAP started losing hundreds of members each month. As neither the German Foreign Office nor the SS supported Clausen while violent assaults on Danish Nazis became commonplace, Clausen realized that he was jeopardizing his position as party leader. In the fall of 1943 he explicitly urged party members to join the Schalburg Corps and announced that he would volunteer to serve on the Eastern Front.[59] After his return in May 1944 Clausen resigned from his position as party leader. Nonetheless, the DNSAP—by then politically irrelevant—survived, under another leader, until the end of the war.

Both the DNSAP and NSDAP-N had hoped to recruit members not only among each other's ranks but also among individuals who were nationally indifferent, the so-called *blakkede* (literally "flecked"). The parties legitimated their attempts by referring to supranational race ideology (which they rejected, for various reasons, in most other contexts) and by emphasizing the overriding regional identity of the Schleswig population. Clausen and Jens Möller shared a consensus that native Schleswigians made up a single group of the same "blood admixture" (*Blutgemischs*). However, while Clausen was convinced that all Schleswigians (including those from south of the Danish border) were, deep down, Danish, German minority leaders were equally certain that they could awaken the slumbering German in every Schleswigian who swore loyalty to Denmark. Both sides chose to ignore the "scientific" contradiction inherent in the idea of a uniform "Schleswigian blood" coexisting with some sort of (variously defined) Danish or German kernel.

As far as SS policies were concerned, both Clausen and Möller played for time. They did not, to be sure, publicly reject SS ambitions for a "Greater Germanic space," yet they tried to delay it until the distant future. For example, the German minority leader Möller maintained in the beginning of 1942 that one must bear in mind that the *Volksdeutsche* could not abandon the "struggle of the peoples" (*Volkstumskampf*) in the borderlands until the Danish majority had started thinking in a "Greater Germanic way," knowing full well that this goal could not be realized within the foreseeable future.[60] Almost simultaneously, Clausen tried to place a similar brake on the "Greater Germanic" efforts of the SS by pointing out that the "fusing of the German lands into a unitary state has taken more than six decades and it would, therefore . . . be a mistake to try, today, to force through in a few months in the North what would have taken years in Germany."[61]

In the end, neither party leader saw fit to erase the distinction between Danish and German in favor of a pan-Germanic consciousness (*gesamtgermanischen Bewusstseins*). Clausen was cautious not to relinquish Danish sovereignty, while Möller was reluctant for fear that this would forestall border revision, effectively leaving North Schleswig a part of Denmark. A further parallel can be found in a position DNSAP and NSDAP-N had adopted toward SS recruitment campaigns. Both parties believed, at least until 1942, that their willing cooperation and the recruitment of the highest possible number of volunteers from their respective ranks would give them a decisive advantage over their rival in the postwar era. For the German minority, such an advantage would entail a border revision; for the DNSAP, an independent Danish state under Clausen's leadership. The SS, however, was silent on this point.

The "Greater Germanic idea" found few followers in Nordschleswig/Sønderjylland, even among local Nazis, not least because the history of the region had been shaped by national conflicts. It did not help that the SS authorities attempted to impose this ideology in blithe disregard of regional peculiarities. On this account, Himmler dramatically overestimated the attractiveness of his vision of a Greater Germanic Reich. He also underestimated the problems that the collaborationist parties' nationalism would eventually pose. In the wake of the occupation both the DNSAP and the NSDAP-N found themselves in a difficult situation. Neither party wanted to abandon its nationalist agenda. However, their ostensible helplessness vis-à-vis the SS prevented them from exerting much pressure on it. As a result, both the NSDAP-N and DNSAP had to recalibrate their strategy. They borrowed those elements of the *völkisch*, or racial Greater Germanic, repertoire that they found useful to their own cause, strengthening them through arguments meant to placate both the SS and their own followers. The resulting—inevitable—contradictions were simply ignored.

As early as the 1930s, the political conflict between the Danish and German Nazis of Sønderjylland broke out into the open. The conflict arose primarily from the border issue, which made any further agreement impossible. Moreover, both parties were competing for the so-called undecided (*blakkede*) while trying to "convert" members of the other camp to their own "nationality." Clausen initially espoused the "Nordic-Germanic idea," as it was promoted by Denmark's southern neighbor. The concept served as legitimation for his attempts to poach recruits from the German minority as well as a bulwark against demands for a border revision. In the early 1930s Clausen's "absorption method" actually showed some success; some pro-German individuals crossed over to the Danish DNSAP. However, to Clausen's dismay, the efforts to recruit members from the local German community had a negative impact on the DNSAP's reputation within Denmark, and it was accused of having become "German." This proved a recurrent problem; the DNSAP was constantly called upon to defend its "national reputation." After 1940 the reinforced emphasis on its nationalist credentials raised the hackles of the German occupation authorities. As a result, most Danes saw the party as overly German, while the Germans accused the party of being excessively Danish. This dilemma had previously limited the party's size; after 1940 it surely contributed to its ultimate decline.

Meanwhile, the German minority in Denmark had already made up its mind about "Nordic-Germanic thought" as promoted in the 1930s south of the border: it was a potential threat to the minority's *völkisch* national identity. Not in a position to openly reject it, the German minority was able to subvert the original meaning of "Nordic-Germanic thought."[62] Furthermore, German minority leaders stressed that, even if Germans and Danes were racially equal, Germans were superior as a *Volk*. Following the occupation by Hitler's Germany, these points were further elaborated in the NSDAP-N's educational dispatches. These internal documents on the one hand provided defensive arguments against unwelcome SS demands and on the other aimed to reassure party members, who were deeply disappointed that a border revision had never taken place.[63]

While the leaders of both the Danish and the German minority Nazi Parties were busy strengthening their members' *völkisch* identity within Denmark, Himmler planned to make the SS into a "Nordic-Germanic" organization. This effort reduced both Danish and *Volksdeutsche* collaboration partners to their supposed Germanic essence, something that was vehemently resisted by both groups. The *Volksdeutsche* protested against the SS "making" them into either Danes or Germanics, which they saw as threatening their *völkisch* identity. Meanwhile, the Danish Nazis insisted on Danish national sovereignty, fearing that Denmark would otherwise end up as a province in a German-ruled "Greater Germanic Reich."

The SS wished to use southern Denmark as Germany's bridge to the north, in disregard of local political and cultural contexts. The "Germanic idea" was supposed to bring together the two "blood-related" peoples who had been fighting over various issues, thus sending a wider message across Scandinavia. This ideal scenario assigned to the German minority the role of a mediator, a role to which it was neither suited nor inclined. The SS's emphasis on the shared blood and history of the "Germanic peoples" would supposedly reawaken the two groups' dormant pan-Germanic consciousness. This reclaimed consciousness was, in turn, to function as a superstructure, creating a common racial identity and simultaneously serving as a demarcation against non-Germans. In the end, though, the concept of a common Germanic ancestry served as nothing but the pretext for gathering all "Germanic" states within the new Greater Germanic Empire.

Notes

Abbreviations Used in the Notes

LAÅ	Landsarkivet for Sønderjylland (Provincial Archives of Southern Jutland, Aabenraa, Denmark)
RA	Rigsarkivet (Danish National Archives, Copenhagen)
PKB	*Den parlamentariske kommissions beretning*, 14 vols. (Copenhagen: J. H. Schultz, 1945–55)
VOMI	Hauptamt Volksdeutsche Mittelstelle (Central Agency for Ethnic Germans)

1. The Nazis considered as "Germanic" the following countries: Denmark, Norway, the Netherlands, the Flemish part of Belgium, Luxemburg, Sweden, Iceland, and Switzerland. The distinction between the terms *Germanic* (*germanisch*) and *German* (*deutsch*) is not obvious in English. *Germanic* usually refers to the ancient past and/or a group of tribes, while *German* refers to contemporary Germany and its people. From the second half of the nineteenth century onward the Germanic tribes were increasingly seen as the ancestors of the contemporary Germans. With Scandinavian mythology incorporated into the nationalist German narrative, Scandinavians were declared "blood-related" peoples (*blutsverwandt*).
2. In 1939 the NSDAP-N received 15.9 percent of the vote, or 15,500 votes, in Sønderjylland.
3. The two different names—North Schleswig in German and South Jutland in Danish—stress the complexity of German-Danish relations in this region.
4. In the 1935 parliamentary elections the DNSAP received 3.6 percent of the total vote in Sønderjylland and 0.6 percent in the rest of the country (sixteen thousand votes in total). Four years later it received 4.3 percent and 1.3 percent of the votes (thirty-one thousand votes in total), respectively. Malene Djursaa, *DNSAP: Danske Nazister 1930–45* (Copenhagen: Gyldendal, 1981), 2:59; John T. Lauridsen, *Dansk Nazisme 1930–45: Og derefter* (Copenhagen: Gyldendal, 2002), 29. In 1939 and 1943 the DNSAP had around five thousand and twenty thousand members, respectively. A total of forty thousand Danes were members of the DNSAP between 1930 and 1945.
5. The term *Race Nordique* was coined in 1900 by the French anthropologist Joseph Deniker and subsequently popularized by the American Madison Grant.
6. Hans-Jürgen Lutzhöft, *Der nordische Gedanke in Deutschland 1920–40* (Stuttgart: Ernst Klett Verlag, 1971), 259; Nazi propaganda was never consistent when talking about the "Aryan," "Nordic," "Germanic," "Indo-Germanic," or "Nordic-Germanic"

races. Within the ss the words *Germanic* and *Nordic* were often used interchangeably, referring to both a race and a people. Allan Lund, *Germanenideologie im Nationalsozialismus: Zur Rezeption der "Germania" des Tacitus im "Dritten Reich"* (Heidelberg: Universitätsverlag C. Winter, 1995), 72, 75.

7. Hans F. K. Günther, *Rassenkunde des deutschen Volkes* (Munich: Lehmann, 1922); Günther, *Der nordische Gedanke unter den Deutschen* (Munich: Lehmann, 1925); Günther, *Herkunft und Rassengeschichte der Germanen* (Munich: Lehmann, 1935).
8. Quoted in Bradley F. Smith and Agnes F. Peterson, eds., *Heinrich Himmler: Geheimreden 1933 bis 1945 und andere Ansprachen* (Berlin: Ullstein, 1974) 38n.
9. Martin Bormann's decree, 12 August 1942, quoted in N. K. C. A. In't Veld, *De ss en Nederland: Documenten uit ss-Archieven 1935–1945*, (The Hague: Martinus Nijhoff, 1976), 1:804.
10. Minutes of a discussion between Himmler and the leader of the Dutch Nazi Movement, Anton Mussert, 13 July 1943, quoted in Josef Ackermann, *Himmler als Ideologe* (Göttingen: Müsterschmidt, 1970), 282; Himmler's speech, 9 June 1942, quoted in Smith and Peterson, *Heinrich Himmler*, 159.
11. Berger to Eggert Reeder, 7 October 1941, quoted in Ackermann, *Himmler als Ideologe*, 182.
12. Bernd Wegner, *Hitler's politischen Soldaten: Die Waffen ss 1933–1945; Studien zu Leitbild, Struktur und Funktion einer nationalsozialistischen Elite* (Paderborn: Schöningh, 1999), 136.
13. Himmler to Berger, 23 April 1940, quoted in Bernd Wegner, "Auf dem Weg zur pangermanischen Armee: Dokumente zur Entstehungsgeschichte des III. ('Germanischen') ss-Panzerkorps," *Militärgeschichtliche Mitteilungen* 2 (1980): 101.
14. Peter Scharff Smith et al., *Under Hagekors og Dannebrog: Danskere i Waffen-ss 1940–1945* (Copenhagen: Aschehoug, 1998).
15. Wegner, "Auf dem Weg," 105.
16. Wegner, "Auf dem Weg," 107.
17. While the Waffen-ss comprised the military units participating in combat operations, the Allgemeine-ss (General ss) was the (paramilitary) noncombat branch mainly deployed in Germany proper. Although separated on paper, personnel continually transferred between the different ss branches, whereas the Germanische ss was also supposed to serve as a reservoir for the Waffen-ss.
18. On the German minority in Denmark, see Sven Tägil, *Deutschland und die deutsche Minderheit in Nordschleswig: Eine Studie zur deutschen Grenzpolitik 1933–1939* (Lund: Svenska Bokförlaget, 1970); Johan Peter Noack, *Det tyske mindretal i Nordslesvig under besættelsen* (Copenhagen: Munksgaard, 1975).
19. Discussion between the Central Agency for Ethnic Germans (VOMI) and German minority leaders, 4–5 December 1940, RA, AA387.

20. Steffen Werther, "Nazi, German and Danish Identities in Danish North Schleswig, 1932–38," in *Bordering the Baltic: Soft and Hard Processes of Boundary-Drawing*, ed. Madeleine Hurd (Berlin: LIT Verlag, 2011), 69–104.
21. Ernst Schröder's report, 6 July 1932, LAÅ, PK303. See also articles in *Flensburger Nachrichten*, 7 April 1933, and *Schleswig Holsteiner*, 14 (1933): 106–7.
22. *Nordschleswigsche Zeitung*, 21 January 1933, 27 January 1933, 9 May 1933, 14 May 1935.
23. Emil Seidelmann (b. 1881) was an active member of various Nazi splinter groups and parties sponsored by the German minority during the years 1933 to 1935, that is, prior to the creation of the umbrella party NSDAP-N. In late April 1935 Seidelmann informed the German minority leadership that he had founded the Standarte Nordschleswig. The following month the Danish Nazi Party's newspaper *Nationalsocialisten* responded to Seidelmann's initiative with a front-page headline: "Pro-German North Schleswigians Call on the DNSAP to Join in the Struggle for a National Socialist Denmark!"
24. See Danish police reports on Seidelmann, LAÅ, PA119/120. To the annoyance of the NSDAP-N's leadership, Seidelmann distributed leaflets claiming that "Frits Clausen wants to settle the Nordic-Germanic brother feud in his homeland." Leaflet prepared by Seidelmann, LAÅ, PK167, T1364.
25. Christian Paysen, *Der Nationalsozialismus als Weltanschauung im Grenzland: Eine Darstellung der Verhältnisse in Nordschleswig* (Tondern: Buchdruckerei Andresen 1941), 26. Paysen served as the NSDAP-N leader in Tønder (Tondern) but was deposed in 1940.
26. Christian Paysen, ed., *Nordschleswigs Sendung im grossgermanischen Reich: Ein Beitrag der Front zur Verständigung zwischen Deutschland und Dänemark* (Tondern: Buchdruckerei Andresen 1942), 9.
27. In this particular case Hitler applied the term *Schicksalsgemeinschaft* only to Germans. Adolf Hitler, "Appell an die Nation," 15 July 1932, accessed 27 September 2010, http://www.dhm.de/lemo/html/dokumente/hitler/index.html.
28. Alfred Rosenberg, "Nordische Schicksalsgemeinschaft," *Norden* 8 (1940): 241–46.
29. Paysen, *Der Nationalsozialismus*, 24.
30. Ewald Albert Lawner to Rolf Kassler, 13 April 1942; Himmler to VOMI, 22 April 1942; Lawner to Cecil von Renthe-Fink, 7 May 1942, in *PKB*, 14:723–24, 726, 729–31.
31. German Foreign Office, note re: SS recruitment in Denmark, 16 May 1939, *PKB*, 14:888.
32. Jens Möller to Himmler, 2 September 1942, *PKB*, 14:929–31.
33. Renthe-Fink's report on Jens Möller, 29 April 1942, *PKB*, 14:727–29.
34. Renthe-Fink's report, 20 May 1940, *PKB*, 14:891.

35. Note on joint session of VOMI and representatives of the German minority, 9 December 1940, *PKB*, 14:696–99.
36. Renthe-Fink's report on Jens Möller, 29 April 1942, *PKB*, 14:727–29.
37. Asmus W. Jürgensen's interrogation records (author of most educational dispatches), 20 March 1946, LAÅ, PK221.
38. NSDAP-N educational dispatch no. 1, "Unser Schleswig und das germanische Werk Adolf Hitlers," fall 1940, LAÅ, PK308.
39. NSDAP-N educational dispatch no. 2, "Das Volkstum als tragende Kraft," [December 1940], LAÅ, PK308.
40. German consulate Apenrade to German legation Copenhagen, 8 October 1941, RA, AA389; NSDAP-N educational dispatch no. 3, "Der Norden und Wir," [February 1941], LAÅ, PK308.
41. Lauridsen, *Dansk Nazisme*, 55.
42. Ole Ravn, *Fører uden folk: Frits Clausen og Danmarks National Socialistiske Arbejder-Parti* (Odense, Denmark: Syddansk Universitetsforlag, 2007), 33.
43. Andreas Monrad Pedersen, *Schalburgkorpset—historien om korpset og dets medlemmer 1943–45* (Odense, Denmark: Syddansk Universitetsforlag, 2000), 24.
44. *National Socialisten*, 11 November 1938.
45. *National Socialisten*, 28 January 1943.
46. Frits Clausen, *Nationalitetsproplemet og nationalsocialismen* (Copenhagen: Landskontor for Folkeoplysning, 1943), 26.
47. Frits Clausen, *Dansk-Tysk, Nationalsocialistiske tanker til Det slesvigske Spørgsmaal* (Fredericia: Forlaget "Landsoldaten," 1934), 24.
48. Lauridsen, *Dansk Nazisme*, 30.
49. On Clausen's political allegiances, see John T. Lauridsen, ed., *Føreren har Ordet! Frits Clausen om sig selv og DNSAP* (Copenhagen: Museum Tusculanums Forlag 2003), 261.
50. *National Socialisten*, 30 April 1938.
51. Clausen to Peter Munch, 22 June 1945, *PKB*, 14:45–46. See also Lauridsen, *Føreren har Ordet!*, 238.
52. Henning Poulsen, *Besaettelsesmagten og de Danske Nazister* (Copenhagen: Udgiverselskab for Danmarks nyeste historie, 1970), 140.
53. Vilhelm la Cour, *Danmark under besaettelsen*, (Copenhagen: Westermann 1947), 2:510–11; *National Socialisten*, 15 March 1941.
54. By giving their consent, the Danish authorities hoped to forestall further German demands, including forced conscription.
55. *National Socialisten*, 11 June 1942.
56. *National Socialisten*, 1 January 1942.
57. Alfred Rosenberg's note, 27 October 1942, *PKB*, 13:747–48.

58. Clausen to DNSAP leaders, 19 April 1943, and Clausen to Himmler, 19 December 1942, quoted in Lauridsen, *Dansk Nazisme*, 718n, 726–27.
59. DNSAP guidelines, 16 September 1943, quoted in Lauridsen, *Dansk Nazisme*, 731; *Fædrelandet*, 1 October 1943; *National Socialisten*, 8 October 1943.
60. Renthe-Fink's report on Jens Möller, 29 April 1942, *PKB*, 14:727–29.
61. Renthe-Fink report on a conversation with Clausen, 2 July 1942, RA, AA201.
62. See the series of articles printed in *Nordschleswigsche Zeitung*, 7 June 1934 and 9 June 1934.
63. German Security Service report, "Allgemeine Stimmung der deutschen wie auch der dänischen Bevölkerung Nordschleswigs, Juli 1940," in *Meldungen aus dem Reich 1938–1945: Die geheimen Lageberichte des Sicherheitsdienstes der* SS, ed. Heinz Boberach (Herrsching, Germany: Pawlak Verlag, 1984), 1370; report by consulate agent Carl Lundberg, 28 June 1940, *PKB*, 14:672.

6

Eugenics into Science

The Nazi Period in Austria, 1938–1945

THOMAS MAYER

The annexation of Austria by Nazi Germany on 12 March 1938, the so-called Anschluss, brought, along with Nazi political and racial persecution, massive changes in administration, especially in science policy. Prior to 1938 eugenics, human heredity, and experimental genetics were not on the curricula of Austrian universities or research institutions, and anthropology was taught only in Vienna. The change in science policy, which emerged as the result of the Nazi program of "hereditary and racial care," led to the establishment of racial and eugenic science as an academic discipline under the headings of *racial hygiene* and *racial biology*. To implement its racial and genetic-health programs, the Nazi regime needed substantial expertise, which scientists were all too eager to provide for their own political, economic, and epistemic benefit.

Research on eugenics in Austria in the period after 1938 tends to emphasize the local context, while scholarship on German eugenics as a rule dwells on Austrian developments only briefly.[1] This chapter presents three case studies that exemplify the process of institutionalization of eugenics at medical faculties in Austria during this period. Making extensive use of university records, it addresses the following four issues: the racial and scientific concepts that were utilized in the establishment of eugenics at Austrian universities during the Nazi period; the degree to which plans to incorporate the study of eugenics into Austrian universities were realized; the extent of involvement of intellectuals and scientists in this process; and what exactly the discipline of racial hygiene and racial biology comprised.

The process leading to the establishment of eugenic institutions in Austria can be described as a struggle for power, resources, and scientific concepts. Ultimately, this chapter argues, it worked in favor of German scholars who had better academic networks, played an active role in science policymaking in Berlin, or started their career in Germany prior to 1938. The establishment of eugenics departments predicated on the concept of racial biology brought eugenics for the first and last time as a scientific discipline to Austrian universities. Due to the broad understanding of the term *racial biology*, the subsequent transfer of scientific knowledge from Germany brought to Vienna first and foremost modern experimental genetics. The comparison of the three Austrian universities demonstrates the different strategies adopted by local scientists, resulting in different outcomes. Overall, the Austrian case exemplifies the second wave of institutionalization of eugenics at German universities after 1938.

After a brief overview of the eugenic education program introduced after 1938, the chapter will examine attempts to develop the discipline of eugenics in pre-1938 Austria. It will then explore the development of racial hygiene and racial science in the years immediately preceding and following the Nazi annexation by mapping how eugenic and racial studies advanced at three different medical faculties in Austria. It will focus on the processes involved in the establishment of eugenics departments, the research that these departments conducted, and the participation of scientists and academics in specific eugenic and racial programs.

The Institutionalization of Eugenics at Austrian Universities: Three Case Studies

Eugenics and human heredity were part of the scientific agenda of a number of physicians and anthropologists in Austria before 1938. Yet from 1920 onward eugenics lectures remained the prerogative of just a handful of hygienists in Vienna and Graz.[2] After 1938 similar lectures became obligatory at Austrian universities. It took another year, however, before eugenics became subject to formal examination in the Reich medical curriculum. This change in medical education in April 1939, which took effect in Austria at the beginning of 1940, was used as an argument for the establishment of new institutions. Preclinical students, for example, had to attend a three-hour lecture on "hereditary science and racial studies" and a one-hour lecture on "population policy," while clinical students had to sit in on a three-hour lecture on "human heredity as the basis for racial hygiene" and a two-hour lecture on "racial hygiene."[3]

As far as the eugenics lectures were concerned, the practice differed at the three medical faculties. In Graz potential instructors were drawn from among those who had been involved in the eugenics movement before 1938. The need for a specially appointed academic chair had been overcome in Graz thanks to lecturers who could step in and offer provisional classes within the discipline. What was meant to be provisional proved to be permanent. In Innsbruck the situation was somewhat different due to the absence of eugenics proponents in that city prior to 1938. Appointed to his position in 1939, the professor of hereditary and racial biology Friedrich Stumpfl was effectively the only academic in Innsbruck able to deliver such lectures. In Vienna, by contrast, there existed a pool of lecturers who had been active in the eugenics movement long before 1942, when the newly established Department of Racial Biology took over. Prior to 1942 most of the lecturers at Vienna were anthropologists. The establishment of the Innsbruck and Vienna chairs fostered the study of modern experimental genetics, with drosophila genetics introduced in Austria from Germany as a teaching subject.

Eugenics and human heredity did not become institutionalized in Austria until after the 1938 takeover due to a combination of lack of political incentive to fund eugenic departments and infrequent as well as ineffectual initiatives by eugenicists. Whenever this issue came to the fore, human heredity was always considered to be part of the discipline. As a result of the failed attempts to establish eugenics at Austrian universities, human heredity was institutionalized as a separate discipline with its own department only in 1938. Before then, the only institution where eugenics, heredity, and biostatistics were taught was a subdivision of social hygiene in the Hygiene Department at the University of Vienna, headed by the hygienist Heinrich Reichel. For two of his students, Friedrich Stumpfl and Karl Thums, it served as a starting point in their career in psychiatric genetics and eugenics. Having worked in Munich under the leading German psychiatrist and eugenicist Ernst Rüdin, both were later appointed directors of eugenic institutions, at Innsbruck and Prague, respectively.[4]

Notably, in the years between 1923 and 1934 there were three failed attempts to establish eugenic institutions at Vienna and Graz. In all three cases anthropologists and physicians—and in one case Viennese municipal officials—endorsed human heredity and eugenics, described as racial hygiene and racial biology, respectively. Twice the Austrian Ministry of Education rejected the initiative due to the tight financial situation in Austria. The final attempt, in 1933–34, was initiated by *völkisch* physicians in Graz, prompted by the political changes in Germany and Austria.[5]

In comparison to Austria, where the first eugenics institution came into existence in 1939, the persistent Nazi interest in eugenics and racial science—and an altogether longer tradition of eugenics, human heredity, and genetics—made the situation in Germany distinct. Despite the Nazis' fixation on race and eugenics, however, not every German university introduced eugenics curricula due to academic infighting, the power struggle between the Schutzstaffel (SS) and the Office of Racial Policy (Rassenpolitisches Amt), or changing political and scientific constellations. The few eugenics departments that had been founded were usually converted from existing departments of anthropology.[6] The process of institutionalization occurred in two phases, from 1933 to 1934 and from 1939 to 1943. The second wave of expansion, which includes the Austrian case, was conditioned by two factors. First, the incorporation of eugenics into the new medical curricula provided a perfect argument for the establishment of new research and teaching institutions. Second, German scholars eagerly availed themselves of the new opportunities that the Nazi annexation of Austria and Czechoslovakia opened up in the field of eugenics. As a consequence, following the Anschluss, medical faculties of the Graz, Innsbruck, and Vienna universities each planned for a department of racial hygiene or racial biology; such departments were eventually inaugurated at Innsbruck and Vienna, both using the term "racial biology."[7]

Lead and Follow: The Hereditary and Racial Biology Department in Innsbruck

In the spring and summer of 1939 the Medical Faculty at the University of Innsbruck became the first such institution in Austria with a eugenics section, the Department of Hereditary and Racial Biology. There are a number of reasons for this. First, the eugenics movement as such was nonexistent in Innsbruck, and therefore the need for university teaching was more urgent there than in Vienna or Graz. Second, the apparent lack of interest in eugenics suggested a lack of internal competition, thus expediting its institutionalization. Third, the Innsbruck Medical Faculty effectively employed a copycat strategy when it adopted the proposal of the University of Vienna Medical Faculty to establish a new eugenics and racial science department at Innsbruck. Indeed, the Innsbruck proposal of July 1938 was strongly influenced by the draft prepared one month earlier by the Viennese faculty, specifically by the dean, Eduard Pernkopf, and the physiologist and state commissioner for education, religion, and national education, Friedrich Plattner. It was Plattner who, in his capacity as head of the Viennese branch of the Reich Ministry of Education in Berlin—the highest authority in Austria with regard to science—sanctioned the Innsbruck department. In April 1939 Plattner appointed Stumpfl head of the Department of Hereditary and Racial Biology and provided him with office space in the Department of General and Experimental Pathology. (Pathology departments as such became dispensable under the new, Nazi-inspired academic curricula. The position of department head had been vacant since the death of its former, Jewish head, who had committed suicide after the Anschluss.)[8] On September 30, Stumpfl was appointed extraordinary professor of hereditary and racial biology and thus officially became head of the newly established department.[9]

Stumpfl was considered one of the leading German experts in the field of criminal biology and psychiatric genetics. His academic training in medicine and anthropology had begun in the 1920s in Vienna and in Freiburg, where he studied under the academic supervision of professors with a profound interest in eugenics and racial science. In 1930 Stumpfl continued his studies at the University of Munich in the Department of Psychiatry, which at that time was headed by Rüdin. Over time Stumpfl became an expert in his field, having conducted research on recidivists and their supposedly inherited proclivity to crime. Following the enactment of the Nazi sterilization law in 1933, Stumpfl began promoting his research by arguing for the social and scientific value of eugenics. In 1936, the same year he defended his professorial thesis (*Habilitation*), he even called for a sterilization law to be applied to recidivists and social deviants, just as Rüdin had done.[10]

In March 1940 Stumpfl presented an ambitious scientific agenda for his new department. He wanted to conduct genetic and racial research in anticipation of a brighter future for eugenics and population studies. Without invoking the term *racial biology*, Stumpfl nevertheless advanced the discipline in the direction of eugenics, human heredity, and anthropology. His research interests led him to focus on the genetic quality of Tyrolean mountain farmers. Stumpfl intended to use this case study as a matrix for a hereditary biological survey aimed at creating a genetic database of the entire population of Austria. Until other eugenics departments were established, his department was also to become the center for twin studies in western Austria. Simultaneously, Stumpfl wanted to continue with his earlier research on the inheritance of psychopathic behavioral traits and participate in the recent discourse over phenogenetics—the study of the phenotypic effects of genetic material.[11] He also focused on anthropological and racial aspects of teaching and research by arguing that smaller university departments of human heredity should perform as departments of anthropology (thereby advancing the arguments of one of the leading German anthropologists, Wilhelm Gieseler). Stumpfl requested two assistants, one trained in medicine and one in anthropology, yet he only received one.[12] Research activity was channeled into two main tracks. Department staff supervised a number of doctoral theses by Stumpfl's students, which examined the inheritance of psychopathic traits, antisocial behavior, and the genetic and social value of students with special needs, thus indirectly advocating eugenic measures.[13] Parallel to this the department launched an anthropological project, ingrained in family research, dealing with the Yeniche people (Jenische, or Travelers) in Tyrol. In 1938 Stumpfl characterized the Travelers as a "racial alien body" and explained their socially deviant way of life as a result of "abnormal" personality traits.[14] From 1941 or 1942 onward Armand Mergen took anthropometric measurements of and collected genealogical and genetic data from 299 members of the Traveler community, publishing the results of this study in 1949, with an emphasis on educational, social, and eugenic aspects of the group.[15]

Stumpfl also contributed to the practice of hereditary and racial care. He wrote at least two hundred expert opinions for the youth welfare offices in Tyrol and Vorarlberg and the juvenile courts of the NSV, the Nazi Welfare Organization. Concurrently, he served as head of an educational counseling bureau at the University of Innsbruck Pediatric Clinic. In spite of his postwar claims that he had wanted to help the juveniles, on at least one occasion Stumpfl recommended forced sterilization, diagnosing imbecility in a fourteen-year-old girl whose personality, he said, was defective, thus making her incapable of performing even basic labor.[16] According to Stumpfl, he wrote at least two expert opinions for the Higher Hereditary Health Court that, however, did not result in sterilization.[17] At this point there is no evidence that Stumpfl ever participated in the Nazi euthanasia program. In addition, Stumpfl issued racial proofs (*Abstammungsgutachten*)—producing at least one such proof to determine Jewish descent of the applicant—and paternity proofs.[18] The two types of expert opinions differed only in their purpose: racial proofs were part of the anti-Jewish persecution campaign, whereas paternity proofs were used to impose alimony obligations.[19] Immediately after the Second World War Stumpfl renamed his department, deleting the word *race*. Nonetheless, he did not lose his belief in the promise of eugenics. Despite the support of the Faculties of Medicine and Law, Stumpfl's department was closed down in 1947 because of its Nazi connection. In the 1950s an academic committee at the University of Innsbruck approved Stumpfl's application to continue with his scientific and forensic career. In so doing the committee not only ignored Stumpfl's role in the Nazi sterilization campaign but also legitimized eugenics as part of the scientific discourse in Austria after 1945.

The Never-Ending Story: Eugenics in Graz

The context for the institutionalization of eugenics in Graz was different from that in Innsbruck. Preexisting eugenic networks and competing interests delayed a proposal for the establishment of a eugenics institution, suggesting a preference for locally trained scholars. This preference, however, never resulted in a research institution formally being established in Graz, in spite of the efforts of certain local academics. The void was eventually filled by one of Graz's leading eugenicists, the dermatologist Rudolf Polland. Polland used to refer to his assignment to teach hereditary biology and eugenics as a "department," even though no other academics were known to have been part of it.[20] On 20 October 1938, at the first Medical Faculty meeting of the semester, the issue of establishing a separate department of racial biology came up in discussion. According to the minutes of the meeting, the name of the department would be racial hygiene rather than racial biology.[21] The change of heart might have been influenced by the Viennese and the Innsbruck draft, which eventually settled on the name *hereditary and racial biology*, reversed from the originally proposed *racial hygiene*. One can speculate whether this decision had anything to do with the presence at the meeting of one of Graz's most influential eugenicists, Heinrich Reichel, and his rejection of the influence of anthropology on eugenics in general. Reichel advocated eugenics as part of hygiene and anthropology as part of racial hygiene.[22] In fact, it was faculty dean Anton Hafferl who combined racial hygiene with racial biology as well as with racial science, and after 1942 with hereditary hygiene, while persistently using *racial hygiene* as part of the name. The struggle over the name of the new department can be interpreted as a struggle between the old and new concepts of eugenics. While the "traditional" racial hygienist Reichel had a preference for the term *racial hygiene*—with the emphasis on the word *hygiene*—*racial biology* was a more recent term that comprised concepts of race and heredity. Primarily used by anthropologists, the term *racial biology* had gradually shifted in the direction of experimental and psychiatric genetics and biostatistics, thanks to Lothar Loeffler.

Hafferl invited eugenics experts who identified with the Munich, *völkisch* branch of the German Eugenics Society to evaluate candidates for the position of head of the proposed Department of Racial Hygiene. Interestingly, in doing so Hafferl ignored the advice given, for example, by the German geneticist and leading eugenicist Fritz Lenz. Lenz had suggested five German scholars, including Friedrich Stumpfl. Lenz considered naming experts a difficult task altogether, on account of similar requests arriving from the faculties in Vienna and Innsbruck and the general lack of qualified scholars. Therefore he recommended requesting the somewhat experienced Reichel to hold the lectures on racial hygiene.[23]

However, Hafferl apparently did not consider Reichel to be the most suitable candidate to head a university research department.[24] In order to implement the new curriculum, on 5 December 1938 Hafferl proposed a chair in racial science and racial hygiene. The office space was supposed to be provided by the Department of General and Experimental Pathology, as was the case previously in Innsbruck and Vienna. Hafferl anticipated that the university would incur considerable expenses for the acquisition of teaching materials and research funds. After complaining about the lack of qualified staff, he discussed seven potential candidates, among them leading eugenicists in Graz (Reichel, Polland, Alfred Pischinger), students of Reichel (Alfred Schinzel, Heinrich Kurzweil), and Heinrich Kranz from the University of Giessen—the only German under consideration. Hafferl suggested three different options. The first option involved the establishment of a separate department, with Polland as preferred candidate for the position of head of department and Kranz, Schinzel, and Stumpfl as shortlisted candidates. The second option envisaged a provisional solution by establishing a subdivision of racial science and racial hygiene within the hygiene department. In that case, Schinzel, Stumpfl, or Kurzweil—all of them students of Reichel—would become director while the lectures would be divided between the director and Reichel. The third option, if the Reich Ministry of Education (Reichserziehungsministerium, or REM) decided to reject the first two, would result in the division of lectures between Reichel, Pischinger, and Polland.[25]

The main objective of the draft was to recruit sufficiently qualified instructors to teach the new curriculum. Hafferl's outline further suggests that he was supportive of a local solution, favoring Polland and Reichel's students. It also demonstrates the influence that the old networks of Polland and Reichel still enjoyed. The idea of a subdivision attached to the hygiene department was a concession to Reichel, who had promoted this idea since 1934. Ultimately the choice fell on Polland, rewarded for his long commitment to the eugenic and hereditary cause.

Regardless of how the ministries in Vienna and Berlin reacted to Hafferl's draft, no significant progress was made in 1939. As the historian Walter Höflechner has speculated, the cause for the delay might have been due to a competition over the position of chair between three faculties.[26] In March 1939 the director of the Department of Criminology, Ernst Seelig, attempted to establish a chair of cultural biology at the Faculty for Law and Political Science.[27] Plausibly, these considerations could have interfered with the plans of the Medical Faculty.

In the years after 1939 the Medical Faculty repeatedly took up the issue of establishing its own research facility, yet it was not until January 1942 that Hafferl launched another serious attempt and sent a new draft proposal to the REM in Berlin, this time suggesting the founding of a chair of hereditary and racial hygiene. Hafferl pressed for the resolution of this issue in light of mounting problems in the faculty. Specifically, he argued that lecturers lacked appropriate teaching materials and received low pay; one of the lecturers, Polland, was over age sixty-five and no substitute candidates were available. Reichel, who was not mentioned in Hafferl's draft, was no longer teaching due to his hospitalization in the summer of 1941.[28]

It is unclear whether Reichel's absence made it easier for Hafferl to relaunch his campaign. The immediate problem seemed to be a lack of space: the office space that had been promised in 1938 was now occupied by the board of trustees (*Kuratorium*).[29] In the meantime, Hafferl suggested Ferdinand Claussen from Cologne University as his first-choice candidate and Günther Just from the University of Berlin as his second choice. As his third preference he suggested Karl Thums of Prague's Charles University, who specialized in psychiatry and therefore, according to Hafferl, was not as suitable as the other candidates for the proposed research center. Hafferl argued that the successful candidate should be well versed in experimental pathology, human heredity, racial science, anthropology, population policy, and racial hygiene. Researchers of human heredity had an edge over experts in anthropology and population policy, who were not considered for the position.[30]

A marked difference from the 1938 draft proposal was that *racial science* was missing from the name of the future center, replaced with *heredity*. As far as potential candidates were concerned, the focus had shifted from Graz to Germany proper, whereas the Austrian Thums—who was not even working in the so-called Ostmark (Austria's name after the Anschluss) at that time—ranked only third on the short list. In contrast to the selection process that had taken place earlier in Vienna, references to experimental genetics were missing whereas psychiatric genetics was considered of less importance, perhaps as an attempt to depart from the influence of Rüdin's Munich school.

It was only in July 1944 that the REM confirmed Hafferl's January 1942 proposal on condition that the issue of office space was solved. Max de Crinis, a key administrator in charge of medical university chairs in the REM, supported Hafferl in his belief in the importance of the eugenics department. However, little happened with respect to the establishment of a eugenic center. Thus, in January 1945, dogged and undeterred, Hafferl launched yet another attempt, now claiming that even though the office space required was still occupied, two or three rooms would do as a provisional solution. Nonetheless, Hafferl conceded that the practical obstacles were still formidable, not least owing to the lack of qualified lecturers.[31] Despite this, no permanent solution was found, even if the REM in Berlin acknowledged the commitment and motivation of the Graz faculty to acquire its own eugenics center.

To ensure a more solid curriculum at the University of Graz, as early as the winter semester of 1938 Polland received a teaching assignment in hereditary biology and eugenics from the Medical Faculty.[32] While similar efforts at Innsbruck came to naught, the Graz faculty was more successful in providing one of Austria's leading eugenicists with a regular income. The dean referred to Polland's argument that he could not depend on the income from his dermatological private practice.[33] As well as academic prestige, political considerations probably played a role in providing Pollard with a teaching position. Having been overlooked for academic positions throughout the 1920s, from 1923 onward Polland had been leader of the eugenics movement in Graz and also a member of the Nazi Party in Styria.

In 1941 Polland began referring to himself as provisional head of a "university department of hereditary science und racial hygiene" (consisting only of himself and probably a secretary), even though the existence of this department was not mentioned in university calendars or any other sources.[34] Most likely Polland just wanted to fill the void. Although in his university personal file Polland was indeed referred to as the "provisional head of the department of hereditary science and racial hygiene," it cannot be determined when and by whom this particular entry was made.[35] The main function of the department was issuing racial, paternity, and sterilization proofs.[36] Since no scientific work by Polland is known to have been published during the Nazi period, it is doubtful whether he conducted research during this time at all. The paternity and racial proofs that he had issued during this time might have generated data for future research, though Polland did not conduct any research or publish after 1945.

Extensive Demands, Resistance, and Genetics: The Vienna Case

Almost immediately after the Anschluss, several Austrian ministries and the Faculties of Philosophy and Medicine at the University of Vienna came up with a plethora of ideas regarding the institutionalization of eugenics and racial science. In April 1938 the Austrian Ministry of Education forwarded to the dean of the Philosophical Faculty a petition originating in the Ministry of Social Administration. In this document the ministry suggested the faculties should establish a department of racial and cultural studies with the prominent German lawyer and legal advisor for the Nazi sterilization law Falk Ruttke at the helm. However, on the basis of negative expert opinions by several Viennese professors of law, Ruttke's candidacy fell through and the chair itself did not materialize.[37]

The initial idea for a separate, nonuniversity research unit came from the dean of the Philosophical Faculty, the orientalist Viktor Christian. He outlined his ideas in an informal letter to Loeffler, in which he proposed a new research department of hereditary biology that would issue paternity and racial proofs.[38]

Christian's true intention was to support his friend, the anthropologist Josef Weninger, whose position was increasingly coming under threat because his wife, the anthropologist Margarethe Weninger, was of Jewish descent. The Weningers and their colleagues in the Department of Physical Anthropology at the University of Vienna had applied Reche's methodology to devise a morphological-genetic paternity proof, starting in the early 1930s.[39] The income generated from those paternity reports provided a financial foundation for *Erbnormalbiologie* (normal human heredity research). Moreover, the introduction of paternity proofs prompted further research programs that explored the problem of inheritance of normal (i.e., nonpathological) morphological traits.

In his reply to Christian, Loeffler subjected the former's plan to criticism. Loeffler, who might have been aware of his impending appointment as head of racial biology in Vienna, rejected Christian's ideas as unrealistic not only because of the alleged racial bias of Weninger (because, in accordance with Nazi laws, his wife was considered Jewish) but also owing to conceptual differences. Loeffler considered Weninger's approach too narrow because it omitted hereditary pathology, which the former regarded as the main advantage of his own approach to racial biology.[40] The term *Erbpathologie* (hereditary pathology) was coined by the German human geneticist and eugenicist Otmar von Verschuer and could be understood to mean medical genetics.[41] In Loeffler's view, hereditary pathology paved the way to racial biology, as it provided the scientific foundation for (negative) eugenic laws. Conversely, Nazi eugenic laws prompted further research in hereditary pathology. As of 1939, Loeffler still favored hereditary pathology over Viennese and German anthropology, arguing for a medical, genetically oriented approach to racial biology.[42] While emphasizing genetic aspects, Loeffler never abandoned a racial perspective. Racial biology was placed at the intersection of eugenics, human heredity, and racial science—an umbrella term for an ultimate solution to the racial problems confronting society. By using a term that had previously been associated with anthropology and racial science, Loeffler effectively incorporated anthropology into racial biology.

The decision by a Medical Faculty committee on 9 May 1938 to establish a department of racial hygiene provided Loeffler with the opportunity to prove his point. During that meeting the initial name was changed to the Department of Racial Biology.[43] The composition of the committee reflected attempts to gain control over university eugenics through various actors within the municipal and state health administration with Nazi Party and SS backgrounds. Apart from the dean, Pernkopf, who was a member of the Sturmabteilung (SA), all the other participants were Austrian academics who had worked in Germany prior to 1938 due to their illegal Nazi Party membership. After the Anschluss they were installed in key positions in the Austrian health system.

The next day a request was sent to the six foremost representatives of German eugenics, genetics, and anthropology: Loeffler, Kranz, Lenz, Verschuer, Rüdin, Eugen Fischer, and the Austrian racial physiologist Robert Stigler. The experts were asked to propose the structure and agenda for the new department and to name qualified candidates for the chair.[44] Stigler was the only one who suggested establishing a separate department of racial physiology with four assistants.[45] The REM, however, wished to postpone the creation of a racial physiology department until a new chair of physiology had been appointed at the university.[46] Soon after this exchange of opinions Plattner was appointed chair of physiology, albeit a separate chair of racial physiology was never established.

The German experts all agreed that racial psychology and racial physiology should not be established as separate departments or subdivisions. Instead, the two disciplines should constitute a subdivision either of the Department of Anthropology or the Department of Psychology and Physiology. With the exception of Loeffler, all of them agreed that the department was to be situated within the Medical Faculty and that it should focus on the study of human heredity, hereditary pathology, and racial hygiene in order to avoid competition with the existing Department of Anthropology. Loeffler was the only one who used the term *racial biology* instead of *racial hygiene,* and he alone favored the inclusion of anthropology as one of six subdivisions of the new department. Verschuer and Loeffler were frequently suggested as the most suitable candidates for the chair. Almost no one, by contrast, proposed Austrian scholars who had worked in Austria before 1938, with the exception of Rüdin, who promoted the cause of Reichel.[47]

On 20 June 1938 Plattner and Dean Pernkopf submitted an official report arguing for the establishment of a department of hereditary and racial biology. Rather unsurprisingly, Loeffler was nominated as the first-choice candidate for the position of director of the new department, along with Gieseler from the University of Tübingen and Kranz from the University of Giessen. The second- and third-choice candidates were two of Rüdin's closest associates, Stumpfl and

Hans Luxenburger. One of the main reasons why Loeffler was considered to be the best candidate was on account of both his experimental genetic research and his administrative experience as head of the Racial Biological Department at the University of Königsberg; no scholar working in Austria was nominated.

The new department gained legitimacy by promoting Nazi hereditary and racial care and thus the health of the whole *Volkskörper* (body politic). The name chosen for the department incorporated the disciplines of general and hereditary biology, racial science, and especially hereditary pathology. The department was meant to be not only a teaching and research facility but also a unit engaged with practical tasks such as issuing racial, paternity, and sterilization proofs. Additionally, it was understood to be an intellectual stronghold of ethnic Germans in Southeastern Europe.[48]

Negotiations between the REM and Loeffler began on a positive note. In November 1938 Loeffler produced a memorandum proposing to create six subdivisions and hire seventeen scientific assistants. Apparently the anthropologist Eberhard Geyer, who had been put in charge of the Department of Physical Anthropology after Josef Weninger had been dismissed, supported Loeffler's proposal. In the winter of 1938–39 he even expressed a desire to be involved in Loeffler's envisaged new department as the head of the planned subdivision of anthropology.[49]

In July 1939 the REM formally approved Loeffler's proposal, allocating a generous budget of approximately RM 835,000 that covered rent, librarian acquisitions, and technical equipment but not the salaries of the seventeen assistants.[50] This amount was significantly larger than that annually received by the Kaiser Wilhelm Institute for Human Heredity Sciences and Eugenics (KWI-A) in Berlin, which employed three subdivision heads and eight assistants—not even half of the staff scheduled for Vienna.[51]

Despite the generous funding, it did not meet Loeffler's expectations entirely since only limited office space was made available to his institution. Furthermore, the room shortage strengthened a counterargument of the anthropologists and their dean, Christian, who feared that the new research unit might marginalize the anthropological discipline within the larger racial biology department. From the summer of 1939 onward Christian and Geyer began to actively resist Loeffler's plans. As the historian Edith Saurer has pointed out, they were driven not so much by ideological differences as by professional and academic considerations: both scientists wanted to prevent the decline of the study of anthropology at

Vienna. Christian, in addition, tried to defend the interests of his faculty and altogether disapproved of Loeffler's ambitions.[52] At the same time he sought to defend the scientific legacy of his friend Weninger and "his" anthropological department.[53] By contrast, Geyer seems to have been motivated by career ambitions.

Evidently power relations worked in Loeffler's favor, since the institutional independence of anthropology as an academic discipline declined in the spring of 1940, supported by the rumor of its imminent incorporation into the Medical Faculty.[54] To prevent this from happening, Christian suggested drastic measures, including restricting anthropology to racial science or even outsourcing the department's main source of income, paternity and racial proofs, to Loeffler's department. The intervention of Geyer, Christian, and university president Fritz Knoll with Max de Crinis in the spring and summer of 1940 bore fruit, and in 1941 Geyer was appointed associate professor.[55] Geyer never gave up his paternity proof work, which remained in demand. As late as 1943 Geyer still considered *Erbnormalbiologie* within the remit of anthropology; according to him, it proved vital for the career development of anthropologists, not least in their capacity as experts in matters related to paternity and racial proofs.[56]

The war mobilization campaign (Loeffler himself was drafted into the German Wehrmacht) caused delays in the institutionalization of racial biology at the University of Vienna. It took until fall 1941 for Loeffler to reopen negotiations with Berlin. For its part, the REM found Loeffler's military service and an inability to find a permanent solution to the office space problem the main obstacles to the progress of negotiations. After protracted negotiations, during which Pernkopf emphasized the pressing need for more teaching staff and urged Loeffler to come to a speedy agreement on accommodations, Loeffler and the REM agreed in principle on part of the former premises of the Department for General and Experimental Pathology as a temporary wartime solution, approximately half of the space negotiated in spring 1939. In June 1942 Georg Gottschewski, who was regarded as one of the most promising younger geneticists in Germany, was appointed lecturer in genetics at Vienna's Medical Faculty. As far as Loeffler's understanding of racial biology went, Gottschewski had to have a position at the Philosophy Faculty, too. Christian, however, resisted Loeffler's attempt to situate the department in the Philosophy Faculty. This opposition provoked considerable friction on the part of Loeffler, who, in a lengthy letter, threatened to withdraw from the project. The REM put the negotiations on hold and asked

Walther von Boeckmann, head of the Viennese branch of the REM, to intervene with Christian.

A meeting between Christian, Pernkopf, Plattner, and his successor, von Boeckmann, still found no solution to the problem. Christian ignored warnings by Plattner, who expressed concerns that the turn of events might lead to possible intervention from the political authorities, thus compromising the faculty's autonomy. Christian gained support from university president Knoll, who pleaded for the retention of the study of anthropology and openly questioned the importance of Loeffler to the university. Von Boeckmann hesitated to interfere in university affairs because overruling the president and the dean would have created a precedent.[57] President Knoll thus proved to be a vital ally for Christian. A solution was finally found: Gottschewski was invited to appear in front of a faculty committee, which made him part of the Philosophy Faculty, a decision supported by zoological and botanical geneticists who appreciated Gottschewski's genetic expertise.[58] Still, Christian was successful in keeping Loeffler out of his faculty. The whole affair had no decisive impact and delayed the establishment of Loeffler's department by only two months. Ultimately, the release of Loeffler from military service proved much more important for the actual launch of the department.

In October 1942 the department was finally inaugurated and Loeffler appointed as head. Due to wartime restrictions only four out of the seventeen assistants initially promised could be hired at any one time.[59] Until June 1943 only four out of six heads of subdivisions had been appointed: anthropology, headed by Hans Ritter; racial hygiene, headed by Loeffler; hereditary psychology, neurology, and psychiatry, headed by Horst Geyer; and experimental genetics, headed by Georg Gottschewski. The subdivisions of hereditary pathology and hereditary statistics were never realized. In the case of the latter, the hereditary statistician Siegfried Koller, who had been considered by Loeffler as early as 1939, was no longer available since he had been appointed head of the Department of Biostatistics in Berlin in 1941.[60] However, due to compulsory military service only two heads, Loeffler and Gottschewski, were working continuously in Vienna.

Loeffler assigned the positions exclusively to scholars from Germany, particularly to colleagues and former students from Königsberg and the KWI-A in Berlin, and this was followed by a transfer of knowledge and genetic material (especially in the case of drosophila and mice genetics, which were of particular interest to Gottschewski). Although the impact of this research on the Austrian scientific community and wider scientific discourse has not been systematically analyzed yet, it is known that the laboratory animals from Vienna were the subjects of the experiments after 1945.

Historians have not yet properly explored the scope of the actual research carried out in the department. Loeffler and Pernkopf worked together on racial and pathological examinations of prisoners of war. Increasingly, Gottschewski began working in the field of cancer research. In 1942 he started collaborating with the Department of Radiation Therapy at Lainz hospital, headed by Emil Maier. This scientific collaboration, which involved experiments on mice for the purpose of cancer research, explored the nature of the gene. At least one doctoral dissertation emerged from this collaboration. In 1947 Oliver Paget, who between October 1943 and April 1945 served as an undergraduate research assistant in Vienna, wrote his thesis on a new method in radiation genetics. Paget used some of Gottschewski's methodological approaches and Emil Maier's radium experiments as well as Gottschewski's remaining drosophila flies for his own experiments.[61] Although the Department of Radiation Therapy was originally a creation of the Social Democratic municipal health official and eugenicist Julius Tandler, during the Nazi period its members were authorized to perform eugenic sterilization with radium irradiation.[62] It is unclear whether Maier and his colleagues carried out forced sterilizations and if they used data gathered from those cases for further research. In any event, German research foundations provided Gottschewski with substantial resources even as late as 1944.[63] Loeffler considered Gottschewski's scientific subdivision of experimental genetics the core unit of the department; without it, he stated, the rest would be only a "torso."[64] As regards Horst Geyer, his wartime studies likely contributed to his 1954 book on stupidity.[65]

Despite lacunae in historical research, the department clearly benefited from Nazi extermination politics through its genetic analysis of the records of the Nazi child euthanasia program.[66] In fact, in 1943 Loeffler encouraged the department's close collaboration with an organization that, as we now know, covered up Nazi child euthanasia, the Reich Committee for the Scientific Registration of Severe Hereditary Ailments (Reichsausschuss zur wissenschaftliche Erfassung von erb- und anlagebedingten schweren Leiden).[67] Loeffler himself was involved in negative eugenics in his capacity as expert at the Higher Hereditary

Health Court in Vienna;[68] by producing racial proofs he effectively participated in racial policy making.

In June 1945 the department was abolished, classified as a Nazi invention. Loeffler and his associates initially fled to Germany proper. Although they chose different career paths after 1945, none of them again received a university position, though Loeffler was lecturing again in the 1950s at the Technical University of Hannover. While Gottschewski resumed his genetic research at various Max Planck Institutes after a ten-year pause, Loeffler, as a member of the German Atomic Commission, became instrumental in the establishment of human genetics as a separate discipline in West Germany in the 1960s.[69]

In contrast to Nazi Germany, no eugenics department existed in Austria prior to 1938, mainly due to the lack of political support for those few initiatives championed by physicians. The institutionalization of eugenics in Austrian universities after 1938 in the form of new departments of human heredity, eugenics, and racial science was driven partly by German policy as well as by the personal and professional ambitions of Austrian academics, scientists, and university administrators. Strengthened by their illegal membership in the Nazi Party prior to 1938, they hoped to advance their careers in the new order. This process of consolidation was defined by three major factors: a desire to maximize one's power and prestige; intrauniversity jockeying for reputation, resources, and power; and conflicting concepts of *racial biology* and *racial hygiene.*

The problem of deciding which scientific disciplines should make up a university department of racial hygiene or racial biology was never definitively solved during the Nazi period. While in Prague by 1942 two departments had been founded at separate faculties—racial hygiene and racial biology—the main protagonists in Vienna opted for Loeffler's ambitious plan, which foresaw six well-staffed and well-funded subdivisions situated within a larger department. Loeffler's concept had also been partially implemented at Innsbruck, though on a much smaller scale.

The different geneses of eugenics departments at Austrian universities can be explained by a number of factors. Due to the nonexistence of a local eugenics network, the Innsbruck faculty adopted a carbon copy of the Vienna draft that blended anthropology with Loeffler's version of racial biology. This move certainly met the wishes of a central protagonist in the story of Austrian science,

Plattner. Conversely, the preexisting eugenics network in Graz prompted competition and ultimately the preeminence of local scholars. Both factors delayed the drafting of an agenda until it became even more difficult due to military mobilization. Moreover, the Graz plans were contrary to Berlin's ambition to install in Austria scholars who adhered to the German scientific tradition. Furthermore, German scholars like Loeffler used their connections in the REM to negotiate their job contracts. These asymmetrical power relations might very well have changed when the Austrian de Crinis acquired a central position in the REM. However, while de Crinis's appointment had an impact on the conflict between medicine and anthropology at Vienna, it did not help the Graz faculty to secure a chair.

The Vienna case is especially interesting for the following three reasons. First, it illustrates the conflict between the Faculties of Medicine and Philosophy and the Department of Anthropology concerning Loeffler's concept of *racial biology*. In many respects, it was a reaction to his ambition for a wide-ranging institutional solution that would incorporate anthropology and thereby might undermine both the autonomy of anthropology and the status of the Faculty of Philosophy. Second, the focus of party, state, and communal officials in Vienna enabled Loeffler to create a huge department comparable to other leading German institutions, in spite of the difficulties caused by a lack of office space and the escalating war efforts. Third, Loeffler introduced the term *racial biology* into Austria as an interdisciplinary admixture of eugenics, human heredity, biostatistics, and experimental genetics by emphasizing hereditary pathology as fundamentally different from anthropology. Nonetheless, Loeffler shared with Austrian anthropology the concept of *race* as the basic building block toward ideological and scientific consensus.

The research carried out at Innsbruck focused on the racial makeup of the local Traveler population and the genetic predetermination of general psychopathic and socially deviant behavior. The department at Vienna, in the meantime, prioritized race-related topics, radiation genetics, and cancer research. By providing expert opinions on eugenic and racial matters, all three medical faculties participated in the Nazi program of hereditary and racial care. From their perspective, however, such participation provided valuable scientific data and thus legitimized the research agendas of their respective departments. The transfer of scientific knowledge from Germany to Vienna was realized by means of experiments on animals within the new discipline of experimental genetics. The envisaged development of medical genetics as hereditary pathology remained at the planning stage, as no scholar capable of doing it could be found in Greater Germany. Ultimately, the conspicuous absence of medical genetics in postwar Austria may provide, at least in part, an explanation for the relatively late introduction of human genetics as a discipline in Austria in the late 1960s.

Notes

Abbreviations Used in the Notes

ÖStA, AVA	Österreichisches Staatsarchiv, Allgemeines Verwaltungsarchiv (Austrian State Archives, General Administration Archives, Vienna)
ÖStA, AdR	Österreichisches Staatsarchiv, Archiv der Republik (Austrian State Archives, Archives of the Republic, Vienna)
PhilF	Philosophische Fakultät (Philosophy Faculty)
RBI	Rassenbiologisches Institut (Racial Biological Institute)
UAG, medF	Universitätsarchiv Graz, Medizinische Fakultät (University of Graz Archives, Faculty of Medicine)
UAI, IERB	Universitätsarchiv Innsbruck, Institut für Erb-und Rassenbiologie (University of Innsbruck Archives, Institute of Hereditary and Racial Biology)
UAW	Universitätsarchiv Wien (University of Vienna Archives)

I would like to thank Mitchell G. Ash, Michael Hubenstorf, and Carola Sachse for commenting on an earlier version of this chapter.

1. On eugenics in Austria, see Michael Hubenstorf, "Medizinische Fakultät 1938–1945," in *Willfährige Wissenschaft: Die Universität Wien 1938–1945*, ed. Gernot Heiss et al. (Vienna: Verlag für Gesellschaftskritik, 1989), 244, 261, 263, 269; Michael Hubenstorf, "Tote und/oder lebendige Wissenschaft: Die intellektuellen Netzwerke der NS-Patientenmordaktion in Österreich," in *Von der Zwangssterilisierung zur Ermordung*, ed. Eberhard Gabriel and WolfgangNeugebauer (Vienna: Böhlau, 2002), 313–19; Edith Saurer, "Institutsneugründungen 1938–1945," in Heiss et al., *Willfährige Wissenschaft*, 303–29; Marion Amort et al., "Humanwissenschaften als Säulen der 'Vernichtung unwerten Lebens': Biopolitik und Faschismus am Beispiel des Rassehygieneinstituts in Innsbruck," *Erziehung heute* 1 (1999), accessed July 29, 2010, http://bidok.uibk.ac.at/library/ralser-unwert.html; Thomas Mayer, "Eugenische Forschung als 'eine politische nationalsozialistische Tätigkeit': Die

akademische Verbindung von Eugenik, Anthropologie, Kriminalbiologie und Psychiatrie am Beispiel des Karriereverlaufs von Friedrich Stumpfl (1902–1997)," in *Schriftenreihe der Deutschen Gesellschaft für Geschichte der Nervenheilkunde*, ed. Bernd Holdorff and Ekkehardt Kumbier (Würzburg, Germany: K&N, 2009), 15:239–65. On German eugenics, see Peter Weingart et al., *Rasse, Blut und Gene: Geschichte der Eugenik und Rassenhygiene in Deutschland* (Frankfurt: Suhrkamp, 1996), 438–45; Paul J. Weindling, *Health, Race and German Politics between National Unification and Nazism, 1870–1945* (Cambridge: Cambridge University Press, 1989), 515–17.

2. Thomas Mayer, "Familie, Rasse und Genetik: Deutschnationale Eugeniken im Österreich der Zwischenkriegszeit," in *Eugenik in Österreich: Biopolitische Strukturen von 1900–1945*, ed. Gerhard Baader, Veronika Hofer, and Thomas Mayer (Vienna: Czernin Verlag, 2007), 162–83.
3. Weindling, *Health, Race and German Politics*, 514.
4. For Stumpfl, see Mayer, "Eugenische Forschung"; for Thums, see Michal Šimůnek, "Ein neues Fach: Die Erb- und Rassenhygiene an der Medizinischen Fakultät der Deutschen Karls-Universität Prag 1939–1945," in *Wissenschaft in den böhmischen Ländern 1939–1945*, ed. Antonín Kostlán(Prague: KLP, 2004), 9:190–316.
5. Verena Pawlowsky, "Quelle aus vielen Stücken: Die Korrespondenz der Anthropologischen Abteilung des Wiener Naturhistorischen Museums bis 1938," in *Vorreiter der Vernichtung? Eugenik, Rassenhygiene und Euthanasie in der österreichischen Diskussion vor 1938*, ed. Heinz Eberhard Gabriel and Wolfgang Neugebauer (Vienna: Böhlau, 2005), 145; Mayer, "Familie, Rasse und Genetik," 179–80.
6. Weindling, *Health, Race and German Politics*, 511–13; Weingart et al., *Rasse, Blut und Gene*, 437–39.
7. See my forthcoming book on the history of the Department of Racial Biology at the University of Vienna, due to be published in 2014.
8. Minister of the Interior and Culture Plattner to President of the University of Innsbruck [name not mentioned], 4 May 1939, UAI, IERB.
9. Reich Minister of Science, Education, and Culture [name not mentioned], Berlin, 30 September 1939, UAI, IERB.
10. Mayer, "Eugenische Forschung," 247.
11. For the establishment and discussion of phenogenetics in Berlin, see Hans-Walter Schmuhl, "Grenzüberschreitungen: Das Kaiser-Wilhelm-Institut für Anthropologie, menschliche Erblehre und Eugenik, 1927–1945," in *Grenzüberschreitungen: Das Kaiser-Wilhelm-Institut für Anthropologie, menschliche Erblehre und Eugenik 1927–1945*, ed. Hans Walter Schmuhl (Göttingen: Wallstein, 2005), 9:319–27.
12. The first assistant was the medically trained Dutch Johannus Antonius van der

Meulen, and the Luxembourgian Armand Mergen came later. The student activist Otto Scrinzi (who worked as a psychiatrist, eugenicist, and right-wing politician after 1945) worked as Stumpfl's undergraduate research assistant from 1940 onward. Mayer, "Eugenische Forschung," 250–52.

13. Mayer, "Eugenische Forschung," 253–54.
14. Friedrich Stumpfl, "Geistige Störungen als Ursache der Entwurzelung von Wanderern," in *Der nichtseßhafte Mensch: Ein Beitrag zur Neugestaltung der Raum- und Menschenordnung im Großdeutschen Reich*, ed. Bayrischer Landesverband für Wanderdienst (Munich: Bayrischer Landesverband für Wanderdienst, 1938), 276.
15. Armand Mergen, *Die Tiroler Karner: Kriminologische und kriminalbiologische Studie an Landfahrern (Jenischen)* (Mainz, Germany: Internationaler Universum Verlag, 1949), 8, 13, 22–25, 173–74.
16. Stumpfl's CV, 23 January 1946, UAI, IERB; Stefan Lechner,"'Deshalb bitte ich, [. . .] mir dieses Unglück nicht anzutun'" NS-Zwangssterilisationen," in *Tirol und Vorarlberg in der NS-Zeit*, ed. Sabine Pitscheider and Rolf Steininger (Innsbruck: Studienverlag, 2002), 241–42.
17. Stumpfl's CV, 23 January 1946, UAI, IERB.
18. Stumpfl to Sonderkomission [*sic*], 10 June 1946, UAI, IERB; Stumpfl to Federal Ministry of Education, 25 June 1946, UAI, IERB.
19. For Austria, see Maria Teschler-Nicola, "Aspekte der Erbbiologie und die Entwicklung des rassenkundlichen Gutachtens in Österreich bis 1938," in Gabriel and Neugebauer, *Vorreiter der Vernichtung?*, 99–138; for Germany, see Georg Lilienthal, "Arier oder Jude? Die Geschichte des Erb- und Rassenkundlichen Abstammungsgutachtens," in *Wissenschaft auf Irrwegen: Biologismus—Rassenhygiene—Eugenik*, ed. Peter Propping and Heinz Schott (Bonn: Bouvier, 1992), 66–84; Hans-Peter Kröner, "Von der Vaterschaftsbestimmung zum Rassegutachten: Der erbbiologische Ähnlichkeitsvergleich als 'österreichisch-deutsches Projekt' 1926–1945," *Berichte zur Wissenschaftsgeschichte* 22 (1999): 257–64.
20. Earlier literature supported Polland's view by assuming the existence of a separate department. See, e.g., Weingart et al., *Rasse, Blut und Gene*, 439.
21. Minutes of Graz Faculty Meeting, 20 October 1938, UAG, medF, Zl.1ex1938/39.
22. Thomas Mayer, "'. . . daß die eigentliche Rassenhygiene in der Hauptsache das Werk Reichels ist': Der (Rassen-)Hygieniker Heinrich Reichel (1876–1943) und seine Bedeutung für die eugenische Bewegung in Österreich," in Gabriel and Neugebauer, *Vorreiter der Vernichtung?*, 87.
23. Lenz to Hafferl, 19 October 1938, UAG, medF, Zl.372ex1938/39.
24. Hafferl to Lenz, 25 October 1938, UAG, medF, Zl.372ex1938/39.
25. Dean Hafferl, note re: the chair in racial science and racial hygiene, 5 December 1938, UAG, medF, Zl.372ex1938/39.

26. Walter Höflechner, *Geschichte der Karl-Franzens-Universität Graz: Von den Anfängen bis in das Jahr 2005* (Graz, Austria: Karl-Franzens-Universität Graz, 2006), 197.
27. Personal file of Ernst Seelig, ÖStA, AdR, 03, K10/87.
28. Mayer, "'...daß die eigentliche Rassenhygiene," 96.
29. Hafferl might have meant "Kurator der wissenschaftlichen Hochschulen in Graz und Leoben" (supervisor of universities and colleges of technology in Graz and Leoben).
30. Hafferl to REM, 31 January 1942, German Federal Archives, Berlin, R4901/13156.
31. Hafferl to REM, 15 January 1945, German Federal Archives, Berlin, R4901/13156.
32. Pollands' personal file, UAG, medF, PA Polland.
33. The dean argued that Polland's engagement with eugenics negatively affected his income from private practice. Hafferl to Ministry of the Interior and Culture, 10 January 1940, ÖStA, AVA, Ministry of Education, 5 Graz Med., Fasz.885, Rudolf Polland.
34. Wolf Bauermeister and Maria Küper, "Die erbbiologische Abstammungsprüfung," *Fortschritte der Erbpathologie, Rassenhygiene und ihre Grenzgebiete* 6 (1942): 146–47, 154.
35. Pollands' personal file, UAG, medF, PA Polland.
36. Until July 1942 Polland produced forty-six expert opinions on the subject of ancestry and paternity. Bauermeister and Küper, "Die erbbiologische Abstammungsprüfung," 147, 165, 189. He was a member of the Hereditary Health Court in Graz. Maria Ladinig, "Das Gesundheitswesen, das Erb- und das Blutschutzgesetz, die Vorgaben der NS-Rassenpolitik und ihre Umsetzung im Gau Steiermark," in *Medizin und Nationalsozialismus in der Steiermark*, ed. Wolfgang Freidl et al. (Innsbruck: Studienverlag, 2001), 58–85.
37. Saurer, "Institutsneugründungen 1938–1945," 317–18.
38. Christian to Löffler, 22 March 1938, UAW, PhilF, PA Weninger.
39. Thomas Mayer, "Akademische Netzwerke um die 'Wiener Gesellschaft für Rassenpflege (Rassenhygiene)' von 1924 bis 1948" (MA thesis, University of Vienna, 2004), 203–13; Teschler-Nicola, "Aspekte der Erbbiologie."
40. Löffler to Christian, 26 March 1938, UAW, PhilF, PA Weninger.
41. Schmuhl, "Grenzüberschreitungen," 371.
42. Loeffler to Mika, 12 August 39, ÖStA, AVA, Fasz.763.
43. Medical Faculty committee's meeting protocol, 9 May 1938, UAW, RBI.
44. Sample letter, 10 May 1938, UAW, RBI.
45. Stigler to Dekan, 13 May 1938, UAW, PhilF, PA Loeffler.
46. Bach to Pernkopf, 11 May 1938, UAW, RBI.
47. Loeffler to Pernkopf, 17 May 1938; Kranz to Pernkopf, 18 May 1938; Lenz to Pern-

kopf, 19 May 1938; Verschuer to Pernkopf, 21 May 1938; Fischer to Pernkopf, 23 May 1938; Rüdin to Pernopf, 25 May 1938, UAW, RBI.

48. Plattner and Pernkopf's report, 20 June 1938, UAW, RBI.
49. Geyer to Mika, 8 December 1939, UAW, RBI.
50. Plattner to REM, 4 July 1939, UAW, Rektorat, 755ex1938/39.
51. For the KWI-A, see Schmuhl, "Grenzüberschreitungen," 214.
52. Saurer, "Institutsneugründungen 1938–1945," 318–19.
53. Löffler to Mika, 12 August 1939, ÖStA, AVA, Fasz.763.
54. Christian to Greite, 17 May 1940, UAW, PhilF, 383ex1939/40.
55. De Crinis to Knoll, 7 May 1940, UAW, PhilF, 383ex1939/40.
56. Ebernard Geyer, "Wissenschaft am Scheideweg," *Archiv für Rassen-und Gesellschaftsbiologie* 37 (1943): 2–3, 6.
57. Walther von Boeckmann, note on the debate re: RBI, 16 May 1942, UAW, RBI.
58. Christian to Knoll, 1 June 1942, UAW, Rektorat, 755ex1938/39.
59. The following individuals joined the department as assistants: Hans Kuttelwascher (former assistant at the Department of General and Experimental Pathology); Johann Jungwirth (in the 1960s head of the Department of Anthropology at the Natural History Museum in Vienna); Charlotte Tiemann; and Egon Gast (who was later replaced with Rudolf Harnisch). Universität Wien, *Personal-Verzeichnis, Studienjahr 1944/45* (Vienna: Verlag Universitäts Wien, 1944), 64.
60. Löffler to Pernkopf, ÖStA, AVA, Fasz.76; Weingart et al., *Rasse, Blut und Gene*, 439.
61. Oliver Paget, "Strahlengenetische Untersuchungen mit Gammastrahlen des Radiums an Drosophila melanogaster nach einer erweiterten ClB-Methode," (PhD diss., University of Vienna, 1947).
62. Maria Rentetzi, "Trafficking Materials and Gendered Experimental Practises: Radium Research in Early 20th Century Vienna," Gutenberg-e, accessed July 29, 2010, http://www.gutenberg-e.org/rentetzi/index.html; Herwig Czech, *Erfassung, Selektion und 'Ausmerze': Das Wiener Gesundheitsamt und die Umsetzung der nationalsozialistischen 'Erbgesundheitspolitik' 1938 bis 1945* (Vienna: Deuticke, 2003), 77; Claudia Andrea Spring, *Zwischen Krieg und Euthanasie: Zwangssterilisationen in Wien 1940–1945* (Vienna: Böhlau, 2009), 242.
63. Ute Deichmann, *Biologen unter Hitler: Porträt einer Wissenschaft im NS-Staat* (Frankfurt: Fischer, 1995), 81.
64. Loeffler to the Office for Wartime Economy of the Reich Research Council, 17 February 1943, UAW, medF, PA Loeffler.
65. Horst Geyer, *Über die Dummheit: Ursachen und Wirkungen der intellektuellen Minderleistungen des Menschen* (Göttingen: Musterschmidt, 1954).

66. Hubenstorf, "Medizinische Fakultät 1938–1945," 261.
67. Loeffler to the Office for Wartime Economy of the Reich Research Council, 17 February 1943, UAW, medF, PA Loeffler.
68. Spring, *Zwischen Krieg und Euthanasie*, 168.
69. Hans-Peter Kröner, "Förderung der Genetik und Humangenetik in der Bundesrepublik durch das Ministerium für Atomfragen in den fünfziger Jahren," in *Wissenschaft und Politik: Genetik und Humangenetik in der DDR (1949–1989)*, ed. Karin Weisemann et al. (Münster, Germany: LIT, 1997), 69–82.

7

Biological Racism and Antisemitism as Intellectual Constructions in Italian Fascism

The Case of Telesio Interlandi and *La difesa della razza*

ELISABETTA CASSINA WOLFF

The promulgation of antisemitic and racial laws in Italy in 1938 prompted no discernible public reaction. Almost unanimously, with the exception of Benedetto Croce, silence followed the expulsion of Jewish academics from Italian universities. Even more striking, though, was the fact that Italian antisemitism, Fascist racial laws, and racial discourse were totally expunged from public consciousness in postwar Italy.[1] The fate of Jews under Fascism, from discrimination to persecution, was largely forgotten until *Storia degli ebrei italiani sotto il fascismo* (A history of Italian Jews under the Fascist regime) was published by the historian Renzo De Felice in 1961.[2] De Felice argued that official racism was introduced to Italy after 1937 in the aftermath of the war in Ethiopia and the establishment of an Italian Fascist empire, while antisemitism had become a dominant ideology in the wake of the political alliance with Germany. He maintained that the establishment of an Italian Fascist empire in Africa encouraged the Fascist regime to think comprehensively about racial categorization and the concept of higher and lower races, thus providing a context for the antisemitic laws that followed. At the same time, he emphasized the marginality of popular antisemitism in Italy in comparison to Nazi Germany, detailing the multiple obstacles that racial legislation encountered in its implementation. De Felice's work was a "courageous and important book," as the Marxist historian Delio Cantimori wrote in the preface. Nevertheless, for decades, Mussolini's antisemitism continued to be neglected by scholars and academics. For one thing, it appeared less authentic than Nazi antisemitism. It was interpreted, following the guidelines outlined by De Felice, as an imitation of German racism developed for opportunistic reasons related to foreign policy. Furthermore, a minority of neo-Fascists attributed it to a radicalization of the conservative factions in the Fascist regime and a deformation of an authentic Fascist worldview, which reached its apogee during the Salò Republic. Filippo Focardi, meanwhile, has identified the origins of the official national amnesia toward the persecution of the Italian Jews in the attempts made in the immediate postwar period by the political class to minimize the antisemitism of the Fascist regime and thus partially rehabilitate it.[3] More recently, a younger generation of Italian historians has contended that the historiographical marginalization of this chapter in Italian history is in part due to the resistance of Italian intellectuals across the political spectrum, reluctant to face their own past.[4] In the past decade Italian scholars have begun to challenge the commonly held orthodoxy of Italian racism in late Fascism as little more than an import from Nazi Germany as well as the popular conception of Italian antisemitism as moderate and marginal. Scientific racism existed in Italy, as elsewhere in Europe, long before Hitler came to power, and antisemitism had deep cultural roots, not least in the Catholic tradition.

This chapter analyzes the development of racism and antisemitism in Italy within the narrow intellectual community that gravitated around the journalist Telesio Interlandi and his journals. An insight into the intellectual life of Interlandi may help to map the development of Italian racism and antisemitism from the beginning of the 1920s until the collapse of the Fascist regime. This chapter contends that racism and antisemitism followed different paths in Italy, shaped by divergent influences, political and cultural as well as domestic and international. Interlandi was an intellectual who strongly supported the biological interpretation of racism in Italy, close to Nazi theories, in opposition to numerous other intellectual Fascist interpretations. An examination of the racial ideas and theories articulated in Interlandi's journals and by the circle of intellectuals who published in them enables a broader perspective on the diverse currents in Italian racism and antisemitism, including the development of biological racism and the role of intellectuals in the dissemination of an Italian racial doctrine. It also facilitates a better understanding of the relationship between legislation and propaganda under the regime and the impact of Nazi theories of race on the shaping of that legislation. It argues that such an impact was indeed tangible, but while in some cases it was pervasive and authentic in others it remained marginal as a result of political and ideological factors.

Telesio Interlandi, Racial Theories, and Fascist Policies during the 1920s and 1930s

Telesio Interlandi became the editor-in-chief of the daily *Il Tevere*, founded by Benito Mussolini in December 1924.[5] The appointment marked the beginning of a brilliant career. For twenty years Interlandi could count on Mussolini's protection and benefited from an incredible freedom of expression in a time when censorship was commonplace in the Italian media. Interlandi's interpretation of Fascism was radical and intransigent, centered on a deep-seated hatred of the bourgeoisie, a demonization of political enemies, and a profound belief in conspiracy theory. The antisemitism preached in the 1920s by *Il Tevere* identified Jews with antifascism and Hebraism with Europeanism and internationalism. Hence the term *Jew* was understood to mean an enemy writ large: a *political* enemy as antifascist, internationalist, and Europeanist; and an *economic* enemy as capitalist and banker. At this point there was little discernible association between antisemitism and racism. Jews were instead placed in a wide category of national enemies, along with Communists, Freemasons, democrats, and antifascists in exile—all of them conceivably involved in a joint conspiracy against Fascism.

Although *Il Tevere* proved a confrontational publication, an exclusive conduit for antisemitism it was not. Antisemitism was featured more prominently in the magazine edited by Giovanni Preziosi, *La vita italiana* (1913–43), which had a great intellectual impact in the 1920s Italy; in *Il regime fascista* (1926–45), owned by Roberto Farinacci; and in the Jesuit journal *La civiltà cattolica*. Since the days leading up to the March on Rome, Preziosi—a former priest and convinced nationalist, he was also antisocialist and antidemocratic—had written numerous articles on the alleged connection between Judaism, Bolshevism, plutocracy, and Freemasonry. He was the first person in Italy to translate, in 1921, the pamphlet *The Protocols of the Elders of Zion*. In the 1920s Preziosi was considered to be the most committed and passionate antisemite among Italian Fascists, despite the fact that he showed no interest in racial theories.[6]

While the intellectual independence of Preziosi has never been in dispute, scholars have long debated whether *Il Tevere* was subject to German ideological directives (prompted by subsidies from Hitler's regime) or whether it was an independent organ.[7] Meir Michaelis, for example, has stated that there is no historical evidence that Interlandi had been paid by the Nazis.[8] The fact that as early as 1926 Alfred Rosenberg, chief Nazi ideologist and editor-in-chief of the Nazi Party newspaper *Völkischer Beobachter*, commented positively on *Il Tevere* does not necessarily prove a direct ideological link to Hitler's Germany.[9] Interlandi spoke no German, whereas he was in contact with the French intellectuals of Action Française, Léon Daudet, and other European intellectuals as well as the poet Ezra Pound. Hence his early antisemitism was probably more influenced by the French brand of antisemitism than by the German one.

Meir Michaelis has argued that *Il Tevere* was Mussolini's *unofficial* organ, just as *Il popolo d'Italia* was the official one.[10] According to this interpretation, Interlandi was not only a journalist entirely dependent on the support of his patron but also a sort of spokesman for ideas that the Duce did not dare to express publicly. The question of the real intentions of Mussolini, of his attitude with regard to antisemitism, is rather complicated. Much more promising is to focus on the actual policies implemented by the Fascist government; the official documents, statutes, and laws approved by the Duce; and the treatises written by him.

The only programmatic Fascist text in existence, "Doctrine on Fascism," cowritten by Mussolini and the Fascist philosopher Giovanni Gentile and published in 1932 in the *Enciclopedia Italiana*, contains no references to racial theories and/or antisemitism. On the contrary, the text affirms an idea of nation inspired by the patriot Mazzini, built around the concept of voluntarism, and hence opposed to racial conceptions. The same year Mussolini declared in an interview with the German-Jewish journalist Emil Ludwig that racial antisemitism did not exist in Italy and that German biological racism was a "stupidity."[11] Simultaneously, he suggested that Hitler should gradually adopt measures against the Jews.[12] By that time certain discriminatory acts had already been implemented in Italy, for example, exclusion of Jews from a number of professional positions, including the Italian Academy of Arts and Sciences. Nonetheless, Mussolini was still far from linking antisemitism to racial science and even further from

Nazi concepts of Aryanism. His preoccupation with Jews had to do with both politics and economics.

Recent scholarship on race and eugenics in Italy has focused on the Italian contribution during the 1920s and 1930s to these fields of research, demonstrating how Italian scientific circles prepared the ground, albeit unintentionally, for the political racism of the late 1930s.[13] Italian eugenics was race based, as it was elsewhere in Europe and the United States, at a time when *race* was a neutral political term. Italian scientists believed in the existence of biologically predetermined races and in the possibility of improving the Italian racial stock. At the same time they rejected negative eugenic measures implemented in some Protestant countries of northern Europe, such as euthanasia, the incarceration of the handicapped, compulsory sterilization, premarriage certificates, and birth control (including forced abortion). Instead, they advocated positive eugenics, which advanced hygiene, education, and physical exercise; improved conditions at home and work; and improved economic conditions for the ordinary Italian family. The Italian tradition, both Latin and Catholic, proved an obstacle to the introduction of negative eugenics, which was perceived by Fascist ideologists as an aberration and sign of cultural degeneration.

One of the leading figures within Italian eugenics was the anthropologist Giuseppe Sergi. Sergi was the author of the so-called Mediterranean Thesis, dating back to 1895. He claimed that the Hamitic race, originating in the northeastern part of Africa, had given birth to the three main racial groups in both the African and European continents: the Nordic, the Mediterranean, and the African. According to this theory, the Italian race had its roots in the Mediterranean group. Sergi's theory was in direct conflict with many of the other theories claiming the European-Caucasian origin of the "Aryan" race. In the 1920s Sergi was the leader of a scientific school that was implicitly anti-German and anti-Aryan, and one supported by a majority of Italian thinkers. Among other eminent Fascist intellectuals who shared Sergi's ideas were Nicola Pende, director of the Institute of Special Medical Pathology at the University of Rome; Sabato Visco, director of the Institute of General Physiology of the University of Rome and director of the National Institute of Biology of the National Research Council; and Giacomo Acerbo, Fascist politician and minister, dean of the Faculty of Economy in Rome, director of the International Institute of Agriculture from 1934 until 1943, and member of the Grand Fascist Council (Gran Consiglio).

At the beginning of the 1930s the vast majority of Italian scientists advocated a sort of Latin model of eugenics, a less extreme version of the racial science that was gaining popularity elsewhere in Europe and the United States, in countries and societies terrified by the prospect of social, biological, and moral degradation. Aside from the principled rejection of negative eugenics, there were two key elements that underscored the contrast between Italian and Nordic eugenics, especially German eugenics: the concept of a pure race and the question of birth control. As far as the latter was concerned, the Latin and Catholic culture of Italy protected the central role of the mother within the family. Regarding the former, Italian scientists were keener to use the term *stock* (*stirpe*) instead of *race*. They claimed that in the case of Italy it was more appropriate to speak of several stocks that had molded the Latin, or Mediterranean, race. At the same time, Italian scientists held the belief that the Italian "race," created through the fusion of several stocks, was superior to others.

In 1924 the first Congress on Social Eugenics was held in Italy. Shortly thereafter the neurologist Ettore Levi founded the National Maternity and Infant Agency (Opera nazionale maternità e infanzia), which introduced a series of programs aimed at improving the health of Italians. The Fascist government lent its support to the activities of the agency. One of the most notable achievements of the agency was the establishment of a system of state-financed maternity centers where women and their children had access to free medical care. During the 1920s and the 1930s the regime promoted a campaign to improve the health of the nation and raise the birthrate in order to increase the size of the Italian "stock." Seen from an ideological point of view, health and pronatalist policies were well integrated within the Fascist project of the economic, military, and political strengthening of the nation. Racial questions, by contrast, were reduced to an equation of the numerical strength and the vitality of Italians; neither biological racism generally nor antisemitism specifically played any part in this equation. Likewise, the strengthening of the Italian nation was to be spearheaded by positive eugenics, not negative eugenics. Social bodies such as the National Maternity and Infant Agency aimed to improve the health of expectant mothers and reduce mortality rates; health agencies placed emphasis on preventative health, paying particular attention to the diets and exercise regimes of ordinary Italians; and the Balilla summer camps for children addressed the health needs of youth. A range of new social welfare initiatives and laws served the same purpose: a Charter of Labor aimed to guarantee improved working conditions; subsidies given to large families encouraged married couples to have more children; and regulations providing for more effective sanitation helped to combat disease and increase life expectancy.

The majority of Italian anthropologists, physicians, and demographers had no reason to oppose the government's initiatives in this sphere of public policy; most of them actually supported the so-called national, or Mediterranean, current.[14] Sergi was considered the scientific leader of the national current, while Giacomo Acerbo was considered its political leader. Moreover, as long as the racial question was reduced to health care and demographic policies or to political propaganda against the hedonistic bourgeoisie who did not procreate enough, it could count on the support of the Catholic Church. At the beginning of the 1930s Interlandi's *Il Tevere* was still strongly influenced by this dominant "Mediterranean" discourse. The same could also be said of Interlandi's other publication, the literary weekly *Quadrivio* (1933–43).

Colonial Warfare and the Defense of the Race

During the 1920s and the beginning of the 1930s, Interlandi's intellectual production reflected an Italian reality in which both racial theories and antisemitism had a political and ideological function, though they were independent from each other as well as from Nazi influence. With regard to racism, it was the 1935 military campaign in Ethiopia (Abyssinia) that marked the beginning of a new phase in Fascist politics as well as in Interlandi's intellectual career. The conquest of new African territories forced the regime to take a more systematic approach to racial questions, especially after Mussolini had declared the creation of the "Italian Empire" in East Africa (comprising Ethiopia, Somalia, and Eritrea) on 9 May 1936. That same month the regime introduced racial laws that enforced strict racial separation between Italians and indigenous Africans in the African territories of the new Fascist empire. A state decree of June 1936 barred children from mixed families from legally obtaining Italian citizenship. Thus the racial laws in the Fascist empire increasingly attracted a biological interpretation. While Italian Jews—many Jewish intellectuals had been committed Fascists from early on—continued to support the regime, including its racial policies in Africa, the Duce continuously underlined the difference between Italian racism and German antisemitism.

Paradoxically, it was from this point onward that biological racism and concomitant antisemitism moved from marginality to the mainstream. In 1935–36 *Il*

Tevere launched a polemical campaign against the economic sanctions imposed on Italy, supposedly promoted by England. Telesio Interlandi insisted on the existence of an anti-Italian conspiracy in which Zionist Englishmen were colluding with Masons, Jewish bankers, and internationalists. Simultaneously, Interlandi began to invoke science in order to defend the racial laws in Africa.[15] Biological racism, initially confined to Africa, hereby made its appearance. In 1935 the young jurist and philosophy student Giulio Cogni began to regularly contribute to *Quadrivio*.[16] A Germanophile racist, he had shaped his worldview to a large extent in Germany, at the Italian-Germanic Institute in Hamburg where he was on a scientific exchange. It was there that he had come in contact with one of the preeminent exponents of Nordic racism, the German racial scientist Hans F. K. Günther. That same year Interlandi introduced Cogni to Mussolini. Cogni's articles that appeared in *Quadrivio* and *Il Tevere* were reprinted in two books published in 1937: *Il razzismo* (Racism) and *I valori della stirpe italiana* (The values of the Italian stock), with a preface by Günther. In the former book Cogni related the essential points of Günther's biological racism, while in the latter he focused on the "Nordic" characteristics of the Italian race. Both books referred to Hitler's Nordic mythology and postulated the superiority of the blonde race, opposing the "Aryan man" to the "beastly man" of Ethiopia. They also claimed a common origin for Italians and Germans based on a shared "Aryan/Nordic" race. Furthermore, Cogni declared his intellectual support for German eugenics, in particular the German law on sterilization of 14 July 1933.[17] However, neither of the books mentioned the Jews.

Cogni's books had no success and drew universal criticism. Many Fascists did not like the obvious imitation of German racism. For their part, Jewish intellectuals appreciated that Cogni had recognized the absence of a Jewish Question in Italy. While the Nazis condemned Cogni for his deficient antisemitism, the Catholic Church denounced the influence of the German racial theorist and pagan Alfred Rosenberg, condemned Cogni's own paganism, and in June 1937 placed Cogni's works on its index of prohibited books.[18] By contrast, the 1937 book *Gli Ebrei in Italia* (The Jews in Italy), by Paolo Orano, attracted more attention. The book asserted a form of antisemitism based on political, economic, and religious arguments yet was critical toward the Nordic brand of racism. In spite of Cogni's fiasco, Interlandi continued to show interest in his biologically defined and Nazi-inspired racist theories. Subsequently he engaged two other journalists: Helmut Gasteiner, an Italian citizen from the Alpine region who moved to the Third Reich after the Anschluss, and Giuseppe Pensabene. In 1937 they were put in charge of the column "Il razzismo è all'ordine del giorno" (Racism is on the agenda) in *Quadrivio*. Both Pensabene and Gasteiner extolled Nazi eugenics methods while questioning the effectiveness of the "Latin" kind of eugenics pursed by the Italian regime. In particular, they supported sterilization of alcoholics and the handicapped. At the same time, they also referred to arguments presented by scientists in social democratic Sweden, based on a calculation of costs and benefits when reducing the number of mental hospitals and nursing homes. It was in this context that a connection between antisemitism and biological racism made its first appearance. Pensabene and Gasteiner wrote that the Jewish Question should be addressed not only on a political, economic, or religious level but also on a racial level. They argued that in Italy and throughout the Fascist empire, proper, strict laws should be introduced by the regime in order to forestall marriages between different races.[19]

In the fall of 1937 Interlandi introduced yet another young anthropologist, Guido Landra, to Mussolini.[20] Landra had been the correspondent for *Il Tevere* in Germany and was a lecturer in anthropology at the University of Rome. Although initially Landra supported the idea that the Italian race had its roots in an African and Mediterranean race, he quickly changed his mind, becoming a strong supporter of an ideological rapprochement with German racism based on the purity of the Aryan race.[21] In February 1938 Mussolini instructed Landra to set up a scientific committee to prepare a racial campaign in Italy. In the spring of 1938 Landra probably attended a number of conferences organized in Italy by Dr. Eugen Fischer, the long-serving director of the Kaiser Wilhelm Institute for Human Heredity Sciences and Eugenics in Berlin (Kaiser Wilhelm Institut für Anthropologie, menschliche Erblehre und Eugenik, or KWI-A) on the issue of race. In any case, Landra kept in touch with Fischer over the following years.[22]

On 14 July 1938, a few weeks after Hitler had paid a visit to Italy, *Il giornale d'Italia* published a long article titled "Il fascismo e i problemi della razza" (Fascism and racial problems), which later became known as "Manifesto della razza" (The race manifesto). The main author of the article was Guido Landra.[23] In the printed version, the Mediterranean racial identity of the Italians, which had thus far been defended by the Duce himself, was superseded by an Aryan identity, on the grounds of a biological concept of race. Moreover, the document marked the definitive transformation of cultural antisemitism into biological racism insofar as it stated that the Jews did not belong to the Italian "Aryan" race. Immediately following the publication of the "Race Manifesto," the Grand Fascist Council approved a series of anti-Jewish laws. Jews were no longer allowed to be members of the National Fascist Party, to be state employees, to own firms with more than one hundred employees or more than fifty hectares of land, or to serve in the military. In addition, the laws put further restrictions on mixed marriages and withdrew Italian citizenship from those Jews who had obtained it after 1918.[24]

In a leading article in *Il Tevere,* Interlandi welcomed the "Race Manifesto" as the first expression of a new Italian scientific school that proclaimed "freedom from prejudice and attention to political needs" and was ready to combat the "anachronistic currents within the Italian university."[25] Interlandi considered the Jewish Question an excellent example of the new thinking. Thus he wrote, "Here is the Jewish problem, imposed biologically rather than politically."[26] By then Interlandi and Landra were emerging as two of the leading figures within the so-called biological racist current, or Nordic current.[27] This current acquired identity in opposition to the national current inspired by Sergi's anthropology.

Meanwhile, Minister of Popular Culture Alessandro Pavolini was planning to open a brand-new Racial Office (Ufficio per la Razza), which began operating in August 1938 with the sole purpose of disseminating racial propaganda. The conflict that evolved between the pro-Nordic Fascists who backed the "Race Manifesto" and a biological interpretation of race and the supporters of the Mediterranean Thesis, who defended the idea of Rome and *romanità,* had political repercussions. Landra, whose writings had first announced the transition to a racial-biological definition of Italian national identity, was replaced after only a few months by Sabato Visco as the director of the Racial Office. The exponents of the national current within the Racial Office did everything possible to boycott the Nordic orientation. Pende and Visco exercised political pressure on prominent figures in the regime in order to modify the line expressed by the "Race Manifesto." They opposed the use of the term *Italian race,* considered the term *Aryan* a useful concept denoting a linguistic or cultural group yet having no scientific basis in biology, and rejected out of hand the Nordic orientation of the document. Several Fascist intellectuals and politicians appear to have agreed with them. The Racial Office promoted various projects such as school lessons that informed students about the Mediterranean characteristics of the Italian people while omitting any references to Nordic, Aryan, or Germanic racial traits. The kind of antisemitism promoted by the Racial Office's propaganda was similarly moderate.[28] In 1938 Giacomo Acerbo managed to gain control over an important policymaking institution, the Superior Council on Demographics and Race, working hard to promote the Mediterranean Thesis.[29] The fact remains, though, that the "Race Manifesto" was never repealed.

Despite this, the new racial biological discourse also made its presence felt in the Racial Office. Coincidentally, one of the office's first actions was to support the publication of a new, biweekly popular journal entirely devoted to the issue of race, *La difesa della razza*, edited by Interlandi. Subsidized by the Ministry of Popular Culture, 140,000 copies were distributed free of charge to the main public offices of Italy (the first issue of the journal appeared on 5 August 1938).[30] The popularity of the journal is debatable, however; by 1940 its print run had declined dramatically, falling to 20,000–25,000 copies.[31] The journal presented a vast variety of articles, covering aspects of Fascist racial legislation and its social and political impact, the defense of the race in the empire, the history of Hebraism, and scientific contributions to racial theories grounded in biology, anthropology, paleontology, and geography.[32] During the six years of the journal's existence, Interlandi participated in the intellectual debate by advocating a radical position on matters of racism and antisemitism. As a reflection of his commitment to racial issues, in August 1938 Interlandi published the explicitly antisemitic pamphlet "Contra Judaeos." Aside from Interlandi, Landra was undoubtedly the most influential personality on the editorial board of *La difesa della razza*, supported by a cohort of young racial scholars who had backed him in the drafting of the "Race Manifesto."

Colonialism, Race, and Eugenics in Interlandi's *La difesa della razza*

One of the recurrent themes addressed by the intellectuals of *La difesa della razza* was the relationship between the racial question and the new Italian colonial empire. Francesco Cassata has emphasized the prominent role played by the young anthropologist Lidio Cipriani in shaping intellectual responses to this issue.[33] Cipriani was professor of anthropology at the University of Florence and director of the National Museum of Anthropology and Ethnology. A signatory to the 1938 "Race Manifesto," he began working at the Racial Office and contributing to *La difesa della razza* almost from its inception. In his studies popularized by *La difesa della razza* Cipriani claimed that the Ethiopians were the original stock from which the Hamitic race had derived, while the white race had its origins on the European continent.[34] The qualities of certain African tribal groups such as the Egyptians and the Zulus, he explained, were due to the "superior" racial characteristics of the Ethiopians. Thus Cipriani disavowed Sergi's theory. Instead he embraced ideas promoting the purity of the Aryan race. He took on the task of reconciling, on a theoretical level, Fascist racism, anthropological theories, and colonial policies, effectively offering intellectual and scientific justification for the regime's colonialist propaganda. According to Cipriani, it was not only a moral duty for the white race to bring civilization to Africa and to exploit the continent's natural resources—which otherwise would have remained untouched by Africans—but also a socioeconomic imperative. Although the theories outlined by Cipriani in *La difesa della razza* referred exclusively to the Ethiopians, or Pygmies and Bushmen, they made the Italian public familiar with arguments that for the first time ever introduced both mental and cultural as well as somatic elements in order to prove the inferiority of one race compared to another. More generally, the numerous articles on the variety of races around the globe, rich in scientific references inspired by contemporary racial theories, contributed to popularizing the idea of the necessity of a racial hierarchy and, ultimately, racial segregation.

La difesa della razza also reflected on the relationship between eugenics and race. One of the questions under discussion was whether a race was exclusively determined by heredity and nature—in other words, by genes—or if it was determined by culture and habitat. The writings of Guido Landra that stressed the biological purity of the race and its transmission through blood became the official line of *La difesa della razza*.[35] His articles were mainly inspired by the ideas of scientists from the United States and Germany. The influence of Nazi eugenics was particularly evident in the communication between Landra and Eugen Fischer. In his capacity as director of the Racial Office, in fall 1938 Landra visited several German institutes of anthropology, where he came into contact with many German scientists, such as Otmar von Verschuer, director of the Department of Genetics at KWI-A. In 1941 Landra was again in Germany, visiting research institutes working on questions of racial hygiene: the laboratories of Eugen Fischer and Otmar von Verschuer; the Institute of Anthropology and Ethnology in Breslau, headed by Egon von Eicksted; and the Institute of Racial Improvement, under Heinrich Wilhelm Kranz, professor of medicine at the University of Giessen.[36]

Landra's advocacy of biological racism was supported by other scientists, physicians, and academics, as well as by lawyers who published in *La difesa della razza*. These authors were profoundly skeptical concerning the potential impact of positive eugenics as espoused by established Fascist thinkers in the fields of education and hygiene.[37] Instead of positive eugenics they proposed measures aimed at controlling the quality of the population. The president of the Italian Institute of Statistics, Franco Savorgnan, for example, supported the creation of a register that would list all the biological characteristics of Italians, a kind of national racial archive where the details of all Italian families would be stored in accordance with their genealogy.[38] The most radical proposals emerged in connection with two main issues: the question of cross-breeding and the fate of handicapped people.

Several articles printed in *La difesa della razza* demonstrated the dangerous consequences of racial cross-breeding, which, as far as the journal was concerned, was always detrimental to the physical and intellectual qualities of superior races. The authors referred to international studies and foreign scientists such as the German Eugen Fischer, the American Charles Davenport, the Swede Herman Lundborg, and the Norwegian Jon Alfred Mjøen, among many others.[39] The journal paid significant attention to the theories elaborated in the United States by Madison Grant and Lothrop Stoddard. Cross-breeding was treated not only as a biological danger but also as a moral threat capable of subverting all principles and values in the Western world. The racial discourse of *La difesa della razza* was consequently linked to political ideology. Fascism was presented as the political system that guarded the natural and divine racial order, protecting both social and political stability and security. Political systems based on liberalism and democracy, and thus open to racial chaos, as in France, however, were deemed to descend into social and political crisis.[40]

The fate of people with disabilities was also discussed, again largely with reference to German and American eugenics. Marcello Ricci, for one, supported any measures that could "limit the reproductive activity of individuals who are noxious to the race."[41] He called for the elimination of handicapped people through the eradication over several generations of all hereditary defects by means of abortion and forced sterilization. He also toyed with the idea of a prenuptial certificate that would force all engaged couples to consent to a premarriage physical examination before they could actually get married. Although this proposal was discussed in Italy during the 1920s, it had never been implemented by the Fascist regime because of strong opposition by the Church.

Ricci therefore studied legislation in countries such as the United States, Soviet Union, and Turkey in order to frame his own proposals accordingly.[42]

La difesa della razza gave readers the opportunity to publish commentaries and suggestions. Many such commentaries demonstrated unconditional enthusiasm for the proposals regarding sterilization and compulsory premarriage certificates, supporting methods of both persuasion and coercion being adopted as part of a broader policy for the improvement of the Italian race. When debating the choice between biological constitution and environment as influential factors weighing on racial development, however, the editorial board of the journal often took part in the discussion, warning readers not to deviate from "true and scientific" racism based on hereditary laws and genetic determinism.[43]

Catholic Anti-Judaism and Racial Antisemitism

Another key issue for *La difesa della razza* was antisemitism. Here the discourse followed three main avenues, all aimed at legitimizing and lending intellectual support to the regime's racial policy in the wake of the publication of the "Race Manifesto." The first approach consisted of giving a physical description of Jews that would easily identify them among Italians, marking the juncture between antisemitism and biological racism. In one of the first issues of the journal, Giuseppe Genna, director of the Institute of Anthropology at the University of Rome, provided a detailed description of the archetypal Jew so that Italian and Aryan citizens could recognize the "alien body" among them.[44] Jews were described by their somatic characteristics, according to strict racial biological criteria. The journal also addressed the topic of the supposed congenital criminality and predisposition to prostitution of the Jews.[45] The arguments were borrowed from both national history and Catholic tradition but also employed references to foreign writers such as the German psychiatrist Max Mikorey and the Nazi physician Edeltraut Bieneck.[46] The identification of Jews had to do with the regime's larger project aimed at registering all Jews in Italy.

In truth, the description of the Jews often moved back and forth between a biological and a spiritual definition, at least until 1942. Interlandi used the registration campaign to articulate his fear that the Jewish population might be even more numerically significant than previously believed.[47] In addition to the racial Jew, a number of writers drew attention to the subgroup of so-called invisible Jews, who conceivably posed an even greater threat. Umberto Angeli, for instance, warned against the "clandestine and self-unconscious" Jews hiding among Italians, pestering Italians—enemies of Fascism, traitors, or just the "pious" kind—who were keen on protecting those Jews whose identity was not in doubt.[48] Consequently, the crypto-Jews were not only those who might have had a Jewish ancestor but also those who had a Jewish mentality, attitude, character, or habits. The "Jew" thus remained the archetypical antifascist bourgeois, or the self-centered Italian bourgeois, under the influence of democratic-bourgeois cultural models.[49] As one of the priorities of the Fascist "revolution," *La difesa della razza* identified studying the "moral attitudes of the individual" and making a "distinction between the true Italian and the false, bringing forward the former, pushing back and eliminating the latter."[50]

The intellectuals of *La difesa della razza* firmly believed that the law of 17 November 1938, which provided the legal definition of the "Jewish race," was insufficient. This law exempted from discrimination all those Italians who were Christian and could prove to have at least one Aryan parent. Likewise, they criticized all Fascist norms that permitted the so-called *arianizzazione* (a legal process enabling one to acquire an Aryan identity). The journal unequivocally denied Italian Jewish converts the opportunity to escape the Italian antisemitic laws by means of seeking the protection of the Catholic Church. According to Gino Sottochiesa, the converted Jews were even more dangerous than the Jews who were faithful to the scriptures, as they disguised themselves and thus became virtually invisible.[51] Remarkably, whenever Interlandi's intellectuals crossed swords with the Catholic Church, the arguments they put forward were definitively informed by biological racism. They believed that, since Jews were marked by blood and race, it was impossible for the latter to be assimilated, whereas conversion could be motivated only by opportunism.[52]

The second approach linked Fascist antisemitism to Catholic anti-Judaism. Interlandi and the *La difesa della razza* contributors made substantial use of rhetoric in an attempt to convince ordinary Italians that the racial laws were not only the continuation of an Italian religious tradition but also the obvious outcome of such a tradition. In other words, they played on the prejudices against the Jews that had their origins in Catholic anti-Judaism.[53] Referring to a common theme within Catholic anti-Judaism, they described the Talmud not as a religious text but as a collection of norms that guided Jews through life and inspired their project for "world domination."[54] Within this interpretation, the primordial racism was identified as Jewish racism against Gentiles.[55] Moreover, the description of the so-called *accusa del sangue* (the blood libel) helped to define Jewish rites as typical examples of a culture impossible to assimilate.[56] It also promoted an argument that spoke to the alleged natural disposition of Jews to cruelty and criminality.[57]

The third approach consisted of rewriting Italian history in order to retroactively expunge the Jews from the national body. The journal set out, as an ideological objective, to prove the total incompatibility of Ebraismo, that is, the Jewish way of life, and Italianità, understood as a combination of Italian history, tradition, and spirit. The intellectuals of *La difesa della razza* provided a fresh interpretation of history, rendered in conspiratorial terms with an emphasis on an *ante litteram* antisemitism. Thus they even claimed to have identified a racial attitude toward the Jews among the ancient Romans.[58] They further argued that the history of the Middle Ages offered numerous examples of how alien Judaism had always been to Christian Italy.

As Francesco Cassata has argued, the construction of the Jews as "the eternal enemies of Rome" (here used as a synonym for Italy) permitted the nationalization of the Jewish Question.[59] In other words, the Jewish Question was retrospectively implanted into the nation's history. The antisemitic discourse of *La difesa della razza* effectively separated Italian Fascist racism from any foreign, and therefore alien, influence and thus legitimized all "massacres" and "deportations" that had occurred during previous centuries as historically and culturally inevitable—the "only possible solution of the Jewish problem."[60] An illustrative example of this thinking was provided by the young journalist Giorgio Almirante. Writing in 1938, he claimed that, "with regard to racism and anti-Judaism, the Italians never did need and never will need to learn from anybody."[61]

Likewise, Interlandi initiated a campaign to demonstrate how obstructive and hostile Jews were in the process of Italian unification, simultaneously promoting the theory of a Jewish conspiracy against the Italian nation.[62] In fact, *La difesa della razza* excelled at rewriting entire pages of Italian history in order to present the Jews, alongside the bourgeoisie, as traitors of the Italian people.[63] Italy's participation in the First World War was identified as the most glaring example of the Jewish conspiracy against the Italian nation. Claiming to have identified a link between Judaism and defeatism on the one hand and Judaism and hostility to Fascism on the other, the journal accused the Jews of both military sabotage and national betrayal.[64] Interlandi himself published two articles in fall 1938 with the intent of alerting the nation to the Jewish threat of racial and ideological contamination.[65] In one of the articles, Interlandi argued for "the necessity of a total separation between the pure Italian race and all those racial elements that are hostile and attempt to contaminate it."[66] However, while such rhetoric illustrated the means by which antisemitic intellectuals like Interlandi utilized cultural and historical imagery, paradoxically, it also exposed the rather limited popular support for purely racial and biological antisemitic arguments in Italy.

By 1939 the three main antisemitic approaches advanced in *La difesa della razza* converged. The Jew assumed not only the traits of an "alien" and criminal within Italian society but also the traits of a political enemy that had to be expelled. The journal insisted on a general interpretation of modernity in terms of progressive domination by the Jews, starting with Luther and Protestantism and finishing with Léon Blum and French socialism, via the English Revolution of 1649, the French Revolution of 1789, the Russian Revolution of 1917, American capitalism, and Russian bolshevism.[67] Within this interpretation, Jewish ideals and the accompanying social and political programs eloquently expressed Jews' opposition to Christianity. In the case of Fascism, its economic policy of autarchy and foreign policy of alliance with Germany and Japan, as well as its wider cultural and social goals, could regard the Jews as nothing but enemies.

As the world war progressed, racial discourse was powerfully steered toward solving the Jewish Question. In 1941 the racist and antisemitic diatribes of *La difesa della razza* reached a crescendo. The journalist Umberto Angeli proposed a new science called "Judeoscopia," which was supposed to establish strict criteria in order to find out which Italian citizens were Jewish owing to their spiritual predisposition, and therefore de facto Jews.[68] That same year Giorgio Piceno wrote, in a militant and uncompromising tone, that it was time to override existing legislation and resort to violence to solve the Jewish problem: "Quite a few people are convinced that laws are of little use against the Jews; there is a widespread consensus that violent actions, which individual countries have resorted to, represent the only means to bring the Jews to reason [*ridurre alla ragione gli ebrei*]."[69]

In the meantime resistance persisted within the regime—especially among older Fascists—to the new, racial biological understanding of Fascist Italian identity. During the spring of 1942 Interlandi assumed a definitive position in favor of biological racism, and in March 1942 his journal reprinted the "Race Manifesto," framing it as the only existing document that provided "official directions" for Italian racism.[70] Simultaneously, Interlandi and the young cohort of racial determinists launched a full-scale attack on Giacomo Acerbo and other leading Fascist intellectuals of the "Mediterranean" current, accusing them of being philosemitic, democratic, and anti-German.[71] Acerbo, supported by sympathetic Fascist intellectuals such as Sabato Visco and Nicola Pende, not only continued to defend Latin Italian eugenics but also launched a counterattack. They disparaged uncritical admiration for German anthropology and racial biology as expressed by *La difesa della razza*, which they claimed paid inadmissibly little attention to its specific, Italian, variant. Gradually Acerbo and his allies acquired more influence in ministerial and bureaucratic offices such as the powerful Minculpop, where they continued to sideline governmental directives related to race.

In spite of this resistance, Interlandi's intellectuals pursued their antisemitic campaign undaunted, if anything intensifying their rhetoric. In February 1943, for example, Guido Landra invited all "true Italian racists" to fight against the nation's enemy number one: international Judaism.[72] In spring 1943, just a few weeks before the Fascist regime fell, Interlandi announced the ultimate failure of the racial legislation and claimed that the enemy within, which had penetrated bureaucracy and society at large, was about to reverse the gains achieved by the Fascist revolution. Using metaphors of dirt, filth, and infestation with regard to Jews, Interlandi's polemic revealed the extent to which Italian racists in *La difesa della razza* had adopted the violent language of Nazi racial biology by 1943: "The Jews do this and more, and yet their dirty quarters have not been sprinkled with pesticide" (Gli ebrei fanno questo e altro, e ancora il loro sporco quartiere non è stato cosparso di liquido insetticida).[73]

By 1943 arguments linked to Italian history had long since been abandoned in favor of more direct attacks. In July of that year, according to Interlandi, the Jewish conspiracy had succeeded in infiltrating the Italian military. Through their articles and polemics, the intellectuals of *La difesa della razza* contributed significantly to laying the foundations for Article 7 of the Verona Program of the Salò Republic of November 1943. This notorious article declared the Jews to be enemies of the Italian nation and paved the way for their deportation and ultimate death.

Racial theories in Italy were developed during the 1920s within an academic community that sought to affirm its independence from German, Northern European, and American influences. From early on the Fascist government used the findings of Italian racial science to promote its racial social policies, though within the framework of positive eugenics. Antisemitism remained a largely marginal phenomenon in the 1920s, defining the "Jew" in political, economic, and religious but not racial terms. In the early 1930s the majority of Italian anthropologists, demographers, hygienists, historians, philosophers, sociologists, bishops, and statisticians opposed German biological racism and antisemitism.

This disposition began to change by the time of the military campaign in Ethiopia, which forced the Fascist government to take the consequences of racial intermixing seriously and brought racial theory into everyday life. Simultaneously, the political alliance with Nazi Germany led to a reevaluation of antisemitism in Italy, whereby political, religious, and socioeconomic definitions were increasingly replaced with racial ones. Although the publication of the "Race Manifesto" in 1938 represented an important shift, until 1943 the interpretation of both racism and antisemitism was anything but linear; "traditional" Fascist theories of race competed with Nazi biological concepts of race for supremacy.

The so-called national current, within the framework of which the majority of Italian scientists and anthropologists operated, remained critical of the biological assumptions of the "Race Manifesto" and racial antisemitism in general. Although the exponents of this current never openly rejected the "Race Manifesto," they were able to influence racial and antisemitic policymaking and eventually subvert the document's most drastic implications through their influential governmental and party positions. Moreover, the scientific appeal of Sergi's Mediterranean Thesis and the Catholic background of nearly all the personalities who had been involved in formulating Fascist racial policymaking militated against the dominance of Nazi racial biology. At the same time, though, the cultural role that both mainstream racial science and antisemitism played in paving the way to a racial antisemitism should not be underestimated. In the late 1930s in particular, public opinion to a large extent embraced the biological racism invoked against the black populations in Africa, serving as a kind of Trojan horse for the penetration of racial antisemitism in Italy.[74]

Nazi concepts of racial biology, by contrast, found their most receptive environment in the writing of Telesio Interlandi and the young scholars who contributed to *La difesa della razza*. Although Interlandi and his collaborators were excluded from powerful positions in the Fascist hierarchy, their journals were important organs of propaganda, mirroring contemporaneous political and ideological developments. With Fascist Italy engaged in the construction of an empire, Mussolini gave Interlandi the task of launching a propaganda campaign aimed at legitimizing the colonial regime's racial categorization and segregation of the conquered African populations. Not coincidentally, the evolution of political antisemitism into biological racism in Interlandi's journals paralleled Italy's growing alliance with Germany. Otherwise, the effect of Interlandi's propaganda proved rather modest: Interlandi himself complained that *La difesa della razza* aroused little interest among the general public, a reflection of the fact that biological racism did not naturally fit with Italian culture.

Mussolini oscillated between the two main racist currents without taking a definitive position. Both initiatives received political and financial support, dictated by domestic and foreign political concerns rather than ideological priorities. This setup helps to explain both the support for *La difesa della razza* and the political dominance of officials and ideologists identified with traditional Fascist racial theories. That said, political considerations and priorities changed abruptly with the fall of the regime and the establishment of a new balance of power between Italian Fascism and German Nazism in the Italian Social Republic between 1943 and 1945. The driving force of Italian racism in the Salò Republic, the uncompromising Giovanni Preziosi and Roberto Farinacci, ensured that Italian antisemitism evolved from the denial of equality and rights to the persecution of individuals, culminating in the deportation of the Italian Jews to Auschwitz-Birkenau. Interlandi remained loyal to Mussolini and Fascism until the very end. However, he refused any position in the Salò Republic, convinced as he was that the Nazi-Fascist struggle against "Hebraism" and plutocracy had already been lost.[75]

Notes

1. See Enzo Collotti, *Fascismo e antifascismo: Rimozioni, revisioni, negazioni* (Rome: Laterza, 2000); Gianni Scipione Rossi, *La destra e gli ebrei: Una storia italiana* (Soveria Mannelli: Rubbettino, 2003); Giovanni Belardelli, "Il fascismo nei manuali di storia dell'Italia repubblicana," in *La storia contemporanea tra scuola e università: Manuali, programmi, documenti*, ed. Giuseppe Bosco and Claudia Mantovani (Soveria Mannelli: Rubettino, 2004); Luca La Rovere, *L'eredità del*

fascismo: Gli intellettuali, i giovani e la transizione al postfascismo 1943–1948 (Turin: Bollati Boringhieri, 2008).

2. Renzo De Felice, *Storia degli ebrei italiani sotto il fascismo* (Turin: Giulio Einaudi Editore, 1961). The English translation of the book, *The Jews in Fascist Italy: A History*, was published by Enigma Books in 2001.
3. Filippo Focardi, "Alle origini di una grande rimozione: La questione dell'antisemitismo fascista nell'Italia dell'immediato dopoguerra," *Horizonte: Italianistische Zeitschrift Zür Kulturwissenschaft und Gegenwartsliteratur* 4 (1999): 135–70.
4. See, for example, Roberto Chiarini, *L'intellettuale antisemita* (Venice: Marsilio, 2008).
5. For biographical data on Interlandi, see Mauro Canali, "Interlandi, Telesio," in *Dizionario Biografico degli Italiani* (Rome: Istituto della Enciclopedia Italiana, 2004), 519–21; Michele Sarfatti, "Telesio Interlandi," in *Dizionario del fascismo*, ed. Sergio Luzzatto and Victoria de Grazia (Turin: Einaudi, 2002), 1:673–74; Meir Michaelis, "Mussolini's Unofficial Mouthpiece: Telesio Interlandi—Il Tevere and the Evolution of Mussolini's Anti-Semitism," *Journal of Modern Italian Studies* 3, no. 3 (1998): 217–40.
6. Aldo A. Mola "Preziosi, Giovanni," in Luzzatto and de Grazia, *Dizionario del fascismo*, 2:422–23.
7. De Felice, *Storia degli ebrei italiani*, 140. De Felice's theory is discussed by Meir Michaelis in "Mussolini's Unofficial Mouthpiece."
8. Michaelis, "Mussolini's Unofficial Mouthpiece," 224; Michaelis, *Mussolini e la questione ebraica: Le relazioni italo-tedesche e la politica razziale in Italia* (1978; repr., Milan: Comunità, 1982), 467–70.
9. Michaelis, "Mussolini's Unofficial Mouthpiece," 237–38.
10. Michaelis, "Mussolini's Unofficial Mouthpiece," 235. According to Michaelis, Alfred Rosenberg wrote in 1932 that *Il Tevere* was of great relevance in Italy because it was close to Mussolini and functioned as a semiofficial Fascist organ.
11. Emil Ludwig, *Colloqui con Mussolini* (1932; repr., Milan: Mondadori, 2000), 55.
12. Giorgio Fabre, "Mussolini e gli ebrei alla salita al potere di Hitler," *La rassegna mensile di Israele* 69, no. 1 (2003): 190.
13. Claudio Pogliano, "Scienza e stirpe: L'eugenetica in Italia (1912–1939)," *Passato e presente*, no. 5 (January–June 1984): 61–97; Giorgio Israel and Pietro Nastasi, *Scienza e razza nell'Italia fascista* (Bologna: il Mulino, 1998); Roberto Maiocchi, *Scienza italiana e razzismo fascista* (Florence: La Nuova Italia, 1999); Aaron Gillette, *Racial Theories in Fascist Italy* (London: Routledge, 2002); Claudia Mantovani, *Rigenerare la società: L'eugenetica in Italia dalle origini ottocentesche agli anni Trenta*

(Soveria Mannelli: Rubettino, 2004); Claudio Pogliano, *L'ossessione della razza: Antropologia e genetica nel XX secolo* (Pisa: Scuola Normale Superiore, 2005); Francesco Cassata, *Molti, sani e forti: L'eugenetica in Italia* (Turin: Bollati Boringhieri, 2006); Cassata, *"La difesa della razza": Politica, ideologia e immagine del razzismo fascista* (Turin: Giulio Einaudi Editore, 2008).

14. On the different currents within Italian Fascist racism, see Mauro Raspanti, "I razzismi del fascismo," in *La menzogna della razza: Documenti e immagini del razzismo e dell'antisemitismo fascista,* ed. Centro Studi F. Jesi (Bologna: Grafis, 1994), 73–89.
15. [Telesio Interlandi], "Razza e impero," *Il Tevere,* 11–12 January 1937.
16. Aaron Gillette, "The Origins of the 'Manifesto of Racial Scientists,'" *Journal of Modern Italian Studies* 6, no. 3 (2001): 308.
17. Cassata, *"La difesa della razza,"* 26.
18. Michaelis, "Mussolini's Unofficial Mouthpiece," 232.
19. Cassata, *"La difesa della razza,"* 32–36.
20. Michaelis, "Mussolini's Unofficial Mouthpiece," 232.
21. Gillette, "Origins," 309–13.
22. Gillette, "Origins," 311.
23. For the genesis of the "Race Manifesto," see Israel and Nastasi, *Scienza e razza*; Gillette, "Origins," 305–23; Tommaso dell'Era, *Il Manifesto della razza* (Turin: UTET, 2008). The "Race Manifesto" was presented as a thesis developed by ten Italian scientists and approved by the minister of popular culture, Dino Alfieri, and the party secretary, Achille Starace. See also Franco Cuomo, *I Dieci: Chi erano gli scienziati italiani che firmarono il manifesto della razza* (Milan: Baldini Castoldi Dalai, 2005).
24. On the history of Jewish persecution in Italy, see Michaelis, *Mussolini e la questione ebraica*; Corrado Vivanti, ed., *Gli ebrei in Italia: Dall'emancipazione ad oggi* (Turin: Annali della storia d'Italia Einaudi, 1997); Michele Sarfatti, *The Jews in Mussolini's Italy: From Equality to Persecution* (Madison: University of Wisconsin Press, 2006); Enzo Collotti, *Il fascismo e gli ebrei: Le leggi razziali in Italia* (Rome: Laterza, 2006); Marie-Anne Matard-Bonucci, *L'Italia fascista e la persecuzione degli ebrei* (Bologna: il Mulino, 2008); Joshua D. Zimmerman, ed., *Jews in Italy under Fascist and Nazi Rule, 1922–1945* (New York: Yeshiva University Press, 2009); Francesco Germinario, *Fascismo e antisemitismo* (Rome: Laterza, 2009).
25. [Telesio Interlandi], "Era tempo," *Il Tevere,* 15–16 July 1938.
26. [Interlandi], "Era tempo."
27. Raspanti, "I razzismi del fascismo."
28. Gillette, "Origins," 317.

29. Gillette, "Origins," 317.
30. Cassata, *"La difesa della razza,"* 56.
31. Canali, "Interlandi, Telesio," 520.
32. For a general introduction to the journal's subject matter, see Valentina Pisanty, *"La difesa della razza": Antologia (1938–1943)* (Milan: Bompiani, 2006); Michele Loré, *Antisemitismo e razzismo ne "La difesa della razza" (1938–1943)* (Soveria Mannelli: Rubbettino, 2008).
33. Cassata, *"La difesa della razza,"* 226–45. See also Francesca Cavarocchi, "La propaganda razzista e antisemita di uno 'scienziato' fascista: Il caso di Lidio Cipriani" *Italia contemporanea* 219 (June 2000): 193–225. Cipriani started a promising career as anthropologist at the University of Florence at the age of twenty-six.
34. Lidio Cipriani, *Considerazioni sopra il passato e l'avvenire delle popolazioni africane* (Florence: Bemporad, 1932); Cipriani, "Il passato e l'avvenire degli etiopici secondo l'antropologia," *Gerarchia* 11 (November 1935); Cipriani, "Gli etiopici secondo il razzismo," *La difesa della razza,* 5 October 1938; Cipriani, "Popoli imbelli e guerrieri in Africa," *La difesa della razza,* 5 November 1938; Cipriani, "Il concetto di razza è puramente biologico," *La difesa della razza,* 20 April 1942.
35. Guido Landra, "L'ambiente non snatura la razza," *La difesa della razza,* 5 December 1938.
36. Cassata, *"La difesa della razza,"* 200–201.
37. See, for example, Elio Gasteiner, "Un pericolo per la razza: La decadenza dei ceti superiori," *La difesa della razza,* 20 August 1938.
38. Giuseppe Lucidi, "Gruppi sanguigni e nuclei razziali: Necessità di un censimento del sangue," *La difesa della razza,* 5 January 1939; Giulio Silvestri, "Per un archivio genealogico nazionale," *La difesa della razza,* 5 March 1941.
39. Leone Franzí, "Il meticciato: Insidia contro la salute morale e fisica dei popoli," *La difesa della razza,* 20 September 1938.
40. Cassata, *"La difesa della razza,"* 204.
41. Marcello Ricci, "Ereditarietà ad eugenica," *La difesa della razza,* 5 October 1938.
42. Marcello Ricci, "Eugenica e razzismo," *La difesa della razza,* 20 January 1939.
43. Cassata, *"La difesa della razza,"* 208–12.
44. Giuseppe Genna, "Gli ebrei come razza," *La difesa della razza,* 5 September 1938.
45. G. L. [Guido Landra], "Considerazioni sulla criminalità degli ebrei," *La difesa della razza,* 5 January 1941.
46. Julius Evola, "Psicologia criminale ebraica," *La difesa della razza,* 20 July 1939; Guido Landra, "Considerazioni sulla patologia degli ebrei," *La difesa della razza,* 5 October 1940.
47. [Telesio Interlandi], "Il censimento degli ebrei," *La difesa della razza,* 20 September 1938.

48. Umberto Angeli, "Gli ebrei manifesti e i clandestini," *La difesa della razza*, 5 February 1939; Angeli, "Tipo fisico e carattere morale dei veri e dei falsi Italiani," *La difesa della razza*, 20 May 1939.
49. Massimo Lelj, "Disarmiamo i borghesi," *La difesa della razza*, 5 October 1938; Giuseppe Pensabene, "La borghesia e la razza," *La difesa della razza*, 5 August 1938; Guido Landra, "La razza dei borghesi," *La difesa della razza*, 20 October 1939.
50. Umberto Angeli, "Tipo fisico e carattere morale dei veri e dei falsi Italiani," *La difesa della razza*, 20 May 1939.
51. Gino Sottochiesa, "Ebrei convertiti," *La difesa della razza*, 5 August 1940.
52. Francesco Callari, "L'ebreo non si assimila," *La difesa della razza*, 20 October 1938; Nicola Salvati, "Conversioni giudaiche," *La difesa della razza*, 20 October 1938; Antonio Trizzino, "Battesimi e conversioni di ebrei," *La difesa della razza*, 5 March 1939; Berlindo Giannetti, "I falsi convertiti," *La difesa della razza*, 5 March 1939.
53. Mario de' Bagni, "Paolo VI e la Carta dei giudei," *La difesa della razza*, 20 March 1939; Paolo Guidotti, "Bolle pontificie contro gli ebrei," *La difesa della razza*, 20 June 1939.
54. Armando Tosti, "L'ebraismo non è una religione," *La difesa della razza*, 5 October 1940; Mario de' Bagni, "Cristo e i Cristiani nel Talmud," *La difesa della razza*, 20 May 1939.
55. "Talmud," *La difesa della razza*, 20 September 1938; "Le due bocche d'Israele," *La difesa della razza*, 5 October 1938.
56. See, for example, Bruno Biancini, "Riti e superstizioni degli ebrei," *La difesa della razza*, 20 November 1939.
57. Armando Tosti, "La razza giudaica: Amoralità e criminalità dei giudei," *La difesa della razza*, 5 October 1941.
58. Agostino Gurrieri, "Ariani e semiti nel Mediterraneo," *La difesa della razza*, 20 January 1940; Paolo Guidotti, "Il popolo più antisociale dell'impero romano," *La difesa della razza*, 20 December 1940; Agostino Gurrieri "Il Mediterraneo e la civiltà ariana," *La difesa della razza*, 5 June 1941.
59. Salvatore Costanza, "Gli eterni nemici di Roma," *La difesa della razza*, 20 June 1939; Cassata, *"La difesa della razza,"* 132–33.
60. Fortunato Matarrese, "Gli ebrei in Puglia," *La difesa della razza*, 20 June 1939.
61. Giorgio Almirante, "Roma antica e i giudei," *La difesa della razza*, 5 September 1938.
62. T. I. [Telesio Interlandi], "Conoscere gli ebrei," *La difesa della razza*, 20 August 1938; T. I. [Telesio Interlandi], "Al principio," *La difesa della razza*, 20 September 1938.

63. [Massimo Lelj], "Le carte degli ebrei," *La difesa della razza,* 20 February 1939.
64. [Massimo Lelj], "La manomissione ebraica della nazione italiana," *La difesa della razza,* 5 July 1939; [Lelj], "La borghesia e l'immigrazione," *La difesa della razza,* 5 December 1938; "Disfattismo ebraico," *La difesa della razza,* 20 October 1938.
65. T. I. [Telesio Interlandi], "Premessa," *La difesa della razza,* 20 October 1938; T. I. [Telesio Interlandi], "Eroica," *La difesa della razza,* 5 November 1938.
66. [Interlandi], "Eroica."
67. Angelo Maria De Giglio, "Il giudaismo fomentatore del protestantesimo," *La difesa della razza,* 5 July 1940; Giuseppe dell'Isola [pseudonym of Giuseppe Pensabene], "Somiglianze tra il giudaismo e la religione degli inglesi," *La difesa della razza,* 20 November 1940.
68. Umberto Angeli, "Judeoscopia," *La difesa della razza,* 20 January 1941; Angeli, "Meglio del pogrom," *La difesa della razza,* 20 June 1942.
69. Giorgio Piceno, "Leggi ariane trucchi giudaici," *La difesa della razza,* 5 September 1941.
70. Guido Landra, "Fondamenti biologici del razzismo," *La difesa della razza,* 20 March 1942.
71. In July 1940 the Fascist politician Giacomo Acerbo published the essay *I fondamenti della dottrina fascista della razza* (The foundation of the Fascist racial doctrine) with the aim of weakening the biological interpretation of Italian racism and separating it from German racism. Acerbo was reluctant to mention the Jewish Question.
72. Guido Landra, "Fronte unico del razzismo italiano," *La difesa della razza,* 20 February 1943.
73. [Telesio Interlandi], "Inintelligenza col nemico," *La difesa della razza,* June 5, 1943.
74. Maiocchi, *Scienza italiana e razzismo fascista,* 210.
75. Michaelis, "Mussolini's Unofficial Mouthpiece," 222.

8

Eradicating "Undesired Elements"

National Regeneration and the Ustasha Regime's Program to Purify the Nation, 1941–1945

RORY YEOMANS

In April 1942, on the first anniversary of the founding of the Independent State of Croatia, a group of train passengers took a journey into the heart of Bosnia. This was no ordinary party of travelers. It comprised leading officials from the Ustasha regime—the Fascist movement that had come to power the previous year—and included Foreign Minister Mladen Lorković, the head of the secret police, Eugen Dido Kvaternik, and the commander of the elite Black Legion death squad, Jure Francetić. Nor was it any ordinary journey: the train took them through eastern Bosnia, which the Black Legion had recently "cleansed" of its Serb population through a campaign of mass murder and expulsion. In his subsequent radio address Lorković explained that the train stations they had passed were "new symbols of the order, creation, and struggle of Ustasha Croatia." Traveling through Bosnia, the view from the train window offered a pleasing picture of a "completely harmonious orderly land," in stark contrast to the situation when the Ustasha movement had come to power. Following the campaign of purification by Francetić's legionaries, there was no more of this "Balkan filth and dirt, a picture of complete negligence and chaos that we saw all the time in this region in the era of the former Yugoslavia." With all traces of the Serb presence in the region eradicated, the railway stations were now "hygienic and orderly" and had home guards and armed militias standing in front of them "offering a beautiful picture" of cleanliness and progress.[1]

The account of the journey undertaken by Lorković and other officials symbolized the contradictory nature of the relationship between the Ustasha regime and racial science. The nucleus of what was to become the Croatian Ustasha movement was founded in the early 1930s by Ante Pavelić, a lawyer and right-wing politician, out of militant university and youth groups (even though its rank and file was made up predominantly of peasants, workers, and sailors). During the 1930s the majority of members of the Ustasha movement lived in exile abroad, in terrorist training camps from which they launched armed raids into Yugoslavia. However, the Ustasha movement enjoyed support across the radical-right spectrum: nationalist university and high school students, Catholic clergy, Catholic youth organizations, the separatist intelligentsia, and nationalist workers' unions. The Ustasha movement aimed to establish, by force if necessary, an independent Greater Croatia comprising Croatia, Bosnia, and parts of Serbia. Its chief theoreticians believed that the economic, moral, and social transformation of the future Croatian state could be achieved only after it had been cleansed of all "alien" influences, the sizable Serb population in particular.[2] During the second half of the 1930s, under the influence of National Socialism, the movement also incorporated antisemitism into its program.

The Ustasha movement viewed the eradication of all "unwanted elements" as a path to national regeneration—a collective panacea for the nation's cultural, moral, and social ills. The Ustasha regime came to power in April 1941 following the invasion and division of Yugoslavia by Axis forces. Established on 10 April, the Independent State of Croatia, comprising Croatia and Bosnia, was divided into two zones of occupation, with Dalmatia annexed and administered by Italian occupation forces and the rest controlled by Nazi Germany. Despite this unpromising beginning for a movement founded on the very principle of national liberation, the Ustasha regime immediately set about realizing its vision of a regenerative national revolution. To facilitate the implementation of its agenda, the regime introduced a series of racial laws and edicts drawing heavily on Nazi biological concepts of antisemitism and race. Simultaneously, the Ustasha media launched a campaign of demonization, imbued with the language of scientific racism, against "foreign" influences and "alien" racial and national groups in the new state. This proved the prelude to a campaign of mass murder to purify the Croatian homeland of Serbs, Jews, Gypsies (Roma), and all other "degenerate" and "undesired" elements.

Despite its attempts to mimic Nazi racial policies, the Ustasha regime's vicious anti-Serb campaign and the inchoate propaganda that accompanied it exposed certain contradictions. The regime's use of moral, social, and economic arguments to legitimize its extremist agenda highlighted the marginality of the Nazi concept of race on the one hand and the centrality of older anthropological theories on the other. It also underscored deep divisions within the movement between "moderate" and "hard-line" factions and exterminatory and assimilationist tendencies and, more generally, underscored the Croatian radical right's ambiguous attitudes toward race and racial science. Armed resistance provoked by the Ustasha campaign of mass murder, and subsequent pressure exercised by the Nazi occupation authorities, eventually forced the regime to modify its policies toward the state's Serbs. The policy of mass murder was first substituted with one of forced assimilation through conversion of the Orthodox Serb population to Catholicism and then, when that failed, by limited attempts to reintegrate the Serbs into mainstream society. At the same time, the presence of a significant number of Jews within the upper echelons of both the regime bureaucracy and the party itself led to the introduction of an exception clause allowing some Croatian Jews to claim "Honorary Aryan" status. However, such contingent decisions were never designed to be permanent. In reality, very few Jews were able to evade eventual deportation and death in one of the Ustasha concentration camps; marginalization, terror, and persecution remained the everyday experience for many Serbs. By 1944 the regime had returned to its original program of mass murder. Perhaps the greatest contradiction of all was that a regime under which racial thinking remained peripheral proved to be far more murderous than regimes in German-occupied Europe with a profound commitment to eugenics.

Becoming Aryan: Scientific Antisemitism and the Education of a New Elite

Officially, race and eugenics played a key role in the Ustasha movement's ideology. The seventeenth principle of its program declared that "balanced breeding, the promotion and perfection of these virtues in all branches of national life, is the goal of public welfare and of state authority."[3] From its foundation in the early 1930s, the Ustasha movement's defining aim had been the establishment of an independent Greater Croatia comprising Croatia, Bosnia, and parts of the Vojvodina that would be cleansed of Serbs and other national enemies, with mainstream society purged of those sympathetic to liberalism and Yugoslav ideals. During the second half of the 1930s, under the influence of Nazi antisemitism and racism, Jews and Gypsies were added to the list of national enemies. By contrast, the movement's intellectual cadre portrayed the Muslim population of Bosnia as Croatian blood brothers and Bosnia—as a radical nationalist student manifesto of 1939 proclaimed—as the "racially purest part of the Croatian nation."[4]

After the establishment of the Independent State of Croatia, questions of racial purity came to dominate legislation and public discourse for a while. Immediately following the movement's seizure of power, it introduced a series of laws defining who did or did not belong to the Croatian "race." These statutes were aimed both at defining and protecting a pure Croat race and at the social, legal, and racial exclusion of Jews and Gypsies. The first such law, the Legal Statute of Race Membership, created a clear division between those who belonged to the national community and those who did not by means of categorizing the population either as "Aryan" (i.e., racially Croatian) or "non-Aryan" (i.e., racially impure). The law defined as Aryan only those who were descendants of members of the "European racial community." Citizens of the Independent State of Croatia were required to produce certificates proving that their parents as well as grandparents were of Aryan stock and that they had no Jewish or Gypsy pedigree.[5]

For the Ustasha regime, being Jewish was a matter of race, not culture or religion. The regime committed itself to rooting out all Jews "hiding" under Croatian names. Subsequently, Jews were banned from changing their names to Aryan ones, while all those who had changed their names before 10 April 1941 were instructed to change them back.[6] If there was any doubt whether Jews could become Croats through assimilation, the Law for the Protection of the Aryan Honor and Blood of the Croatian Nation made it clear that they could not: the law criminalized racial miscegenation and banned not only marriage but relationships of any kind between "Aryans" and Jews. Stating that a Croatian Aryan woman would be racially defiled by engaging in sexual intercourse with Jewish men, the law provided for the rape of an Aryan girl by a Jewish male to be punished by death. Jews were further banned from employing maids who were younger than forty-five, on the assumption that younger women were vulnerable to sexual exploitation by the Jews.[7]

Croatia was to be culturally as well as racially cleansed. Jews were forbidden from having "any involvement in influencing the construction of national and Aryan culture." Consequently, they were banned from cultural, sports, and youth organizations and from contributing to literature, journalism, the visual and musical arts, architecture, theater, and film. Mobility restrictions confined Jews to ghettos and imposed curfews, preventing them from using money to "stimulate disorder and rebellion in the territory of our homeland."[8] A statute of June 1941 banned Jewish businesses from operating between seven in the morning and seven in the evening. Jews were also barred from displaying Croatian flags and other insignia on their shops and business premises, which from then onward had to be marked with a "Jewish" sign. Eventually Jewish businesses were nationalized by the regime, which appointed morally upstanding commissars to oversee their operation. Jews above the age of fourteen were required to wear a large patch on their chest with the letter *Ž* for *Židov* (Jew), and another on their back, effectively marking them as distinct from members of the national and blood community.[9]

In his speech in the Croatian parliament on 10 February 1942, Interior Minister Andrija Artuković declared that the Independent State of Croatia, finding itself in a state of self-defense against "these insatiable and poisonous parasites," had solved the "so-called Jewish Question" with "healthy and decisive action."[10] Ustasha antisemitic policy was certainly decisive and in some respects even more radical than the antisemitism of Nazi Germany. For example, while Jews were legally required to wear a Jewish sign on their chests in many Nazi-occupied states, the Ustasha authorities insisted that they wear a large sign on their backs, too; some local Ustasha councils stipulated that babies and small children should also be required to wear such signs, not just Jews above the age of fourteen. Public sentiment was hostile to such zealotry, as a confidential report of 26 May 1941 by the Ustasha secret police made clear: "In the wider public there is the feeling that it is not necessary for Jews to be identified on both their front and back, but only on their front with the Star of David." Of course, the very fact that members of the public agreed that they should be identified as different from other Croatian citizens suggests that popular opinion was not necessarily particularly sympathetic to the Jews either: at best, it seems, they were often viewed as just one of a hierarchy of hostile elements threatening the survival of the nation, even if they were not considered the most dangerous. In fact, as the report went on to explain, some citizens proposed that, in addition to the Jews, such signs should be worn by those traitorous members of the Aryan national and blood community who had acted as the most "faithful servants" of Jews and Freemasons, "to the detriment of the Croatian people." These traitors were "worse even than the Jews who, by race, are not Croats" and consequently they should be "more clearly marked" than the latter. The report noted that German officials, too, had complained "that on the subject of Jewish signs, we have gone too far." In consequence, the police ordered that Jews should cease wearing signs on their backs, with immediate effect.[11]

Despite the substantial number of ordinary citizens' letters to the Jewish Sections of the Interior Ministry or the Ustasha police denouncing Jewish colleagues, friends, neighbors, and even relatives, popular participation in and support for the regime's antisemitic policy remained marginal. Grassroots Ustasha activists, as well as high-ranking officials, became increasingly aware of how unpopular that policy actually was. Veteran soldiers and militia members, in particular, frequently complained about their chastening public encounters. For instance, when Stjepan Tomljenović, a father of two small children, moved into an "Aryan" apartment building in early 1942, the landlady's children treated him with barely concealed hostility. Her son accused Tomljenović of taking over the property illegally and threatened him with eviction. In his letter of complaint to the Interior Ministry, Tomljenović contrasted the sympathy the landlady's daughter had expressed for the former Jewish tenants with her callous attitude toward his own family. She "helped the Yids carry things onto the street in a wheelbarrow, cried and, with tear-stained eyes, looked with compassion on the Yids. But, by contrast, not a second thought for warriors! Without any pity and any mercy, she would throw me on the street."[12] The fact that the antisemitic statutes had contributed to the death of twenty-seven thousand out of the estimated thirty-four thousand Jews living in the Independent State of Croatia by 1945 owed far more to politics than popular mobilization.[13]

The regime was anxious to give its antisemitic, racist policy a scientific and intellectual framework. In 1941 the Ministry for Education and Religion (in 1942 renamed the Ministry for National Education) established a Racial Political Commissariat and entrusted it with developing plans for the legal reform of "racial biology, racial politics, and racial hygiene or eugenics"; ensuring that legal statutes were in accordance with the principles of racial politics; and advising the regime on questions of racial politics and eugenics. The commissariat was also to arbitrate on all cases concerning racial membership—a decisive factor when it came to state employment and military service. The main function of the commissariat, however, was collecting racial and demographic statistics and "educating the nation" on issues of race and eugenics. The commissariat was composed of a Racial Political Council and a Racial Political Bureau and was staffed by experts on medicine, biology, and education as well as the national economy. All members of the commissariat and their spouses had to prove their Aryan credentials since the posts were explicitly political, requiring appointees to take part in the work of the Interior Ministry.[14]

Among other means by which officials in the Ministry for National Education attempted to introduce the nation to the principles of racial science was a series of academic courses. Starting in winter 1941 the ministry offered to state officials and members of the armed forces racial biology lessons and "psychometric and anthropological" examinations in racial biology. As an advertisement in one Croatian pedagogical journal pointed out, upon completion of this course racial experts would be placed in the army, where they could fall back on their training to reject applications from racially suspect candidates. This particular service would sustain a racially pure army—one of the essential pillars of the new state.[15] Officials in the security police, the Ustasha Supervisory Service, and various other agencies of state repression such as the Bureau for the Eradication of Criminality were all required to attend. In so doing the regime made explicit not just the link between these courses and state repression but also the connection between race and criminality.[16]

In addition, the ministry developed a series of courses to educate teachers, medical students, and doctors, with the purpose of "highlighting the importance of the laws of heredity" and equipping students to supervise "anthropological and psychometrical examinations to be implemented on the entire territory of the Independent State of Croatia."[17] The courses included several modules by the eugenicist Professor Boris Zarnik covering various aspects of heredity, racial and familial genetics, blood groups and physiological differences, and physiognomy and physical characteristics of the main racial groups, Jews and Gypsies in particular. Zarnik's colleague Đuro Vranešić concentrated on spiritual rather than biological aspects of race, including modules on race, nation, and "the spiritual culture of the nation and its connection to racial elements."[18]

The regime employed modern scientific concepts not just to educate the public about race but also to warn it about the consequences of high birthrates among racially alien groups. In the case of the Jews, demographic studies—such as the one undertaken by the State Statistical Department in the summer of 1941 to ascertain how many Jews were living in Croatia—were disseminated in the state media, playing on wider anxieties about the large numbers of Jewish refugees who had arrived from Nazi-occupied Europe during the late 1930s. Articles describing how towns in Croatia had been demographically "colonized" by Jews routinely appeared in the press. In June 1941, for example, a regional newspaper wrote about the formerly "pure Croatian town" of Ilok, which in the course of two decades had become dominated by Jews, both demographically and commercially.[19]

The culmination of the Ustasha regime's anti-Jewish propaganda was the antisemitic exhibition held in Zagreb in May 1942, chronicling the "destructive and exploitative" work of Croatia's Jews before 1941. The exhibition brochure, produced by officials of the State Investigative and Propaganda Bureau, pointed out that the majority of the exhibits had been collected by racial experts, demographers, scientists, and ordinary citizens, as well as the bureau's investigators. Among the many illustrations, the brochure featured graphics depicting the growth rate and colonization of Croatia by Jews and statistics demonstrating how the Jews dominated commerce in comparison to their share of the country's population. Although the statistics were hardly credible, the authors of the brochure repeatedly stressed the rigorous science behind the analysis. The exhibition's promotional poster embodied racial bias. It depicted a naked, muscle-bound Aryan warrior, armed with a sword and shield adorned with the Croatian coat of arms, fighting a poisonous Jewish snake; thus the new Ustasha man was purifying Croatia's body politic of the racially degenerate Jew.[20] While exposing the economic injustices and abuse to which honest Croatian workers were subjected under Jewish hegemony in Yugoslavia, the brochure related the sexual mistreatment of young Croatian girls at the hands of Jews. In the exhibition's promotional film, a narrator explained how Jewish pimps had picked up vulnerable young Croatian women, seducing them and selling them into a sordid life of brothels, prostitution, and white slavery.

The Ustasha regime's antisemitism was in part grounded in its suspicion of the city and the cosmopolitan values it represented. The regime not only associated Jews with the urban, liberal values of interwar Yugoslavia (the "Jewish

Eldorado") but more generally depicted the atomized, alienating city as a cesspit of Jewish cosmopolitanism, sexual license, and racial miscegenation. The fact that these ideas were imbued with racial biological terminology illustrated the symbiotic relationship between moral and racial conceptions. Arguing that only peasant values could decontaminate the city, state propaganda encouraged Croatian peasants to migrate into the city to improve the biological composition of the urban population, "in whose veins flows ever more foreign, Jewish blood." One newspaper thus wrote in the wake of the introduction of the Aryan laws in May 1941, "The Croatian nation must be like any other ethical movement if it wants to protect its racial purity, to protect the purity of its blood, if it wants to accomplish its historical mission. We have to protect our blood from Jewish, Gypsy and non-Aryan manifestations because this is one of the basic prerequisites of the construction of a new Croatia."[21]

Purifying the Nation: Racial Anthropology and the Mass Murder of the Serbs

Since the mid-1930s the radical right in Croatia had been struck by a fear: the declining demographics of the Croatian nation in relation to "foreign" and "immigrant" populations. In 1939 Mladen Lorković published a study in which he argued that the Croatian nation was dying out, squeezed out by fast-breeding, aggressive foreigners, specifically the Serbs. According to him, the Serbs descended from Vlach nomadic populations brought to Croatian lands, primarily Bosnia, in the sixteenth and seventeenth centuries by the invading Ottomans. Upon arrival they had burned down the houses of indigenous Croats, killing and expelling many of them. In their wake the Serbian Orthodox Church had arrived, leading to the disappearance of the Catholic Church in some regions and the conversion of the Vlachs to Orthodoxy. Simultaneously, many Croatian Catholics were forcibly converted to Orthodoxy, thereby losing their Croatian consciousness and becoming Serbs. Although the book aimed to provide a racial history of Croatian "lands," it mainly focused on current demographics. According to Lorković, the Serb population had grown exponentially, especially in the period after 1918 when many Serb settlers had received farmsteads and land plots in Croatia and Bosnia from the Yugoslav regime. By contrast, the Croatian population was declining, weakened through low birthrates and mass emigration.[22]

The idea that the Serbs of Bosnia were a foreign element had its roots in the racial theories of a group of Croatian anthropologists and ethnologists writing in the 1900s, especially the director of the Agricultural Museum in Sarajevo, Ćiro Truhelka. In a 1904 treatise he argued that the foreign origin of the Serb population in Bosnia was visible in its racial makeup. Unlike blond-haired and blue-eyed Croats and Muslims, but like the Jews, they were dark-skinned, dark-eyed and pigeon-chested, representing "a black-skinned, overwhelmingly dark, physically degenerate type." Linking biological traits to character, among the values he associated with the Serb population were craftiness, greed, and a talent for moneylending.[23] Alongside polemical pamphlets, Truhelka frequently wrote scholarly articles on the "biologically recessive" racial composition of the Serbs.[24]

Notions of Serb racial inferiority and immigrant origins intensified among the separatist and ultranationalist intelligentsia in the 1930s in response to political events. Following the establishment of a semiautonomous Croatian Banovina in 1939, some Serb community leaders began calling for political autonomy for areas of the Banovina where the Serbs constituted a majority. In consequence, they became a target of radical newspapers and journals, which attacked them as "emigrant Vlachs, Serbs, and other Balkan Greek Easterners" who should either keep quiet or emigrate.[25] Some radical younger nationalists suggested an even more drastic solution to the Serb "problem." In a book review in 1940 one young writer, Emil Medvedović, cited the mass deportation of the Greek minority by the Turkish government in 1922 as "most inspiring for us."[26]

In early 1942 the aged Truhelka published his magnum opus on the racial composition of the Balkans, *Studije o podrijetlu* (Studies of origins). Among other things he argued that prior to the arrival of the Turks there had been no Serb presence in Bosnia. Instead, they had arrived as nomadic Vlach "colonists," penetrating ever farther into Croatian lands. Settling in urban areas, they had become economically powerful, accumulating wealth through usury and "any other business in which they could make large profits." However, while the Serbs had become influential in Bosnia's economic, social, and political life, they remained immigrants at heart, unlike the "indigenous" Catholic and Muslim populations: as immigrants, they had no right, he argued, to "covet the lands" or deny the existence of the Croats, the "native" population.[27]

When Truhelka died in the summer of 1942, Hasan Šuljak, a twenty-five year old Muslim Ustasha journalist, called him the "apostle of Croatian Bosnia," making a direct link between Truhelka's ideas about the Serbs as "a foreign element" and the Ustasha regime's ideology.[28] As Šuljak's obituary made clear, the significance of Truhelka's scholarly and anthropological work went far beyond the academy: numerous Ustasha intellectuals cited him as an important influence on their own racial thinking about both Muslims and Serbs.[29] Whether his racial theories reflected ideas ingrained in extreme nationalist ideology that the Ustasha regime had appropriated or directly inspired the campaign of mass murder against the Serb minority embarked on by the Ustasha regime in the summer of 1941, there is little doubt that they provided intellectual and scientific context for both. In the spring of 1941, drawing on theories in Truhelka's studies, in newspaper interviews, articles, and speeches at rallies, party officials and ideologists publicly denounced the Serbs as a racially "alien" and dangerous presence in the beating heart of the Independent State of Croatia. For example, in an interview with the newspaper *Vihor*, Mladen Lorković explained that before Serbian immigrants were settled in Croatia by the Yugoslav government after 1918 there had been few real Serbs in Croatia. Those who now called themselves Serbs were mostly "the remnants of Balkan-Romanian and Gypsy half-breeds" who had arrived with the Ottomans. "Although racially they are neither Serbs nor Croats," he continued, "they represent an unstable element open to foreign influences, which, because they belonged to the Serbian Orthodox Church, succumbed to political Serbianization."[30] Writing in the Ustasha weekly *Hrvatska gruda* in August 1941, Mirko Košutić, an education policy adviser, linked the Serbs' supposedly degenerate moral values to their racial origins. The Serbs, he argued, were a nation that "always caused evil and misery" and were "inclined to material possessions, full of selfishness, wanting to impose their will and domination on others" so that they could live "an easier and better life at the expense of others." This was at least in part, Košutić reasoned, because they possessed "significant amounts of Gypsy blood, Vlach blood, Greek, Turkish and Cincar blood—largely nomadic blood."[31] In a speech at Varaždin, meanwhile, Mile Budak, the minister for education and religion, declared that the Serbs had come to Croatia as "Balkan trash," brought from the east by the Turks, "who used them as vassals and servants, water carriers and beggars who fell on the deserted hearths of Croats like locusts." Now it was time for them to leave: the "Vlach-Serbs" would be driven out, by force if necessary, to the other side of the River Drina to protect the racial purity of Croatia.[32]

Although the anti-Serb propaganda was often rooted in racism and century-old stereotypes, it was not as racially "scientific" as the regime's antisemitism was. Rather, ideologists spoke about the Serbs in the language of degeneration and disease. Father Ivo Guberina, a radical priest and official in the Ministry for National Education, compared the Serbs to a poison in the national body and Ustasha militias to physicians who had to "surgically remove this poison" from the organism. "The Ustasha movement has been given the task of carrying out this work using the tools which serve every surgeon who is carrying out an operation," he wrote. "Where necessary, he makes the incision."[33] Ustasha theorists also linked racial ideas about the Serbs with notions of moral disease, arguing that the racial purification of the nation would lead to its moral regeneration. The journalist Josip Frajtić maintained that the Serbs had left the Croatians "a repulsive, disgusting, and completely undesired inheritance," a consequence of their "many disgusting, shameful, and sinful habits and curses." All "honorable and upstanding" Croats would see the need to purify themselves from the inside and "shake off all those unwanted Serb inheritances." In the Independent State of Croatia, ruled by the Poglavnik and the Ustashas, all this "evil and ugliness . . . filth and shame" would disappear.[34]

Unlike in the case of the Jews, the Ustasha regime never introduced any racial legislation against the Serbs. There were no legal attempts to prevent Croats from marrying Serbs, even though the media spoke at length about the danger of miscegenation to the "purity of the race."[35] Nevertheless, while the regime did not legally define the racial characteristics of Serbs, Ustasha rhetoric frequently linked Serb racial characteristics to those of Jews, Gypsies, and Armenians and the Serb "problem" to the Jewish Question. In a speech on 27 July in Donji Miholjac, Lorković explicitly stated that the overriding mission of the Ustasha movement was to create a Croatia purified of Serbs and Jews. The Croatian nation, he announced, must "cleanse itself from all those elements which are the misfortune of the nation, which are foreign and alien, which drain healthy forces in our nation, which through decades and centuries went from one evil to the next. These are our Serbs and Jews." This campaign, he vowed, would be carried out to its "logical conclusion."[36]

In the late spring and early summer of 1941, both regime officials and regional party leaders gave inflammatory speeches imbued with violent imagery and the language of destruction, calling for the eradication of the Serb population. These official speeches proved to be the prelude to the systematic campaign of mass murder launched by the Ustasha regime in 1941. Whether the Ustasha leadership had decided upon such a campaign prior to 1941 is open to debate, though the postwar testimonies of some Ustasha officials suggest that they had. At their 1946 trial both Ante Brkan and Ljubo Miloš, high-ranking personnel at the Jasenovac concentration camp established in the autumn of 1941, claimed that plans for concentration camps and the systematic expulsion and mass murder of Serbs, Jews, and Gypsies had been developed years before the regime came to power. Moreover, given how rapidly after coming to power the Ustasha regime established the first concentration camps and began the organized persecution of the Serbs and the Jews, it is highly likely that at least a rudimentary plan of extermination was developed by the Ustasha leadership during the decade they lived in exile.[37] In any case, once the Ustasha state had been established its leading cadres became fairly outspoken about the regime's plans for the Serbs. In a speech on 6 June, for example, Milovan Žanić made it quite plain that the establishment of the Independent State of Croatia meant that the Serbs, who had come as "immigrants" and "spread like hedgehogs," would have to leave. "This is a Croat land and nobody else's and there are no methods that we would not use to make this land truly Croat and cleanse it of all Serbs who have endangered us for centuries and who would do so again at the first opportunity," he declared.[38] At a Ustasha rally on 28 May, Viktor Gutić, party chief in Bosanska Krajina, was even more explicit. He stated with pride that he had begun "the grandiose work of cleansing the Bosnian Krajina of all undesired elements." These elements would be destroyed so that "soon not even the seeds of them will remain, and all that will be left will be the evil memory of them." Two days later he told another gathering in Sanski Most that he had "published drastic laws for [the Serbs'] complete economic destruction, and new ones will follow for their total extermination." He called upon his followers to "destroy them wherever they may be found" so that they would disappear from Croatia forever.[39] Newspapers were also open about the new state's program for the Serbs. In fact, as early as 14 April the leading Ustasha daily, *Hrvatski narod,* proclaimed that the "resurrection" of the Croatian state after eight and a half centuries was contingent upon "bloodily confronting our eternal enemies, our native Serbs."[40]

A series of laws introduced in the spring of 1941 aimed at destroying the Serb community. The main cultural institution of the Serbs, the Serbian Orthodox Church, was outlawed. A series of newspaper articles detailing the historical misdeeds and crimes of the Serbian Orthodox Church in Bosnia and Croatia—

including "Greater Serbian nationalism . . . deceit, craftiness, lack of principles, and avarice"—heralded a parallel campaign of persecution against the Serb Orthodox clergy. Along with the destruction and forced closure of hundreds of Serbian Orthodox churches and cathedrals across the country, hundreds of priests and patriarchs were imprisoned, murdered, or expelled to Serbia, and the property of the church confiscated.[41] The very fact of Serb identity was criminalized. The Cyrillic script was banned; Serb cultural and educational organizations were closed down and their property nationalized; streets, villages, and towns with Serb names were renamed.[42] A law of July 1941 censored the use in the press and official documents of the expression "Serb Orthodox." Henceforth Serbs were to be referred to as "former Serbs," "Greek Easterners," or "Vlachs."[43] According to an editorial by Mirko Košutić in *Hrvatska gruda* on 24 May 1941, all those in the Ustasha state who called themselves Serbs were merely "Orthodox Vlach, Cincar, Greek, Morloch, Romanian, and numerous other races [who had] settled in our towns" and adopted a Serb identity under pressure from the Serbian Orthodox Church. The Serbian Orthodox Church's propaganda had prevented the assimilation of "this conglomerate of various Balkan races," which had instead attempted to build a "new homeland" in Croatia, thus becoming the Croatian nation's worst enemies.[44] The Ustasha Supervisory Service and police expelled Serbs into ghettos along with Jews, barring them from leaving the city limits under any circumstances.[45] In some regions they were required to wear insignia identifying them as Serbs.[46] Simultaneously, the regime purged Serbs from government departments and other bureaucratic agencies as well as from private businesses and cultural institutions; Serbs' businesses, like those of Jews, were nationalized in the name of the state and placed under the administration of specially appointed commissioners.

In June 1941 the regime established a special department for the mass deportation of Serbs, the State Directorate for Regeneration.[47] The directorate aimed to achieve social mobility and national regeneration through the expulsion of 170,000, largely educated or affluent, Serbs to Serbia proper, targeting in particular those who had arrived as "colonists" at the end of the First World War. Their property, land, and assets were subsequently to be expropriated for the benefit of impoverished Croat peasants and workers. While many Serbs did not survive the brutal deportations, newspaper articles described in idyllic terms the "kind and humane manner" in which the nation's "enemies and former oppressors" had been treated: according to one highly unrealistic account in *Hrvatska krajina*, they had sung, eaten, and drunk with gusto as they traveled in comfortable train compartments to their destination.[48] It was not only propaganda outlets that heralded the removal of a privileged segment of the Serb population with enthusiasm. Following the directorate's establishment, it was inundated with applications from idealistic young anthropologists, social scientists, and civil servants eager to participate in what they believed was the social transformation of the Croatian countryside and the transfer of property from a "colonialist" employer class to landless peasants and workers.[49] However, although some of the expropriated property and land was distributed among Croatian peasants or used for other socioeconomic purposes, most of it ended up in the hands of local Ustasha officials, their relatives, and supporters, thus contributing little to national "regeneration."[50]

Simultaneously, Ustasha militias and death squads armed with axes, knives, guns, and mallets embarked on a campaign of mass murder against Serb communities in the Croatian and Bosnian countryside. In some cases villagers were tied together with wire, loaded onto carts and trucks, and pushed off cliffs, locked inside churches that were set alight, or lined up and thrown down pits after their throats had been slit or their skulls crushed.[51] The regime also established a network of concentration camps for the incarceration of Serb, Jewish, and Gypsy "national enemies" as well as Muslim and Croat "political enemies." Furthermore, the Ustasha movement was the only fascist regime to erect concentration camps specifically for the incarceration of infants and children.[52] Estimates put the number of Serb civilians who died at the hands of Ustasha death squads or in the regime's concentration camps at anywhere between 200,000 and 350,000 out of the nearly 2 million Serbs residing in the Independent State of Croatia, although it is possible that as many as 500,000 perished.[53]

While the campaign of mass murder and deportation was underway, Croatian demographers and racial anthropologists were busy carrying out studies of the Serb population in an attempt to ground the regime's policy of extermination in statistical science. A 1941 study by the demographers Petar Bašić and Franjo Lačen used not just the statistics but the earlier arguments of Mladen Lorković, who had attempted to demonstrate the extent to which the indigenous Croatian population had declined on account of "foreigners." Another demographer, Zvonimir Dugački, used population censuses from the mid-eighteenth century to the late 1930s to show how the nation's enemies had intentionally encouraged the colonization of Croatian lands by racially alien minority groups in an attempt to destroy it. His research highlighted the exponential growth of the Jewish population from the late nineteenth century onward, drawing a direct link between low birthrates among Croatians in Slavonia and the settlement of "foreign elements" in the region, Serbs in particular.[54]

Despite the viciousness of the regime's anti-Serb policy, younger members of the party's intellectual elite sought to legitimize the campaign of extermination on ethical grounds. In a series of articles in youth journals marking the publication of Truhelka's 1941 landmark study, they enthusiastically embraced Ustasha racial policies, acknowledging the influential role Truhelka had played in the evolution of their own radicalism. One of these young acolytes, an undergraduate student named Vatroslav Murvar, wrote that the Ustasha ideology had taken hold among his generation of university youth in the late 1930s as "an intellectual instinct" and a "moral idea." He specifically recalled the influence of Truhelka's "Vlachian" scientific theories on his thinking. By 1940, when national liberation remained just a utopian dream, students had enthusiastically embraced the idea that they were the original inhabitants of Bosnia. They found comfort in the belief that they were cheated out of the heart of the Croatian homeland by the deception of "Greek Easterners—the obstacle to our liberation and a menace to our independence—who settled here as part of the Turkish occupation of our land."[55] Meanwhile, in an article printed in 1942 in *Ustaški godišnjak* Murvar asked rhetorically what should be done with those "foreigners from the Balkans." Unlike most other countries that had successfully assimilated minorities, in Croatia alone "this colonizing Vlach population always betrayed and spat on the land and people who had accepted them." Murvar concluded, "It would be hard to have this enormous alien population living in the heart of Croatia. Therefore, we should be prepared for all solutions and not discount any of them."[56]

"Honorary Aryans" and "Orthodox Croats": Contingency, Catholicism, and Racial Ideology

Many young intellectuals echoed Murvar's sentiments. In contrast to Murvar's threatening rhetoric, though, their writings signaled an apparent change in the regime's public discourse, if not policy, regarding the Serbs. This trend was exemplified by an article written in November 1941 by the Ustasha youth leader

Jere Jareb in the party's central youth journal, *Ustaška mladež*. He noted that one part of the "Greek Eastern" population was derived from "remnants of various nations" and "an anational Vlach-nomadic element" brought to the region as Turkish border guards. However, he also argued that the "overwhelming majority of Greek Easterners on our territory" were really of the "purest Croatian and Catholic origins": they were Catholic Croats who had been converted to the "Greek Eastern faith" under Ottoman pressure and had begun to identify as Serb only in the 1840s, "thanks to Serb propaganda." Jure Boroje, a student at the University of Zagreb, agreed. Writing in the intellectual youth journal *Plava revija* in fall 1941, he reiterated the idea that until the arrival of the Ottomans Herzegovina had been "purely Catholic and Croat," with the "Cincars, Vlachs, and others who later began feeling Serb" appearing in the region only after 1463. Variations on Jareb and Boroje's thesis began increasingly appearing in newspaper articles and in the speeches of Ustasha officials, as the original, racial discourse faded into the background.[57]

Such ideas reflected a growing power struggle within both the regime and the movement. By the late summer of 1941 much of the countryside was engulfed in armed insurrection against the regime and Ustasha authority was crumbling. The tough, working-class émigré wing, which had formed the nucleus of the prewar organization and had dominated the regime's policy making in the first six months, was increasingly sidelined in favor of a less extreme, more intellectual faction. Later gathered around the party's cultural and intellectual journal *Spremnost*, some of its leading exponents had never even been members of the movement or had joined only after April 1941. Consequently, the émigrés viewed this competing faction with suspicion and loathing, considering them weak, opportunistic, inauthentic, and potentially harmful. In the long run, this battle for influence was to play a decisive role in the resolution of the Serb "problem." By the fall of 1941 many regime officials had concluded that due to the large number of Serbs and the ongoing armed rebellion, killing or expelling them en masse would be an impossible task. Many, though by no means all, of the leading figures in the Nazi occupation administration disapproved of the campaign of mass murder, mainly for strategic reasons. Some officials, most notably the German plenipotentiary general in Croatia, Edmund Glaise von Horstenau, however, were genuinely shocked by the cruelty and sadism of Ustasha militias.[58] The ascendancy of "moderate" Ustasha officials combined with the chaotic conditions in the state forced the regime to change direction.

The regime did not suddenly abandon its campaign of mass murder and deportation. Nonetheless, it did begin to seriously think about the systematic implementation of a second favored policy option regarding the "Serb problem": forced assimilation. Party officials and ideologists, as well as the regime leadership, believed that a program of compulsory conversion to Catholicism would transform the apostate Serbs into the Croats that they had once supposedly been. It seems likely that Ustasha officials had been thinking about a policy of mass conversion since at least the beginning of the summer. In fact, in some parts of the new state conversions of the local Serb population had already been carried out by regional Ustasha councils and parish priests, many of whom were members of these councils. In addition, the Ministry of Education and Religion had issued guidelines on conversions as early as May 1941. Nonetheless, it was probably only in the fall of that year that a systematic campaign was put into effect.[59] The framework for the mass conversion program was developed by Andrija Radoslav Glavaš, a priest and official in the Ministry for Education and Religion. His policy framework documents emphasized that conversion was as much an attempt to eradicate the Serb identity in Croatia as the campaign of mass murder and expulsion had been. According to a circular that he sent to local authorities in July 1941, educated Serbs, affluent peasants, the intelligentsia, middle-class Serbs, teachers, and Serb Orthodox clergy would not be allowed to convert to Catholicism since their identity was supposedly too strong. They were to be killed. Without their influence as community leaders, it was believed, the process of making ordinary Serbs into Croats would be easier.[60]

The attitude of the Catholic hierarchy to forced conversion was ambiguous. In a letter of 16 July 1941 Josip Lach, secretary to Archbishop Alozije Stepinac, conveyed to the Ministry for Education and Religion that the church understood the state's "justified concern" to protect itself against those elements that sought to use conversion to "sneak into the Croatian national organism" and "work destructively"; likewise, the church wanted to protect the Catholic faith from being "profaned and insulted" by opportunists. Nevertheless, the church was opposed to a total ban on conversion in the case of the Serb intelligentsia, since Christ had come into the world to save all souls, Lach pointed out. Instead, he argued unsuccessfully that the church should accept middle-class "schismatics" whenever "it has been clearly established that their intention is to enter the Catholic Church with sincerity and honesty."[61]

The Catholic theory of Serb origins was not new: it had been articulated in the 1930s in a massive German-language history of Catholics in Bosnia by the priest and church historian Krunoslav Draganović. In his book Draganović had emphasized the aggressive proselytizing of the Serbian Orthodox Church under Ottoman occupation. After the arrival of the Turks in 1463, he wrote, "Serb monasteries on Bosnian land grew like mushrooms after the rain and an immense mass of Orthodox Christians appeared on all sides." The Catholics either perished or were sold into slavery; an even greater number of them settled in safer Catholic regions, leaving their homeland in the thousands due to "persecution, violence, and threats of arson."[62] At the height of the campaign of mass conversion in early 1942, a translation of Draganović's study appeared in a Catholic journal in Sarajevo. His ideas not only provided a scholarly context to the campaign of conversion but also legitimized the campaign of mass murder that had preceded it. Many of the Serb settlements that Draganović's book had identified as "aggressive" outposts of Orthodox colonization were razed to the ground and their inhabitants exterminated by Ustasha militias in the summer of 1941. One of the most gruesome cases was that of the monastery town of Žitomislići, whose inhabitants, including women, children, and the elderly, were herded onto trucks and driven into the countryside, where a local Ustasha unit subsequently slaughtered them.[63]

Along with chronicling the history of the Catholic Church in Bosnia from the medieval era onward, the book exposed state-sanctioned apostasy in interwar Yugoslavia. According to its translator, Draganović's study would "open the eyes" of Croats to the Catholic Church's difficult history, in particular the "organized attack by Saint Sava Orthodox Serbs on the old Christian faith and Roman Catholic Croats" during the 1920s and 1930s.[64] Other clerical supporters of the regime backed these claims. Thus in an essay written in 1943 Ivo Guberina recalled the injustices inflicted on the Catholic Church in the interwar period. According to Guberina, during the twenty years of the Yugoslav state Catholic areas of Croatia had been "systematically colonized by the Orthodox [Serbs]," conversion to Orthodoxy "was more or less openly favored," and mixed marriages "were skillfully encouraged." As a result, the Catholic Church lost almost two hundred thousand members of its congregation through "apostasy and mixed marriages."[65]

Beginning in fall 1941 the head of the Religious Section of the State Directorate for Regeneration, a radical young Franciscan and Ustasha militia member, Dionizije Juričev, dispatched zealous Catholic "priest-missionaries" into the

Croatian and Bosnian countryside to convert the Serb peasants to Catholicism in mass baptismal ceremonies attended by local officials and overseen by armed Ustasha militias before large, makeshift altars. The religious ritual was accompanied by the symbols of statehood: choirs singing national and party hymns, Croatian and Ustasha flags, life-sized portraits of the Poglavnik, and an official banquet in the local Ustasha party hall—all of which emphasized the secular, national aims of the religious conversions.[66] Officially, the Serb masses welcomed the opportunity to convert to Catholicism. According to newspapers as well as an official publication promoting the conversion program, joyous delegations of peasants sought an audience with the Poglavnik to express their joy at reclaiming their Catholic origins.[67] Telegrams of thanks were also sent to the Poglavnik from Serb communities, thanking him for allowing the Orthodox population to return to the "faith of their forefathers."[68] In reality, after an initial wave of conversions the campaign began to falter. Although an estimated two hundred thousand Serbs (or 10 percent) had converted by the beginning of 1942, this was still far fewer than the Ministry of Religion had envisaged. Much depended on the local context: while in some regions a significant proportion of the Serb population converted to Catholicism, often at their own instigation, in many others the conversion campaign met with little or no success.[69]

Despite the initial dynamism of the conversion campaign, officials such as Glavaš and Juričev began expressing concerns about a wide range of issues: the long-term viability of the campaign, the perceived financial and political marginalization of the Religious Section and its programs, and the sincerity of those Serbs who had converted to Catholicism. Juričev, in particular, notwithstanding his own ideological extremism, began voicing misgivings about the independent actions and conduct of some of the more radical missionaries. His concerns ran against their lionization in the regime press as self-sacrificing, iron-willed warrior-priests and "martyrs for the faith and Catholic Croatia."[70] Echoing the sentiments of regional party chiefs, he complained in a report from December 1941 about the irresponsible behavior of some of the missionaries, who he said went beyond their duties as priests. Unfit to participate in the conversion program, their involvement actually had a counterproductive effect. Highlighting the tensions between the secular aims of the regime and the religious impulses of many of the young missionaries, Juričev, speaking in his capacity as a state official, stressed that the groundwork for the conversions had to be laid by the secular authorities, not religious ones.[71]

Ultimately, however, the fate of the program was most decisively influenced from within the regime: by the ambivalent and suspicious attitude of many Ustasha officials and activists who continued to view the Serbs as a racial enemy who should be exterminated. In a testy circular that Juričev sent to the Directorate for Public Order and Security (Ravnateljstvo za javni red i sigurnosti, or RAVSIGUR) and all regional party chiefs at the height of the program in December 1941 he reminded them that the Religious Section had been given exclusive authority over all religious questions relating to the Serbs. Contrary to an earlier order from the Interior Ministry that guaranteed security and protection for all Serbs who had either applied to or had already converted to the Catholic faith, he complained that some regional Ustasha authorities were using "various forms of violence" against them, including incarcerating entire families. "It has been noted," he continued, "that some of your regional authorities and agencies are actively hindering the religious conversion of Greek Easterners to the Roman Catholic faith and ridicule and denigrate these conversions, with some attempting to prevent conversion ceremonies taking place through violent means. This has to stop once and for all. It has gone so far in some individual regions that your cohorts are publicly stating: 'What use is it for the Vlachs to convert? We should kill them all.'"[72]

In fact, the deeply hostile attitude of some party cadres was present from the beginning: as early as May 1941, months before the campaign of mass conversion was implemented, Catholic newspapers were sharply criticizing the attitude of Ustasha bureaucrats as well as members of party militias, pointing out that Ustasha "ruffians" and "drunks" continued to terrorize, imprison, and kill the Serb population, even after the latter had converted to Catholicism.[73] In fact, in some regions invitations to attend conversion ceremonies were used by local Ustasha authorities simply as a pretext to massacre whole Serbs communities, often with the active support and involvement of village priests.[74] Finally, some regional Ustasha officials, such as the party chiefs in Travnik and Garešnica, confirmed that the violent nature of the conversion program, combined with the openly hostile attitude of many local officials, had served only to discredit it in the eyes of many ordinary Serbs.[75]

When the Serb masses realized that converting to Catholicism would not save them from persecution or death, active resistance to conversion increased. The resistance took different forms. Young men escaped into the woods and joined resistance groups. Meanwhile, the widespread mistrust and fear among

Serbs regarding the intentions of local authorities, parish priests, and missionaries meant that, increasingly, they had to be forcibly brought to conversion ceremonies under armed guard; intensified resistance prompted ever more draconian measures by Ustasha officials to compel the Serb masses to covert. When Vladimir Sabolić, the county head of Posavje, was informed in early 1942 that Serbs in the Bosanski Brod region had applied en masse for certificates of conversion, but had never actually presented themselves at their parish priest's office, he sent a strongly worded reprimand to the chief Ustasha administrator for Bosanski Brod. Out of the ten thousand Serbs in Bosanski Brod, seven thousand had applied for certificates. However, he pointed out that only a few of these Serbs had actually contacted their parish priest about baptism. Consequently, he instructed village Ustasha councils to "most severely punish" all those who had been issued certificates of confirmation but had failed to report to the parish office at the designated time. For all those who actively obstructed the conversion program he suggested incarceration in concentration camps. In his reply the chief administrator emphasized that the authorities were using every opportunity to communicate to the Serbs that it was "in their own interests to convert to the Roman Catholic faith." Simultaneously, he pointed out that there were now fewer than ten thousand Serbs in the region, as a significant number had already "resettled" with their families in Serbia, while many others had long since been deported to concentration camps or to Germany as prisoners of war.[76]

In spring 1942 the Ustasha regime began to move away from the idea of mass killing, deportation, and forced assimilation, instead raising the possibility of the reintegration of the Serb minority into mainstream society through the establishment of a Croatian Orthodox Church.[77] Serbs would be nationally, racially, and legally recognized as Croats yet could retain their Orthodox faith. Thus culture appeared to have triumphed over race and religion. A Croatian Orthodox Church was officially established in February 1942. Consequently, Orthodox churches were reopened and Serbs were admitted back into public life; special Orthodox labor and military battalions were created; Orthodox representatives were appointed to parliament and admitted to the military. On the surface, many Serbs were able to return to some semblance of normality: the children of mixed Serb and Croat couples—earlier labeled by Ustasha propaganda as an outcome of "racial miscegenation"—were admitted to the Ustasha Youth organization; Serbs were appointed to positions in the local bureaucracy; some Serbs even applied to officially join the Ustasha movement and sought permission for their children to join the Ustasha Youth.[78] In reality, however, few Serbs embraced the newly established church and even fewer of those Serb Orthodox priests who had survived the mass killings of 1941 agreed to serve in the Croatian Orthodox church. Despite the propaganda hailing the reintegration of Serbs into mainstream society, those Serbs who joined the state bureaucracy were a tiny minority. Despite the limited measures to readmit Serbs into public life, marginalization, persecution, and terror remained an everyday experience for many of them, especially for educated and middle-class Serbs, whom the Ustasha regime consistently saw as an enemy.

The assimilation campaign was not designed as a permanent answer to the Serb "problem" but rather as a temporary tactic imposed on the regime by rebellion in the countryside. In the long run, the regime appeared intent on completing the Final Solution it had begun in 1941. This intention transpired in a conversation of September 1942 between Edmund Glaise von Horstenau and Eugen Dido Kvaternik. According to Horstenau, Kvaternik confided that "in a certain period of time he would kill the remaining one and a half million Serbs, including women and children." Shortly afterward, the Poglavnik dismissed Kvaternik, whom the leadership subsequently made into a scapegoat (along with a number of other hard-line officials who had been replaced at the same time in a regime purge). The fact that a number of Kvaternik's allies in the regime's inner circle escaped the purge, however, suggests not only that some factions within the regime remained committed to the Final Solution of the Serb "problem" but that the attitude of the leadership itself was far more ambiguous than its public pronouncements and policy initiatives might have indicated.[79]

For one thing, the racial stereotyping of the Serbs never really stopped; it persisted long after religion- and culture-based definitions had become the dominant public discourse and even intermittently experienced a revival. For instance, in a 1942 study looking at the racial composition of Croatia, student leader Milivoj Karamarko alleged that at least 15 percent of Serbs had visible traces of "non-Aryan and far-eastern" blood and another 40 percent Armenian blood. Foreign blood accounted for their short noses and dark complexions as well as for the "craftiness, fickleness, deception, and insincere flattery" of the Serb bourgeoisie, with its "mercantilist" spirit. Like Jews, Serbs had racially contaminated Croatia's cities with "very obvious racially alien influences." By contrast, the Croats were marked by the high percentage of Nordic and especially Dinaric blood in their racial makeup. For Karamarko, the Dinaric race embodied all the best qualities of the Croatian nation: tall and physically strong, muscular and healthy, well built and blond.[80]

Moreover, party intellectuals and commentators continued to express nervousness and ambivalence about the identity and status of the Serb masses, who, they insisted, remained foreign and alien, despite their legal classification as Orthodox Croats. As late as July 1943, for instance, the historian and priest Kerubin Šegvić was asking, "Are the Orthodox in Croatia Serbs?" Like many state intellectuals and commentators, he insisted that there were numerous Orthodox Croats who had identified themselves as Croats and who had even led Croatia's national revival. Likewise, he attributed the Serb identity of much of the Orthodox population to the proselytizing of the Serbian Orthodox Church in the nineteenth century, encouraged by the Austrian authorities in order to divide and weaken the Croatian nation. In this sense, the establishment of the Independent State of Croatia and the seizure of power by the Ustasha movement had put an end to that process and reunited the long-divided Croatian population in "love and sacrifice" for the homeland. Yet he was also troubled by the tendency of the Orthodox population in the Independent State of Croatia to retain its Serb identity. Foreign visitors could not understand why these "so-called Serbs . . . , these Orthodox who want to call themselves Serbs" had ever wished to settle in Croatia. The answer was to be found, he argued, in the racially alien origins of one segment of the Orthodox population in Croatia who had settled in Croatian lands as émigré "shepherds" and "Gypsies" fleeing the advance of the Ottoman forces. While their descendants would play an important role in Croatia, in few of these immigrant households was Croatian spoken or a Croatian identity adopted. As a consequence, they always remained apart, in contrast to the well-assimilated immigrant communities elsewhere in Europe.[81]

Although by the spring of 1942 the regime had, at least publicly, abandoned the systematic policy of mass murder and the violent rhetoric that accompanied it, the militias continued to intermittently visit mass terror on the Serb minority. For example, in July 1942, that is, five months after the creation of the Croatian Orthodox Church, the Black Legion, with the assistance of German troops, launched an assault on the Serb population in the Kozara Mountains who supposedly constituted the nucleus of a violent Communist insurrection that had committed barbaric depredations against innocent Croatian civilians. Thousands of men were shot while others were despatched to Germany as slave laborers; the women and children were deported to concentration camps, where many of them perished.[82] Although the Ustasha press portrayed the Serb peasant population as victims of Communist rabble-rousing, the Kozara operation was nonetheless understood in racial, not ideological, terms. The newspapers "explained" the savage behavior of the Serb peasants by their degenerate racial composition, filthy lives, and nomadic nature.[83]

In any case, by September 1944, with the radical émigré faction once again in the ascendancy, explicitly racist rhetoric about the "former" Serbs returned. The Serbs were demonized as racially degenerate "Balkan-Asiatic-Gypsy-Cincar-Vlachs" who should be "shown the door."[84] As biological conceptions of nation became fashionable again, the regime announced the establishment of a State Anthropological Institute. In his letter of 9 July 1944 to the Ministry for National Education, professor of medicine Franjo Ivaniček gave some indication of the repressive uses to which the intended institute might be put. The institute was supposed to conduct anthropological and paleontological research to uncover the "racial makeup of the Croats" and the "ethnogenesis and racial history of the Croats." At the same time Ivaniček wrote that the institute, as a symbol of the modernity of the Croatian state, should be informed by modern racial and eugenic principles. To that end it should be working closely with the Institute for Human Genetics and Eugenics, the Criminological Institute, and the State Bureau for Settlement and Colonization (an agency of the State Directorate for Regeneration).[85]

Ivaniček outlined his ideas in more detail in an article published in *Spremnost*. Among other things, he wrote that one of the institute's primary tasks should be educating citizens about the "biological constitution of the nation, in other words, the racial system of the nation" in order to cultivate a racially and biologically conscious nation. Another task, according to Ivaniček, would be training a new cohort of physicians as "practical eugenicists" who would eradicate hereditary diseases for the benefit of future generations.[86] In any event, the movement's militant émigré faction was hardly interested in intellectual disquisition. Under their influence the regime's original plans for mass murder were revived: in autumn 1944 two new elite paramilitary units were created for this purpose. The resurgent dominance of the hard-liners seeped into every facet of life, including culture and public discourse. The popular press assumed the same kind of fanatical language that had defined the first six months of Ustasha rule. The growing hysteria culminated in an apocalyptic Easter 1945 editorial marking the state's fourth anniversary. The journal *Ustaša* called for a legion to cross the River Drina and "devastate and butcher" the Serbs, as they had failed to do in their "stupidity" in 1941.[87] One month later the Independent State of Croatia collapsed; the remaining inmates at Jasenovac and other Ustasha concentration camps were murdered prior to Ustasha personnel destroying the camp installations.

In the case of Jews, too, there was, technically, a way out of racial categorization. A number of the Ustasha movement's leaders, activists, and intellectuals either were married to Jewish women or were of Jewish origin themselves. Mixed marriages between Ustasha officials and Jewish women started at the apex of the Ustasha leadership—both Slavko Kvaternik and Ante Pavelić married women of Jewish origin—and went right down to the grassroots level. The Ustasha regime also contained a number of prominent Jewish members, including Eugen Dido Kvaternik and Vladimir Singer, a high-ranking member of the Ustasha Supervisory Service, prewar Ustasha student activist, and idol of the Ustasha Youth. Jews were also represented at the regional and local levels of power, including in the Ustasha Youth. Prominent Jewish intellectuals within the movement included the Ustasha Youth ideologist Ivo Korsky and the journalist Stipe Mosner. In recognition of this fact, the Legal Statute on Race Membership made an exception for anyone who had proven "their service to the Croatian nation, especially to its liberation, as well as their spouses with whom they were joined in matrimony" prior to 10 April 1941. Such people could apply for or be granted "honorary Aryan" status, which would exempt them from the antisemitic laws and bestow upon them "all the rights that belong to people of Aryan origin." However, the decision about whether to grant honorary Aryan status was taken on ideological grounds: being an assimilated Jew, for example, was not enough. Individuals on active military duty and those with Aryan spouses and children baptized before 10 April 1941 could be exempted from the law requiring them to wear the Star of David on their chest.[88]

It also appears that even before the legislation regarding honorary Aryan status was published, some regional Ustasha authorities made autonomous decisions to grant Aryan status to some prominent local Jews. Thus on 20 April 1941 the Osijek newspaper *Hrvatski list* announced that, on account of his longstanding Croatian nationalism, a small businessman, Gustav Kraus, would not have to place a Star of David on his shop front. It distinguished his "honorable and respectful attitude" toward the Croatian liberation struggle from that of the majority of local Jews. They had not only served the Yugoslav regime but had tried to destroy his business, "astonished" at his ideological sympathies.[89] In the spring and summer of 1941 the Jewish Section of the Interior Ministry was deluged with letters from Jewish citizens seeking honorary Aryan status. Many of them stressed their loyalty to both the new state and Ustasha ideals, their years of struggle for Croatia's liberation from Serb oppression, and their hostility to "international Jewry." One of the first Jews to apply for and be granted honorary Aryan citizenship was Vladimir Sachs-Petrović, a veteran separatist activist who had spent twenty-one years of his life in exile following the creation of the Yugoslav state in 1918. In May 1941 he wrote to the Jewish Section pointing out his fifty years of sacrifices and persecution "for the sake of the Croatian nation and especially for its liberation." He further declared in his application that his wife would "commit suicide if she had to wear the shameful sign of international Jewry against which we have both determinedly fought."[90] Many of those writing to the Jewish Section were baptised Jews or were married to Gentiles (some of them to Ustasha activists and officials), while others were the offspring of mixed marriages. While the Jewish Section received many letters from ordinary Croatians alleging deception or mistreatment at the hands of Jews during the Yugoslav era, many petitions from Jews were accompanied by letters of support from neighbors, employers, and friends, including members of the Ustasha movement, high-ranking regime officials, or iconic figures in the interwar separatist movement.

For example, in an application of May 1941 on behalf of their mother and three sisters, Robert Neumann and his brother Slavko wrote to the Jewish Section to request honorary Aryan status. Robert requested the status on the basis that he was married to the daughter of a "pure-blood Croat peasant family" as well as on account of his political activities. Not only had he been arrested by the Zagreb police during nationalist student demonstrations in 1933, but in 1936 he also had given a public lecture in which he had called for "all native-born Jews to fight shoulder to shoulder with the entire Croatian nation for their freedom and independence, putting this in the context of examples of brave Croatian Jewish Ustasha warriors such as [Vladimir] Singer, [Ljubomir] Krešimir, etc."[91] Meanwhile, Celine Kohn, a sixty-two-year-old widow, pointed out that she had sheltered the family of the celebrated Ustasha martyr Matija Soldin in the 1920s and 1930s and as a result was "subject to police searches and interrogations at all times of the day and night."[92]

The substantial number of applications for hononary Aryan status seems to have surprised the Interior Ministry; it also provoked sharp divisions in the party itself as well as resistance from regional party branches. In party journals, commentators and activists expressed apprehension about the potential of the honorary Aryan clause to allow an escape mechanism to far more Jews than was originally intended. In Bjelovar, for instance, *Nezavisna Hrvatska* openly ridiculed the letters that Jews had written to the Jewish Section. It commented sardonically on those Jews "yearning to attain 'honorary Aryanism'" and collecting signatures to attest to "their loyal behavior in the days of the struggle of the Croatian nation for liberation." At the same time it sought to identify those Jews who had taken "Aryanized" names after baptism to evade the state's antisemitic laws, and it publicly named and shamed those supposedly upstanding Aryan citizens—doctors, lawyers, businessmen, and housewives—who had acted as godfathers and godmothers at their christening ceremonies.[93]

Eventually, the party's central organizations also began voicing concern about the clause and the danger that it might be misinterpreted by some officials. In an article of September 1941 *Ustaša* attempted to set the record straight. The discussion of the types of intolerable behavior by Ustasha officials and members prominently featured fraternizing with Jews. Some activists, it warned, were "still constantly sitting in corners with Jews. Your accounts are not the cleanest."[94]

As it turned out, they should not have been so worried. Only a few of these applications—including those submitted by members of the Ustasha movement—were successful; even then, it often proved only a temporary reprieve. Although in October 1941 the Kohn family did receive honorary Aryan status, ultimately the entire family was deported to Jasenovac—in spite of a plea from Soldin's mother Katarina, one of the heroines of the Ustasha movement. Similarly, the law exempting state officials and those in military service from wearing the Star of David had no power of deferment and no "bearing on the right to exemption from other orders which are related to racial membership."[95] In other words, honorary Aryan status afforded little protection from the possibility of future persecution. Unlike Serbs, who in certain cases were able to join state and Ustasha agencies (e.g., high school, local bureaucracy, or youth and cultural organizations), only those Jews who were fortunate enough to be classified as honorary Aryans could avail themselves of that option. The fact that so few Jews were able to gain this status, and the precarious existence of many of those who did, betrays its true purpose—protecting the Ustasha movement's honor and ideological cohesion and the lives of some of its more important leaders and intellectuals. In the end, not even those few individuals could always feel safe. Ironically, in late 1941 Vladimir Singer, Pavelić's faithful follower, was arrested by the police on suspicion of aiding the Communist resistance. He was then interrogated, deported to the Jasenovac concentration camp, and subsequently executed.

Can racial science be built on contradiction? The Ustasha regime certainly thought so. Driven by the aspiration for independence, whereby the Croatian nation would be purified of all foreign elements and influences, the regime presided over a state that adopted Nazi (i.e., foreign) biological concepts of race and nation. The Ustasha regime publicly aimed at the moral regeneration of the nation but inaugurated the most brutal campaign of mass murder against a civilian population that Southeastern Europe has ever witnessed. While the regime attempted to remake society in the spirit of Ustasha ideology, it frequently had to be modified in the face of widespread popular resistance and the displeasure of the occupation authorities. The prevailing ideology was also undermined by conflict between rival factions within the regime, pulling it in different directions.

The campaign of mass murder and deportation against the Serb population was initially justified on racial scientific principles. As this campaign proved to be unviable, it was later abandoned on religious and historical grounds in favor of a policy of forced assimilation and then limited reintegration into society. From the very beginning, Ustasha anti-Serb policies had been driven as much by culture as by race. While Nazi racial theories largely guided the regime's antisemitism, the propaganda used against the Serbs was dictated by an older tradition that drew upon anthropological, and thus cultural, concepts. Similarly, initial racial biological propaganda against the Jews was supplemented, or even superseded, by ethical criteria decrying the economic exploitation, social destruction, and moral degeneration inflicted upon the Croatian nation by the Jews. The fact that the regime's antisemitic and anti-Serb propaganda was linked to a wider set of socioeconomic, cultural, and moral questions demonstrates the extent to which racial politics was perceived not as an end in itself but the motor for societal and national transformation.

By mid-1942 the campaign of mass murder against the Serbs largely came to an end, though smaller-scale killings continued. In many parts of the Independent State of Croatia Serbs were readmitted into mainstream society. There also existed a small, protected category of honorary Aryan Jews. However, these policy changes should not be exaggerated. Few Jews were able to enter this privileged category; even some of those who did were later murdered by the regime. For their part, many Serbs continued to experience hostility, persecution, and terror on a daily basis. The Ustasha regime's concessions to the Serb minority were transient, willfully ignored by many of its paramilitary agencies. To a large degree the fate of the Serbs was dependent on whichever Ustasha faction happened to be in the ascendancy. Under the pressure of regime collapse in 1944, the hard-line faction was once more in the ascendancy. The project of extermination was relaunched, and so was the racial categorization that had originally defined it. Once again, national regeneration was to be realized through racial purification and the violent eradication of "undesired elements."

Notes

Abbreviations Used in the Notes

AJ	Arhiv Jugoslavije (Archives of Yugoslavia)
ANDH	Arhiv Nezavisne Državne Hrvatske (Archive of the Independent State of Croatia)
AVII	Arhiv vojnoistorijskog instituta (Archive of the Military-Historical Institute, Belgrade, Serbia)
DRP	Državno ravnateljstvo za ponovu
HDA	Hrvatski Državni Arhiv (Croatian State Archives, Zagreb, Croatia)
JTSRH, OP-A	Javno tužilaštvo Socijalističke Republike Hrvatske, Optužnica Pavelić-Artuković
MNP	Ministarstvo narodne prosvjete
MPB	Ministarstvo prosvjete i bogoslužje
MUP	Ministarstvo unutrašnjih poslova
NARA	National Archives and Records Administration, Washington DC
NDH	Nezavisna Država Hrvatska
OS	Opći spisi
RSUP	Republički sekretarijat unutrašnjih poslova

SRH Socijalistička Republika Hrvatske
VO Vjerski odsjek
ŽO Židovski odsjek

1. Mladen Lorković, "Ustaška straža na Drini," *Hrvatski narod*, 26 April 1942.
2. See Ladislaus Hory and Martin Broszat, *Der Kroatische Ustascha-Staat, 1941–1945* (Stuttgart: Deutsches Verlags Anstalt, 1965); Fikreta Jelić-Butić, *Ustaše i Nezavisna Država Hrvatska* (Zagreb: Liber, 1977); Bogdan Krizman, *Ante Pavelić i Ustaše* (Zagreb: Globus, 1978); Krizman, *Ustase i Treći Reich* (Zagreb: Globus, 1983).
3. Danijel Crljen, *Načela hrvatskog ustaškog pokreta* (Zagreb: Matica Hrvatska, 1942), 99–100.
4. Grga Ereš et al., "Ne damo Bosnu!" *Hrvatski narod*, 3 June 1939.
5. "Zakonska odredba o rasnoj pripadnosti," *Narodne novine*, 30 April 1941.
6. "Naredba o promjeni židovskih prezimena i označavanju židova i židovskih tvrtka," *Narodne novine*, 4 June 1941.
7. "Zakonska odredba o zaštiti arijske krvi i časti Hrvatskog naroda," *Narodne novine*, 30 April 1941.
8. "Zakonska odredba zaštiti narodne i arijske kulture Hrvatskog naroda," *Narodne novine*, 4 June 1941.
9. "Naredba o promjeni židovskih prezimena"; "Dužnosti povjerenika u židovskom i srbskim podužecima," *Nezavisna Hrvatska*, 12 July 1941.
10. "Hrvatski državni sabor-brzopisni zapisnik III sjednice hrvatskog sabora Nezavisne Države Hrvatske dne 24 veljača 1942," *Narodne novine*, 25 February 1942.
11. Report of the Directorate for Public Order and Security, 26 May 1941, HDA, NDH, MUP, ŽO, 227/1941.
12. Letter from Stjepan Tomljenović to the Jewish Section of the Interior Ministry, 18 March 1942, HDA, NDH, MUP, ŽO, 2027.42.
13. Vladimir Žerjavić, *Gubici stanovništva Jugoslavije u Drugom svetskom ratu* (Zagreb: Jugoslovensko viktimološko društvo, 1989); Bogolub Kočović, *Žrtve Drugog svetskog rata u Jugoslaviji* (London: Naše Delo, 1985); Tomislav Dulić, "Mass Killing in the Independent State of Croatia, 1941–1945: A Case for Comparative Research," *Journal of Genocide Research* 8, no. 3 (2006): 271–72.
14. "Naredba o ustrojstvu i delokrugu rada 'Rasnopolitičkog povjerenstva,'" *Narodne novine*, 4 June 1941.
15. "Rasno-biologijski tečaj u Zagrebu," *Napredak* 83, no. 5–6 (1942): 185–86.
16. See, for example, the letter from the Ministry for National Education to Ivan Esih, the head of the State Bureau for the Eradication of Criminality, 28 April 1942, HDA, NDH, MNP, 7.146/189.

17. Letter to Luka Degoricija, inspector for Zagreb's schools, from the adjutant (name illegible) of the General Department of the Ministry for National Education, 30 February 1942, HDA, NDH, MNP, 648/1942.
18. Syllabus for the racial-biological course, 30 February 1942, HDA, NDH, MNP, 648/1942.
19. "Židovi su bili osvojili Ilok," *Novi list*, 27 June 1941.
20. *Židovi—izložba o razvoju Židovstva i njihovog rušilačkog rada u Hrvatskoj prije 10.iv.1941: Rješenje židovskog pitanja u Nezavisnoj Državi Hrvatskoj* (Zagreb: Državni i promičbeni ured, 1942), 5–6, 22–5.
21. "Povijestna važnost zakonskih odredba o zaštiti arijske krvi," *Hrvatski branik*, 10 May 1941.
22. Mladen Lorković, *Narod i zemlja Hrvata* (Zagreb: Matica Hrvatska, 1939), 83–86, 141–63, 261–70.
23. Ćiro Truhelka, *Hrvatska Bosna: [Mi i oni tamo]* (Sarajevo: Hrvatski dnevnik, 1907), 2–6, 12–13, 16, 27–29.
24. Ćiro Truhelka, "O porijeklu bosanskih Muslimana," *Hrvatska smotra* 2, no. 7 (July 1934): 257.
25. Luka Grbić, "Još o Srbo-Cincaro-Vlasima," *Nezavsina Hrvatska Država*, 4 November 1939; M.O., "Vlasi i ne Srbi," *Nezavisna Hrvatska Država*, 1 June 1940; M.O., "Dosljenje Srba u Hrvatsku i turska politika Svetozara Pribićevića," *Hrvatski narod*, 13 October 1939.
26. Emil Medvedović, "Turska—danas i sutra," *Hrvatska smotra* 8, no. 7–8 (July–August 1940): 434.
27. Ćiro Truhelka, "O podrijetlu žiteljstva grčkoistočne vjeroispovijest u Bosni i Hercegovine," in *Studije o podrijetlu: etnološka razmatranja iz Bosne i Hercegovine* (Zagreb: Matica Hrvatska, 1941), 29–43.
28. Hasan Šuljak, "Apostol Hrvatske Bosne," *Nova Hrvatska*, 22 September 1942.
29. See, for example, Ladislav Vlašić, "Mi i bosansko-hercegovački Muslimani," *Ustaški godišnjak* 1 (1942): 124–37.
30. "O Srbima i četničkom bandama," *Vihor*, 7 September 1941.
31. Mirko Košutić, "Srbi su narod koji uvijek donosi zlo i nesreću," *Hrvatska gruda*, 28 August 1941.
32. "Doglavnik dr. Budak o seljačkog politici Nezavisnoj Državi Hrvatskoj," *Novi list*, 4 August 1941.
33. Ivo Guberina, "Ustaštvo i katolizam," *Hrvatska smotra* 9, no. 10–11 (July–October 1943): 439–40.
34. Josip Frajtić, "Neželjena baština," *Hrvatska gruda*, 26 September 1941.
35. "Čistoća rase," *Sarajevski novi list*, 24 March 1943.
36. "Značajan politički govor ministra dra Lorkovića," *Hrvatski narod*, 28 July 1941.

37. Testimony of Ante Brkan to the Republic Secretariat for Internal Affairs, 15 May 1946, HDA, SRH, RSUP, 013.041; testimony of Ljubo Miloš, 20 May 1946, HDA, SRH, RSUP, 013.044. See also the testimony of Vladimir Zidovec, a prewar Ustasha regional organizer and later diplomat, to the Countrywide Commission for the Establishment of War Crimes, HDA, Narodna Republika Hrvatske, Zemaljska komisija za utvrdivanje ratne zločine, 24 May 1947, 013.056.
38. "Uredit ćemo ova država kako propisuju ustaška načela," *Hrvatski narod,* 6 June 1941.
39. "Stožernik Viktor Gutić dobio je naročite pohvale sa Najvišeg mjesta za svoj dosadašnji rad," *Hrvatska krajina,* 28 May 1941; "Triumfalan put stožernika dr. Viktora Gutića u Sanski Most," *Hrvatska krajina,* 30 May 1941.
40. "Poslije osam i pol stoljeća uskrsla je nova Hrvatska," *Hrvatski narod,* 14 April 1941.
41. "Fatalna uloga srpsko-pravoslavne crkve u životi srpskog naroda," *Novi list,* 4 May 1941.
42. "Naredba o promjeni imena sela i katastralne obćine Srpsko polje i Srpska kapela u Hrvatsko polje i Hrvatska kapela," *Narodne novine,* 4 October 1941; "Nova imena ulica i trgova u Mitrovica," *Hrvatski narod,* 17 May 1941; "Zakonska odredba o preuzimanju imovine 'srbskih zavoda i ustanovima' u Hrvatskim Karlovacima u vlastničtvo Nezavisne Države Hrvatske," *Narodne novine,* 19 September 1941.
43. "Ministarstva odredba o nazivu grcko-istočnje vjere," *Narodne novine,* 19 July 1941; Circular of Mijo Bzik to the Ministry of Education and Religion, 15 July 1941, HDA, NDH, MPB, 31.218/119/1941.
44. Kš., "U državi Hrvatskoj nema niti može biti srpskog naroda," *Hrvatska gruda,* 24 May 1941.
45. "Evakuacija Srba u Zagrebu koji stanju u odrednjim svjernim djelovima grada," *Hrvatski narod,* 10 May 1941; "Židovi i Srbi moraju za 8 dana napuštiti svjerni dio Zagrebu," *Hrvatski narod,* 10 May 1941.
46. See, for example, the order of the Ustasha headquarters in Požega from 13 May 1941 ordering Serb Orthodox citizens to wear white armbands with the word "Orthodox" on them on their left arm. AJ, AVII, ANDH, 313.55, 540/1941.
47. "Zakonska odredba o osnutku Državnoga ravnateljstva za ponovu," *Narodne novine,* 24 June 1941.
48. "Prvi transport iseljenih srba napustio je Banju Luku," *Hrvatska krajina,* 9 July 1941. By the time the process was halted in September 1941 by the German occupation authorities, 120,000 Serbs had been expelled.
49. See, for example, letter of application from Krunoslav Stjepan Soldo to the State Directorate for Regeneration, 7 July 1941, HDA, NDH, DRP, VO, OS, 447/141/127/41; letter of application from Zvonimir Maričić to the State Directorate for Regen-

eration, 30 May 1941, HDA, NDH, DRP, VO, OS, 447/141/34/41; and reply dated 11 June 1941, HDA, NDH, DRP, VO, OS, 447/141/222/41.

50. See, for example, the confidential report by the Ustasha Supervisory Service from early 1942, HDA, NDH, Ustaška nadzorna služba, 12 February 1942, 1.248/VT8/42.

51. Among the best accounts of the Ustasha regime's campaign of mass murder are Nikola Živković and Petar Kačavenda, eds., *Srbi u Nezavisnoj Državi Hrvatskoj* (Belgrade: Institut za savremenu istoriju, 1998); Tomislav Dulić, *Utopias of Nation: Mass Killing in Bosnia and Hercegovina, 1941–1942* (Uppsala, Sweden: University of Uppsala Press, 2005).

52. The standard works on Ustasha concentration camps are Antun Miletić, *Koncentracioni logor Jasenovac, 1941–1945: Dokumenti*, 3 vols. (Belgrade: Narodna knjiga, 1986–87); Vladimir Dedijer, *Jasenovac: Dokumenti* (Belgrade: Rad, 1987).

53. See Žerjavić, *Gubici stanovništva*; Kočović, *Žrtve Drugog svetskog rata*; Dulić, "Mass Killing," 271–72.

54. Petar Bašić and Franjo Lačen, "Demografija," in *Naša domovina*, ed. Filip Lukas (Zagreb: Glavni Ustaski Stan, 1943), 1:113–22; Zvonimir Dugački, "Demografijske i narodne prilike," in *Zemljopis Hrvatske*, ed. Dugački (Zagreb: Matica Hrvatska, 1942), 619–37. See also Filip Lukas, "Zemljopisni i geopolitički položaj," in Dugački, *Zemljopis Hrvatske*, 20–21.

55. Vatroslav Murvar, "O podrijetlu pučanstva bosanske Hrvatske," *Plava revija* 2, no. 4 (January 1942): 128–32.

56. Vatroslav Murvar, "Ustaška vjera," *Ustaški godišnjak* 1 (1942): 82–85.

57. J[ere]. Jareb, "Grko-istočnjaci u Hrvatskoj," *Ustaška mladež*, 30 November 1941, 3; Jure Boroje, "Prodor grčko-istočnjaštvo u Hercegovini," *Plava revija* 11, no. 2–3 (November–December 1941): 93.

58. The attitude of the Italian authorities was even more negative. Contemporary accounts of German and Italian reactions to the Ustasha regime's campaign of mass murder include Peter Broucek, ed., *Ein General im Zwielicht: Die Lebenserinnerungen Edmund Glaises von Horsteneau* (Vienna: Verlags Anstalt, 1980), 3:108, 167–68; Alfio Russo, *Rivoluzione in Jugoslavia* (Rome: Donatello de Luigi, 1944). Cf. Jonathan Steinberg, *All or Nothing: The Axis and the Holocaust* (Oxford: Oxford University Press, 1998); Jonathan E. Gumz, "Wehrmacht Perceptions of Mass Violence in Croatia, 1941–1942," *Historical Journal* 44, no. 4 (2001): 1015–38.

59. "Zakonska odredba o prelazu s jedne vjere na drugu," *Narodne novine*, 5 May 1941; "Uputu o prelazu jedne vjere u drugu," *Narodne novine*, 27 May 1941. For a useful discussion of the conversion program, see Mark Biondich, "Religion and Nation in Wartime Croatia: Reflections on the Ustaša Policy of Forced Religious

Conversions, 1941–1942," *Slavonic and East European Review* 83, no.1 (January 2005): 71–116.

60. Directive from Andrija R. Glavaš to regional Ustasha authorities, 14 July 1941, HDA, NDH, MPB, 42.678/B-41.
61. Letter from Josip Lach to the Ministry of Education and Religion, 16 July 1941, HDA, NDH, DRP, Srpski odsjek, OS, 2/155/1941.
62. Krunoslav Draganović, *Massenübertritte von Katoliken zür 'Orthodoxie' im kroatische Sprachgebiet zür Zeit der türken Herrschaft* (Rome: Pontifacium Institutum Orientalium Studiorium, 1937), 5–12.
63. Report on the massacre from the Hum County mobile Ustasha unit commander to the Adriatic divisional region commander, 16 August 1941, AJ, AVII, ANDH, 85.46.
64. Alozije Budzinski, "Masovni prelazi katolika na 'pravoslavlje' na hrvatskom jezičniom područuju za vrijeme vladavine Turaka," *Vrhbosna* 56, no. 1 (January 1942): 25–29.
65. Ivo Guberina, "La formazzione Cattolica della Croazia," in *Croazia sacra*, ed. Krunoslav Draganović (Rome: Tipografija angostiana, 1943), 15–21.
66. See, for example, a description of one such ceremony in a report by the local Ustasha leader in Osijek of 10 November 1941, HDA, NDH, DRP, VO, OS, 2/383/41/1941.
67. "Poglavnik je primio prelaznike s grcko-istočne vjere iz velike župe Baranja," *Hrvatski narod*, 19 November 1941; *Povratak vjera otaca* (no place or date of publication: probably Zagreb, 1941).
68. Telegram to the Poglavnik of the Independent State of Croatia from convertees from Belišće, Baranja, no. 214–29.xi, 29 November 1941, HDA, NDH, Presedništvo vlade, Glavno ravnateljstvo za promičbu, 4.234.
69. Cf. report of 21 November 1941 from Petar Klein to the State Directorate for Regeneration with report of 18 November 1941 from the Party chief in Garešnica to the State Directorate for Regeneration, HDA, NDH, DRP, VO, OS, 2.155/1941 and 2.8646/1941.
70. The missionaries most often glorified in the official media were precisely those about whom Juričev and others had the gravest concerns; newspaper articles often presented their violent conduct as a virtue. One of the most well-known priests was Sidonije Scholz, a young Franciscan notorious for his violent methods of conversion. After he was killed in the village of Našiče by Partisans while returning from a conversion ceremony, the clerical press idealized his "martyred death." Našičanin, "Otac Sidonije Scholz," *Katolički list*, 4 May 1942; "Sprovod ustaše O. Sidonija Šolca," *Hrvatski narod*, 30 May 1942.
71. See report of Juričev, 31 December 1941, and report of Party chief in Osijek, 17 December 1941, HDA, NDH, DRP, VO, OS, 2/470/41/1941 and 2/155/1941.

72. Circular from Juričev to RAVSIGUR and regional party chiefs, 16 December 1941 HDA, NDH, DRP, VO, OS, 584/1076/603/41.

73. See, for example, the stern criticism of the conversion process in the commentary "Ljudi bez kvalifikacija," *Katolički tjednik*, 18 May 1941, 1.

74. This appears to have been especially common in Herzegovina. In one particularly notorious case in early 1942, Serbs from Štikada in the parish of Gračac were summoned by Gračac's parish priest, Morber, to attend a reconversion ceremony, where they were subsequently massacred by a Ustasha militia from neighboring Gudura. See the testimony of Jovan Trbojević and others before the Supreme Court of the People's Republic of Croatia, 23 September 1946, HDA, SRH, JTSRH, OP-A, 121.1421, 232, 1946.

75. See, for example, report from the Travnik Party chief to the Religious Section of the State Directorate for Regeneration, fall 1941, HDA, NDH, DRP, VO, OS, 2/8646-1941/603/41, and similar complaints to the directorate outlined in report of the Garešnica Party chief, 18 November 1941, HDA, NDH, DRP, VO, OS, 2/8646/1941.

76. Report from Vladimir Sabolić to the Bosanski Brod regional authority, 22 January 1942, HDA, SRH, JTSRH, OP-A, 124.1421, 221/42; unsigned reply from Bosanski Brod local authority chief administrator to Sabolić, 10 February 1942 (unreadable signature), HDA, SRH, JTSRH, OP-A, 124.1421, 30/1.

77. "Položaj pravoslavne crkve u Hrvatskoj," *Glas pravoslavlja*, Easter 1944; "Položaj pravoslavlja u Hrvatskoj," *Pravoslavni kalendar za 1943 godine*, 85–88.

78. See, for example, an official report concerning Ustasha propaganda work in the Slavonia region, 15 June 1942, HDA, NDH, MNP, 572/1942; report of officials from the Propaganda Department of the Main Ustasha Headquarters in Zagreb concerning a visit to Ustasha Youth organisations in the Stara Pazova region, 29 March 1942, HDA, NDH, Glavni Ustaški stan, Ured za promičbu, 3.249; "U službi domovine," *Glas pravoslavlja*, 15 May 1944.

79. Reports from Glaise von Horstenau, 14 and 27 September 1943, HDA, NDH, NARA, T-501, 264/968; HDA, NDH, NARA, T-501, 268/82.

80. Milivoj Karamarko, "Dinarska rasa i Hrvata," *Spremnost*, 22 November 1942.

81. Kerubin Šegvić, "Jesu li pravoslavci u Hrvatskoj Srbi?" *Hrvatski krugoval*, 25 July 1943, 13.

82. Regarding Kozara, see Dragoje Lukić, *Rat i djeca Kozare* (Belgrade: Književne novine, 1980); Petar Stanivuković, *Dece logorima smrti* (Belgrade: Rad, 1988); Zora Delić Skiba with Jovan Kesar and Dragoje Lukić, *Djetinstvo moje ukrađeno* (Prijedor: Nacionalni Park Kozare, 1983).

83. "Vidio sam te odmetnike . . . ," *Hrvatski narod*, 26 July 1942.

84. Miško Petrić, "Zablude o Balkan," *Ustaša*, 3 December 1944, 4–7; "Duh Vidovdana i Kajmakčalana," *Ustaša*, 3 December 1944, 14.

85. Letter from Franjo Ivaniček to the Ministry of National Education, 9 July 1944, HDA, NDH, MNP, 7.33/42652/1943.
86. Franjo Ivaniček, "Naša antropologija," *Spremnost*, 23 April 1944. Ironically, *Spremnost* generally represented the perspectives of the moderate faction of the Ustasha regime.
87. "Uskrsnuće i mir," *Ustaša*, Easter 1945, 1.
88. "Zakonska odredba o rasnoj pripadnosti"; "Naredba o promjeni židovskih preizmena."
89. "Priznanje jednom čestitom osječkom trgovcu," *Hrvatski list*, 20 April 1941.
90. Letter from Vladimir Sachs-Petrović to the Jewish Section of the Interior Ministry, 25 May 1941, HDA, NDH, MUP, ŽO, 194/41.
91. Application of Robert and Slavko Neumann to the Jewish Section of the Interior Ministry, 30 May 1941, HDA, NDH, MUP, ŽO, 33/41.
92. Letter from Celine Kohn to the Jewish Section of the Interior Ministry, 7 May 1941, HDA, NDH, MUP, ŽO, 68.6/41.
93. Sp., "Zasluge vrlo zaslužnih," *Nezavisna Hrvatska*, 28 June 1941; "Židovi skupljaju podpise . . . ," *Nezavisna Hrvatska*, 28 June 1941; "Pokršteni Židovi," *Nezavisna Hrvatska*, 28 June 1941.
94. "Naše slabe i 'jake' strane . . . ," *Ustaša*, 7 September 1941, 2.
95. Letter from the Interior Ministry to Celine Kohn, 27 October 1941, HDA, NDH, MUP, ŽO, 48.175/41; letter of Katarina Soldin to the Jewish Section of the Interior Ministry, 14 January 1942, HDA, NDH, MUP, ŽO, unnumbered; "Naredba o promjeni židovskih preizmena."

9

"If Our Race Did Not Exist, It Would Have to Be Created"

Racial Science in Hungary, 1940–1944

MARIUS TURDA

The development of racial science in Hungary during the Second World War remains a largely unexplored subject.[1] This is evident when one attempts to understand the expansion of racial ideas into the public and political spheres and the formalization of race as a normative category of Hungarian national identity. That race was a fluid concept at the time has long been acknowledged, but a detailed analysis of racial sciences in Hungary, their internal dynamics and functions, still awaits its historian.[2]

Defining the nation in racial terms in Hungary between 1940 and 1944 expressed a broader attempt to build a new national state adapted to the unique conditions resulting from the territorial re-creation of Greater Hungary after 1938. While Hungary's political alliance with Nazi Germany during the Second World War undoubtedly played a role in the wide acceptance of biological concepts of identity by political and cultural elites, native racial thinking played a far greater and more influential role.[3] In Hungary, the concepts of race and nation had been in a symbiotic relationship since the late nineteenth century, but it was during the 1940s that they acquired the quality of distinctive constituents of political and cultural discourse.[4] Race, like nation, was an instrument of identity construction.

One can approach the relationship between race and nation in Hungary between 1940 and 1944 in two ways. The first is to examine the political culture of race, that is, to determine whether racial arguments so frequently encountered in public speeches and in parliamentary and scholarly debates in Hungary during the 1940s were indeed expressions of a pervasive racial culture, as was the case in Nazi Germany, or just enunciations of a figurative language employed when speaking and writing about the nation. The second is to survey the racial texts intended for the general public but written by experts. Even before it became common in the 1940s, anthropologists, sociologists, and biologists had often used race to explain ethnic differences. During the Second World War racial science acquired renewed importance in the public imagination, as the nationalist and antisemitic press carried an increasing number of articles exploring the meaning of race and its importance for Hungarian society. To explore this tendency, following a short overview of contemporary interpretations of race, this chapter examines how these descriptions informed racial science and antisemitic discourses in Hungary between 1940 and 1944.

Interpreting Race

Racial science in Hungary was not monolithic but rather a contested field with competing theories. Advancements in anthropological research notwithstanding, one particular question persisted well into the 1940s: "Is there a Hungarian race, or at least, do the Hungarian people have any racial characteristics that are different, independent, and distinguishable from those of neighboring peoples?"[5]

This question brought together professionals from various academic disciplines. Lajos Bartucz, the most important Hungarian anthropologist during the 1940s, was one author who dedicated much of his creative energy to discussions of the "Magyar racial type" and the "Magyar race."[6] Bartucz claimed that centuries of ethnic migration and social interaction, augmented by geographical peculiarities, generated a racial fusion in Hungary. He rejected theories of racial purity and relevant ideas of racial superiority.

Although Bartucz did not deny that a sense of belonging to a racial community sustained the country's national character, he insisted that, in strict, scientific terms (i.e., subspecies), one could not speak of a "Magyar race." Bartucz proposed instead the term "Magyar (Alföld) type" (or the type of the Hungarian plain).[7] However, if race were to be defined as a "biological symbiosis" with a "special racial structure," then those living on any given territory for a thousand years constituted a "harmonious race" (read: "Magyar race"), both physically and spiritually. "The Hungarian national character," Bartucz argued, "was formed from three essential sources: a special physical and spiritual racial structure of the national body; the biological [structure]; and the reproductive community, created by both history and the millenary exposure to our country's environment."[8] The Hungarian race was as much a product of heredity as of historical forces and was an uninterrupted living presence in the Carpathian basin.[9]

The law professor István Csekey reinforced this view when he insisted, "Race is therefore something constant, but the people and the nation vary frequently. Race is heredity."[10] Csekey also rejected the idea of racial purity, insisting that the Hungarian race was racially heterogeneous: "It is precisely in her particular racial composition, a consequence of the quantity and the quality of the racial elements represented in her, that this Hungarian nation differs from all other nations of the world. And as such, one can indeed call it 'a unique and solitary branch.' In this sense a Hungarian race exists. It is a mosaic, a mixture that itself cannot be found anywhere in the world."[11]

Another feature of this particular trend in Hungarian racial science relates to the way in which race (*faj*) was viewed when compared to people (*nép*) and nation (*nemzet*). The dominant framework for understanding *race* was, broadly speaking, hereditarianism, using the term in its biological function, while *people* referred to a myriad of cultural and linguistic factors, determined by environment and history. This was more than just a matter of semantics, as illustrated by the 1936 exhibition on Hungarian anthropology organized at the Natural History Museum in Vienna.[12] On behalf of the host institution, the Austrian anthropologist Viktor Lebzelter suggested "The History of Races in Hungary" as the title of the exhibition, but Bálint Hóman, the Hungarian minister of religion and education, opposed the use of the word *race* in the title, suggesting instead "The History of the Hungarian Nation."[13] Despite the strong racial component, in the end the exhibition was titled simply *The Anthropology of Hungary*.

Despite these tensions, racial scientists continued to insist on the biological as well as cultural components of national identity. The anthropologist Béla Balogh separated them thus: "Race is a biological concept, whereas people is linguistic and cultural." There was an additional corollary to this argument. "We cannot speak," Balogh further maintained, "of a homogeneous Hungarian race, any more than we can speak of a German or Slav one." To accept the idea of racial intermixing meant coming to terms with one of the main tenets of racial sciences—biological worth. "Were races different in terms of their biological value?" Balogh asked. Furthermore, if Hungarians were racially mixed, which of the existing racial strands would then need to be protected and cultivated in order to ensure "the nation's biological future"?[14] According to Balogh, as far as racial and eugenic value was concerned, two racial strands were particularly worthy of strengthening, namely the Turanic and the East Baltic.[15]

Although Balogh accepted the existence of races, he questioned the obsession with racial protectionism (*fajvédelem*). Strengthening the body politic through eugenics and racial policies was deemed necessary, but with one important caveat. Would the process of "making the national body biologically stronger," Balogh pondered, "also contribute to the emergence of a more unitary Hungarian spirituality?"[16] Balogh spelled out his position at the end of this particular article, emphasizing the negative effects of racism on the legitimate attempts to identify and explain Hungarian racial particularity and its corresponding historical trajectory.

Opinions such as these represented one direction in Hungarian racial science. Another was represented by authors who were less interested in physical anthropology and advanced religious and psychological arguments about race and racial essentialism instead.[17] The philosopher József Somogyi was one such author; openly indebted to Christian morality, he delivered a scathing criticism of racial mythologies. His 1940 book, *A faj* (The race), grew out of his belief that religion and science could be merged in a scientific theology based on faith and Darwinism.[18] He demonstrated a continuing commitment to the objective nature of science, a strategy he detailed in his studies on eugenics and later used to dismiss racist arguments.[19] In his writings Somogyi warned of the dangers of confusing race with people and nation, taking the view that the former was the sum of different physical characteristics, while the latter embodied psychological, religious, and cultural qualities. Mendel's theories of heredity, Somogyi argued, had transformed the study of man and yielded new insights into the history of human groups. The crude taxonomy of race proposed by "racial fanatics" like Arthur de Gobineau, Houston Stewart Chamberlain, and Hans F. K. Günther, however, contributed to a racialized, indeed "fictitious," interpretation of ethnicity.

Somogyi was equally disapproving of the "Turanic-Magyar myth," which he claimed was an imitation of the "Aryan myth."[20] Turanists were particularly keen to emphasize the connection between the "Magyar" and other "Aryan" races, but racial insights garnered from ethnography and linguistics did not impress Somogyi.[21] While "recognizing the significance of blood and race" for defining one's ethnic identity, he insisted on relating these features to a cultural framework based on "intellectual lineage, heritage, traditions, and values."[22] Race thus emerged as a synthesis between nature and nurture; this synthesis, in turn, incubated both Somogyi's nationalism and his corresponding Christian philosophy.

Bartucz and Somogyi's ideas of racial symbiosis had their roots in the intellectual environment of pre–First World War Hungary, defined by ethnic diversity and the awareness that racial science could contribute to national assimilation.[23] During the 1940s, however, these ideas encountered opposition, particularly from the radical right and various antisemitic circles.[24] The racial protection of the Hungarian nation, or lack of it, became a dominant topic. Some authors—who believed that the assimilation of ethnic minorities, particularly the Jews, was manifestly incompatible with the logic of Hungarian racial identity—advocated mandatory racial politics. Far from identifying assimilated Jews as "pure" Hungarians, those authors claimed that the Jews' religious conversion and social integration could not supplant a "Magyar racial affinity" based on heredity and blood. The antisemitic implications of this strand of racial science are obvious.

Lajos Méhely, the head of the Department of Anthropology at the University of Budapest, exemplified this trend.[25] In 1926 Méhely became a member of the German Society for Blood Group Research (Deutsche Gesellschaft für Blutgruppenforschung). It was also during the 1920s that Méhely made a name for himself as an implacable racial biologist, a status he carefully cultivated in his antisemitic journal, *A Cél* (The target), and a number of his writings. In his 1936 book, *Fajvédelem és fajnemesítés* (Race protection and race improvement), for example, he outlined a violent program of racial antisemitism and evoked eugenics to propose a serological definition of racial identity.[26]

Méhely's racial awakening took shape during the First World War through a powerful and eclectic range of intellectual influences, most notably German.[27] His hereditarianism was extreme: Hungarian national character could be expressed only through race and blood, he believed. As remarked by the prominent right-wing journalist István Milotay in his overview of Hungarian racial science written in 1940 for the Italian journal *La difesa della razza*, for Méhely "the fate of the nation was governed by biological forces; the nation's grandeur, or its decline, was due to biological causes. The most precious treasure of the state was not its economic wealth but its racially healthy population."[28] To be sure, such an emphasis on biological determinism was not exclusive to Méhely and reflected a growing political interest in racial differentiation between ethnic groups, as illustrated by the anti-Jewish laws introduced in Hungary after 1938.[29]

In 1940 Méhely published *Vér és faj* (Blood and race), a book that synthesized his ideas on race, eugenics, racial biology, and racial protectionism.[30] Méhely drew from similar frameworks in Fascist Italy and Nazi Germany. Notably, he emphasized the duty of the state to promote racial awareness, including "racial biology education," thus encouraging "early marriages." Very much in line with the argument regarding strengthening Hungarian "racial values" was his proclivity for eugenically informed social reforms. He argued in favor of financial aid for poor yet racially valuable members of the Hungarian community so that they could marry, state support for large families, the introduction of a bachelor tax, and the outlawing of celibacy.[31] Like many of his fellow antisemites, during the 1940s Méhely's racial discourse centered on the exclusion of Jews from the Hungarian political and cultural mainstream.

In the name of racial protection, Méhely categorically rejected any notion of racial mixing, which might have given other ethnic minorities, Jews in particular, a leading role in Hungarian society. Such practices, he claimed, would not only destroy Hungarian racial supremacy but also potentially lay the foundations for the worst form of political despotism, namely Communism.[32] Unsurprisingly, the Hungarian political right and some influential cultural and religious leaders shared this view. Ferenc Szálasi, the leader of the pro-Nazi Arrow Cross Party, maintained that the Jews were "a foreign race that has nothing to do with the Hungarians."[33] The Protestant bishop László Ravasz declared in the same vein that "Judaism is a race with strong racial characteristics that prevent its assimilation. Although the Jews mingle with people of other races, Judaism [stubbornly] continues to maintain those racial characteristics."[34] Finally, the writer László Németh described the Jews as "a race incomparably denser and tougher than the Hungarians."[35] These opinions increasingly became the mainstream in the early 1940s, encouraging racial scientists like Méhely and their racial antisemitic views to assume an enhanced political significance.

Throughout his career as Hungary's foremost racial biologist, Méhely was unequivocally committed to the idea that the bonds of Hungarian race transcended the narrowly conceived territorial, social, and institutional boundaries of post-Trianon Hungary. This racial version of the idea of national belonging was closely linked to his crude social Darwinism and in many ways directly flowed from it. Thus in all his writings, Méhely subscribed to the belief that human history was but a continuous struggle for existence, whereby one race always dominated the others.[36] Modern Hungarians may have been the product of centuries of racial metamorphoses, but it was the Turanic racial elements that had most noticeably contributed to fashioning Hungary's much-celebrated historical past. By asserting a superior political role for their nation in Central Europe, Hungarian racial thinkers claimed not only lost territories but also the lost vitality of the ancient "Magyar race."

As mentioned earlier, Méhely moved in antisemitic and racist circles, though he remained affiliated with other Hungarian anthropologists, such as Mihály Malán and János Gáspár.[37] Like Méhely, Malán was immersed in German eugenics and racial science, having completed his studies in 1936 at the Kaiser Wilhelm Institute for Anthropology, Human Heredity and Eugenics in Berlin (Kaiser Wilhelm Institut für Anthropologie, menschliche Erblehre und Eugenik, or KWI-A) with the German racial hygienist Eugen Fischer.[38]

Yet compared to Méhely, Malán was cautious when employing eugenics and racial typologies for political purposes, particularly antisemitism.[39] In his 1940 article devoted to contemporary developments in Hungarian racial research, Malán expressed his anxiety over the fact that Hungarian anthropology remained a fragmented and unfocused discipline: "For 28 years, since the death of professor Aurél Török, who ended up in a cul-de-sac with his skull-measurement approach, [this discipline] has had no university chair; [moreover] we have no anthropological museum or an institute for racial biology."[40] Malán's account has important and wide-ranging implications for the understanding of racial science during this period. But it was not merely a matter of restating the importance of anthropology for Hungarian racial politics. What Malán argued for, ultimately, was an "institutional solution . . . that would bring us closer to a better understanding of our own race."[41]

Institutionalizing Race

Malán may have slightly exaggerated. To be sure, there was no consensus on race among Hungarian scientists, but Hungary did not completely lack institutions and centers of racial research. The Department of Anthropology at the University of Budapest had existed since 1881, and in 1922 Lajos Bartucz established an anthropological section within the Hungarian Ethnological Society. Malán himself set up the Anthropological Laboratory at the College of Physical Education in Budapest in 1930, which he supervised until 1942. But Malán could not have voiced his concerns at a more propitious moment. It was only in 1940 that the Hungarian government established the Institute for Anthropology and Racial Biology at Horthy Miklós University in Szeged under the leadership of Bartucz.

The territorial expansion of the early 1940s did much to encourage the creation of new institutions devoted to racial science. The government officially recognized the importance of racial research for the new national politics formulated after the recovery of borderlands. Thus, when Királyi Ferenc József University reopened in Kolozsvár (Romanian: Cluj) in 1940, one chair in anthropology was created for Malán and another in racial biology and heredity was given to the geneticist Lajos Csík.[42]

Demands for the creation of an institute of racial biology had emerged as early as the 1920s. At the time, Hungarian political and cultural elites were gripped by a sense of intellectual urgency as they struggled to come to terms with the country's territorial dismemberment. Kunó Klebelsberg was one such dedicated politician who, as minister of religion and public instruction between 1922 and 1931, advanced an array of cultural and educational policies, consistently promoting state-sponsored scientific research in Hungary and abroad. Klebelsberg was an enthusiastic supporter of the idea of "Hungarian cultural supremacy" because he believed that only culture could inform national values.[43]

Such insistence on culture, however, prompted leading racial scientists like Méhely to assert that hereditary knowledge in general and eugenics in particular were likewise important in the quest for the nation's rejuvenation.[44] "Racial elements constituting the country's population," Méhely argued in 1927, "have to be studied systematically, for both their internal and external racial characteristics, so that a racial biological inventory can be realized, one that could be used as the background for all sorts of decisions by the intellectual, national, and economic leaders of the country."[45] A sense of historical urgency underlined Méhely's plea for a nationwide eugenic inventory. "For us Hungarians," he maintained, "the achievement of this objective has a dual importance: it is not only about establishing the scientific anthropology of our race, which is dangerously declining, but also . . . about taking stock of those racial strengths and environmental influences, which we can use in order to prepare the way for the physical and spiritual renewal of the Magyar race."[46]

Following the model offered by the State Institute for Racial Biology established by the Swedish eugenicist Herman Lundborg in Uppsala in 1922, a year later Méhely proposed the creation of a similar institution in Hungary at Pázmány Péter University in Budapest. Although Méhely's proposal was initially endorsed by the Department of Arts and Humanities, the university's council rejected it. Zoltán Vámossy, a member of the council and the editor of the influential periodical *Orvosi Hetilap* (Medical journal), invoked financial constraints (a lack of state funds) and structural problems (the faculty of medicine had long wanted, but did not get, an institute for experimental biology) to explain the negative decision. To this financial rationale—to be expected in view of the economic situation of the 1920s—Vámossy added yet another: the university already had "anatomical, zoological, anthropological, and physiological institutes driving hereditary and racial-hygienic research."[47] In other words, there was no need for an institute exclusively devoted to racial biology.

Méhely was undeterred. Taken in its broader national sense, he retorted, the institute would serve as the principal conduit of eugenic propaganda, taking the lead in applying racial biology to Hungarian realities. He substantiated his claim by proposing the following divisions within the prospective institute: genealogy and medical demography; anthropology, including criminal biology; experimental biology; and pathology, along with a museum devoted to heredity and genetics.[48]

The aim of this detailed plan was to urge Hungarians to remain faithful to national values. The true enemies of Hungarian national identity, from this perspective, were those who did not see the importance of racial protectionism:

> All over Europe, both in the victorious and the defeated states, a great stirring storms through the souls, a movement that *regards nursing the racial sentiment and maintaining racial self-awareness as its most sacred responsibilities!* We would be foolish if we did not understand this [movement] and did not follow suit by trying to strengthen the basic pillars of our national existence, and surrendered to the languid spirit of decay instead. I still believe that we have not degenerated too much just yet, and we still feel the significance of that innate great truth: *the race is sacred, and to it we must erect an altar in our heart!* [Emphasis in the original][49]

At a time when Hungarian political and cultural elites struggled to reach a national consensus, tenets of racial biology and eugenics enabled some members of the political and cultural establishment to publicly assert their nationalism and antisemitism.[50]

In 1938 Hungary annexed southern Slovakia from Czechoslovakia and two years later, northern Transylvania from Romania. When Nazi Germany invaded Yugoslavia in 1941, Hungary incorporated Muraköz, Muravidék, and western Vojvodina.[51] This territorial expansion gave a new impetus to racial science.[52] The anthropological theories of the 1920s and 1930s were recast as ideological arguments in disputes over the newly acquired territories.[53] Yet it did not prompt a racist euphoria among physical anthropologists, nor was it confined to Hungary. Anthropologists cherished the new forms of racial science developing in Europe at the time, contributing their expertise and guidance to governmental policies on population control and management. The state, in its turn, promised the restoration of the family and a healthy body politic.

The 1940s witnessed a number of major state initiatives to institutionalize racial science and policies of population control in Hungary. The most important of these initiatives was the Hungarian Institute of National Biology (Magyar Nemzetbiológiai Intézet), officially inaugurated on 31 May 1940.[54] For eugenicists and racial scientists concerned with the health and protection of the Hungarian national body, the institute offered not only government funds for research but also a potential source of ethnic empowerment. To this end, national biology was defined as a summative science combining eugenics with anthropology, demography, and public health, with the aim of cultivating a stronger sense of national belonging in the enlarged Hungarian state.[55]

This form of politicized eugenics received its codification in a theory of "biologism" proposed by the institute's director, the physician Lajos Antal. Formulated as a universalist system of thought, biologism strove "for a better biological future of mankind." Although modeled after Fascist and Nazi eugenic narratives, Antal's biologism nevertheless retained its nationalist undertones: its practical personification was Hungarian biopolitics, whose aim was "a better biological future for the Hungarians." Biologism, biopolitics, and the Institute for National Biology circumscribed Hungary's national regeneration within the new geopolitical boundaries of Europe. According to Antal, Hungarians—like the Germans and the Italians—were now in the vanguard of the biological revolution sweeping Europe and the world, maintaining that Hungarians were essential "for the biological balance of Europe." All this was written with a strong belief in Hungarian racial superiority. Antal concluded his argument as follows: "If our race did not exist, it would have to be created."[56]

In addition to guarding the nation's biological worth, the Institute of National Biology was responsible for the physical and moral education of the population.[57] The institute consisted of ten departments, each dealing with a particular area of research: new perspectives on national life (Lajos Antal); statistics (Alajos Kovács); biology and Hungarian history (Miklós Asztalos); national education (Géza Féja); population growth (János Hidvégi); the biology of large families (János Néveri); heredity, racial biology, and eugenics (János Gáspár); national nutrition (Harald Tangl); science of labor (György Gortvay); and national psychology and public opinion research (Ferenc Rajniss). In comparison to other, similar institutions existing in Europe at the time, including the KWI-A and the French Foundation for the Study of Human Problems (founded in 1941), the Hungarian Institute of National Biology offered a holistic vision of society based on a correlation between social and racial science.[58]

This vision is clearly expressed in the program Gáspár compiled for the Department of Heredity, Racial Biology, and Eugenics. If the institute as a whole dealt with understanding the nation's *external* conditions—be they social, economic, national, or cultural—his department, Gáspár remarked, concentrated on "*internal* biological factors [emphasis in the original]." The strategic purpose of this endeavor was to understand "which national elements were harmful, and thus subject to elimination, and which useful to Hungarians." Eugenics focused simultaneously on the individual's genetic worth and the racial quality of the collective body. To ignore the complex interplay of hereditary factors shaping the national body was to misconceive its complicated biological existence and endanger not only its physical but also its intellectual future. "Ultimately," Gáspár insisted, "a deeper understanding of heredity and racial biology was meant to eugenically enhance the national spirit."[59] Far from being concerned with just the biological transformation of society, eugenics mediated the emergence of a new national ontology and a corresponding epistemology of belonging, one that expressed and reinforced the normative qualities of the Hungarian race.

Complementing these efforts aimed at qualitative and quantitative racial improvement were specific attempts to unify what the psychiatrist László Benedek termed in his 1941 lecture to the Institute of National Biology as "empirical hereditary prognoses." His suggestion was the creation of a Central Institute of Hereditary Biology and Population Policy (Központi Öröklésbiológiai és Népesedéspolitikai Intézet).[60] It was yet another example of how the growth of research on heredity endorsed the eugenicists' ambition to control welfare institutions in Hungary, in a period when the evaluation of the population according to racial categories became central to public health programs.[61]

In fact, eugenicists perceived the nation's racial improvement as crucial to state social policy. Lajos Antal, for instance, endeavored to assist the government by offering a conceptual synthesis whereby the emphasis was not on race but on population. He termed it "population biology" (*népesedésbiológia*), hoping thus to transcend the limitations of classical demography and eugenics. In his view population biology combined biology, sociology, history, and statistics; it aimed at examining and understanding the role of "biological laws in determining the growth of the Hungarian population." Biology was not merely a scientific discipline but an active agent in shaping the future of the nation. Antal ultimately defined population policy as "applied population biology."[62]

Antal thus spelled out what he perceived as the main scientific rationale underpinning his theory of biologism. Racial science was for him the key to the emergence of this new nationalist morality; awareness of one's racial importance represented a duty before both individuals and the collectivity. Between 1940 and 1944 racial scientists constantly emphasized the biological definition of the nation as a core component of Hungarian nationalism. Gábor Doros, a prominent member of the Hungarian Union for the Protection of the Family (Magyar Családvédelmi Szövetség), was one author who consistently tried to develop a form of national eugenics whose objective went beyond merely making visible improvements in the physical health of the population.[63]

It was not merely a matter of emphasizing a biologized narrative of national identity against competitive interpretations of the nation.[64] Racial scholars like Antal and Doros also aimed at educating the younger generations in the latest theories of eugenics and biological racism. In December 1943 the Hungarian National Defense Association (Magyar Országos Véderő Egyesület, or MOVE) established the first university college (Fajvédelmi Főiskola) purposely devoted to racial education. The new college offered public lectures in the history of race, population policy, eugenics, social politics, Turanism, and so on. Antal, for example, spoke on population biology, while Doros lectured on eugenics.[65]

Racial science was thus effectively used to communicate to the general public ongoing discussions about social and biological engineering.

Yet there was an additional component to the racial discourse promoted by organizations like MOVE, one that linked race to antisemitism. The ultimate aim was, as the former subprefect of Pest County, László Endre, noted, "to protect the Hungarian race against everyone else, but primarily against the harm caused by the Jews."[66] The broadening of racial propaganda by the beginning of 1944 was not only an indication of growing antisemitism but, most important, of the erosion of Hungary's hitherto successful refusal to assist Nazi Germany in the Final Solution of the Jewish Question.

Race and Antisemitism

"The protection of the race was not confined exclusively to the Jewish Question."[67] While this statement from the editorial team of *A Cél* was undoubtedly written to assuage general anxieties over the journal's racial antisemitism, it was obvious to just about anyone that the Jews were at the receiving end of racial science. Although many intellectuals and politicians were reluctant to enunciate radical racist statements, they did accept Hungary's emerging national hierarchy, one in which the division between those deemed "racially Hungarian" and others—especially the Jews—was enforced by law. Even if the regime's constitutional framework, at least until 1944, did not allow for the physical annihilation of the Jews, it did little to prevent the spread of antisemitism among the general population, particularly after the adoption in 1939 of the so-called Second Jewish Law that included a biological definition of the "Jew" based on race.[68]

The increased presence of antisemitism in Hungarian politics invited attempts to create an infrastructure that would support racial research into the Jewish Question.[69] Inspired by similar research centers in Germany and Nazi-occupied Europe, particularly Alfred Rosenberg's Institute for Research of the Jewish Question (Institut zur Erforschung der Judenfrage), racial antisemites in Hungary endeavored to establish an institute of their own: the Hungarian Institute for Research into the Jewish Question (A Zsidókérdést Kutató Magyar Intézet).[70] In a programmatic statement written for *A Cél* in 1942, the institute's future director Zoltán Bosnyák outlined the following areas of research: *history*, dealing with the history of the Jews in Hungary; *social sciences*, focusing on

Jewish social life as well as the Jews' social stratification; *racial biology*, concentrating on the anthropological and serological research of the Jews; *statistics*; *culture*, recording the Jewish impact on film, literature, the arts, and the general press; and finally, *racial protectionism*, concentrating on the activities of various antisemitic groups in Hungary. The institute was endowed with a library, a museum, and a publishing house of its own.[71] This structure was a modified version of Mehély's idea for an institute of racial biology, but one tailored to reflect Bosnyák's antisemitism.[72]

The racial policies advocated by Bosnyák aimed "to purify the nation," that is, to eliminate all those categorized as being "alien" and "degenerate." Like Méhely, Bosnyák viewed racial hygiene as a means to justify the exclusion of various groups from what was deemed the Hungarian nation and race. Several population groups were identified as posing a threat to the health of the nation, including medical categories, such as the physically and mentally disabled; social categories, such as prostitutes; and ethnic categories, such as the Roma.[73] Ideas of racial purity became the norm rather than the exception in Hungarian political life.

In a series of publications during the 1940s, Bosnyák persistently equated the protection of Hungarian racial values with the elimination of the Jews.[74] Indeed, once the Institute for Research into the Jewish Question became informally operational on 1 January 1943, Bosnyák's racial ideology and antisemitism were declared the very incarnation of a new form of Hungarian nationalism.[75] When László Endre, in his new position as the state secretary in charge of Jewish affairs in the Ministry of Interior, formally inaugurated the institute on 11 May 1944, he announced that the government had decided "to bring about a final solution to the Jewish Question within the shortest possible time."[76] This determination was confirmed by Gábor Vajna, the minister of the interior in the Szálasi government, on 23 February 1945: "I have started the complete and, if necessary, draconian settling of the Jewish and Gypsy question necessitated by the behavior of these two alien antinational races."[77] Indeed, in a short period of time during the second half of 1944, more than half of Hungary's Jewish population—nearly 440,000—were deported to Auschwitz-Birkenau.[78]

While scholars continue to debate whether this sheer brutality against the Jews was the result of the long history of Hungarian antisemitism, the German occupation, or both, it is obvious that scientific and popular writings about the "Magyar race" also played a crucial role; racial science produced a multifarious biological discourse that amplified mainstream culture and politics in defining the Jews as inferior, less worthy, and ultimately dangerous to the body politic.

The comprehensive history of racial science in Hungary during the Second World War has yet to be written. This chapter has sought to capture some aspects of this history. First, it is important to remember that eliminationist policies deriving from racial science can be identified not only in Hungary but across Europe.[79] Second, in Hungary, antisemitism was only one symptom of a wider process of racial appropriation in which the alleged biological particularity of an ethnic group—the Roma or the Jews—was reinvented for nationalist and political purposes.

The introduction of antisemitic laws after 1938 both built upon and modified the interpretations of race discussed in this chapter. Albeit in different ways, Hungarian authors discussed in this chapter had all attempted to endow the nation not only with new racial foundations but also with a new racial consciousness. Defining *race* was a primary concern during the war years, despite the lack of consensus as to what a "Magyar race" really constituted. Anthropologists like Bartucz and philosophers like Somogyi never joined the radical right and questioned the racist rhetoric that racial biologists like Méhely and Bosnyák inveighed against the Jews.

Yet anthropology did much to legitimize Hungarian racism. The proliferation of anthropological writings within racial science during the 1940s reinforced the belief about the existence of different races that shared specific hereditary characteristics. As elsewhere in Europe at the time, particularly in Nazi Germany, Hungarian racial scientists lent their support to new developments in genetics and eugenics, accepting that heredity determined the transmission of pathological and racial characteristics.[80] The nation's uniqueness was accordingly embodied in an ideal racial type, one that needed to be protected and defended from internal and external enemies. This exclusivist definition of the nation reflected both an indigenous tradition of race and the pressure exercised by the Nazi regime and its supporters in Hungary.

The glorification of Hungarian racial values gradually morphed into a virulent antisemitism aimed at segregating the Jews from the Hungarian nation. In contrast to the idea of racial symbiosis advocated by anthropologists like Bartucz, racial antisemites like Bosnyák promoted an alternative image of the future, namely that of a Hungary without Jews and other ethnic groups deemed "inferior." With the

Arrow Cross's seizure of power on 15 October 1944, antisemitism reached a new level. Racists like Endre grew more assured, basing their actions on a political agreement that attested to an accepted racial difference between "racially pure Magyars" and the Jews.[81] In this context, race was exercised not as a scientific concept but as the main component of an antisemitic policy that ultimately provided legitimacy to the deportation of the Jews to Nazi extermination camps.

Notes

1. No historical study has directly addressed this topic, although attempts have been made, particularly in the 1960s, to discuss Hungarian racial science within the context of Nazi racism. See, for example, Rudolfné Dósa, "A fajbiológia a magyar fasizmus szolgálatában," *Tudományos Közlemények* 1 (1967): 1–22. More recently, it is primarily historians of antisemitism who have discussed the impact of race on Hungarian political culture. A notable example is János Gyurgyák, *A zsidókérdés Magyarországon* (Budapest: Osiris, 2001). See also his *Ezzé lett magyar hazátok: A magyar nemzeteszme és nacionalizmus története* (Budapest: Osiris, 2007) and especially his *Magyar fajvédők* (Budapest: Osiris, 2012).
2. On race as a fluid concept, see, for example, the arguments put forward by Max Weber in *Economy and Society: An Outline of Interpretative Sociology*, ed. Guenther Roth and Claus Wittich (Berkeley: University of California Press, 1978). For marginal discussions of racial sciences in Hungary, see Péter Sipos, *Imrédy Béla és a Magyar Megújulás Pártja* (Budapest: Akadémiai kiadó, 1970); Nicholas M. Nagy-Talavera, *The Green Shirts and the Others: A History of Fascism in Hungary and Rumania* (Stanford CA: Stanford University Press, 1970); Gyula Juhász, *Uralkodó eszmék Magyarországon, 1939–1944* (Budapest: Kossuth könyvkiadó, 1983); Margit Szöllösi-Janze, *Die Pfeilkreuzlerbewegung in Ungarn: Historiker Kontext, Entwicklung, und Herrschaft* (Munich: Oldenbourg, 1989).
3. József Vonyó, *Gömbös Gyula és a jobboldali radikalizmus: Tanulmányok* (Pécs: Pro Pannónia, 2001); Ignác Romsics, ed., *A magyar jobboldali hagyomány, 1900–1948* (Budapest: Osiris Kiadó, 2009).
4. See Marius Turda, *The Idea of National Superiority in Central Europe, 1880–1918* (New York: Edwin Mellen, 2005); Turda, "Race, Politics and Nationalist Darwinism in Hungary, 1880–1918," *Ab Imperio* 1 (2007): 139–64.
5. Mihály Malán, "A magyar fajkutatás problémái," *A Cél* 30, no. 7 (1940): 1.
6. Lajos Bartucz, *A magyarság antropológiája* (Budapest: Királyi Magyar Egyetemi nyomda, 1938); Bartucz, "La composition raciale du people hongrois," *Journal de la Société Hongroise de Statistique* 17, nos. 1–2 (1939): 32–55.

7. For a general discussion, see Marius Turda, "Entangled Traditions of Race: Physical Anthropology in Hungary and Romania, 1900–1940," *Focaal* 58 (2010): 32–46. In 1943 the Austrian anthropologist Ämilian Kloiber reviewed the debate between Lajos Bartucz and Viktor Lebzelter over the "Alföld" race and type for the *Archiv für Rassen-und Gesellschaftsbiologie*. As an anthropologist working for the German army, Kloiber justified the importance of his review as a contribution to the growing body of Nazi "scientific knowledge" about Eastern Europe. See Ämilian Kloiber, "Zur Frage der 'Alföld-Rasse,'" *Archiv für Rassen-und Gesellschaftsbiologie* 36, no. 5 (1943): 358–75.
8. Lajos Bartucz, *Fajkérdes, fajkutatás* (Budapest: Királyi Magyar Egyetemi nyomda, 1940), 318.
9. Lajos Bartucz, *Magyar ember, típus, faj* (Budapest: Athenaeum, 1940).
10. István Csekey, "Race et nation," *Nouvelle revue de Hongrie* 32 (1939): 111.
11. Csekey, "Race et nation," 114.
12. "Die Ausstellung Anthropologie der Ungarn," *Neue Freie Presse* no. 25960 (17 December 1936): 8. The exhibition was held on 14 and 15 December 1936.
13. "Sonderausstellung Rassengeschichte Ungarns" (report), 26 November 1936, Austrian State Archives, Vienna, 15 B 1 Naturhistorischen Museum Wien, Faszikel 3206/40027. I would like to thank Dr. Margit Berner for her help in locating this document.
14. Béla Balogh, "A magyar fajiság," *Természettudományi közlöny* 71, no. 1095 (1939): 273–74, 281–82.
15. The consensus was that modern Hungarians emerged as a mixture of the following "races": Nordic, Mediterranean, Alpine, Dinaric, East Baltic, Taurid, and Turanic ("Alföld"). See Lajos Bartucz, "Die Rassenelemente des ungarischen Volkskörpers," *Ungarische Jahrbücher* 19, nos. 2–3 (1939): 255–80.
16. Balogh, "A magyar fajiság," 282.
17. See, for example, István Boda, "A 'magyarság' mint lélektani kérdes," *Magyar Psychologiai Szemle* 13, nos. 1–4 (1940): 56–66.
18. József Somogyi, *A faj* (Budapest: Az Athenaeum kiadása, 1940).
19. See, for example, József Somogyi, *Tehetség és eugenika* (Budapest: Eggenberger, 1934).
20. Somogyi, *A faj*, 73–76. As early as 1839, Hungarian scholars used the word *Turan*, an ancient Iranian name for the country to the northeast of Persia, to describe the Turkish lands of Central and Southeast Asia. Subsequently the term *Turanian* was applied to a group of people and languages comprising Turkish and Mongol as well as Finnish and Hungarian. See Lajos Sassi Nagy, *A Turanizmus, mint nemzeti, faji és világeszme* (Budapest: Pátria, 1918).

21. Vámbéry, for example, declared, "The Hungarian nation arose from the successful union of Asia and Europe. Turanian heroism was united with Aryan diligence and Aryan tenacity." See Arminius Vámbéry, "On the Hungarian Race," in *The Millennial Realm of Hungary: Its Past and Present*, ed. John Horowitz (Budapest: Carl Louis Posner, 1896), 38. In the late 1930s there were other attempts to relate Central and East European nations to the then-fashionable "Nordic-Aryan" pedigree. The Croat priest Kerubin Šegvić suggested that the Croats, unlike the Serbs, were not of Slavic but German origin. The Bulgarian anthropologist Stefan Konsulov put forward a similar argument in his 1937 *The Racial Appearance of the Bulgarians*, arguing that Bulgarians were not Slavic but Nordic by origin.
22. Somogyi, *A faj*, 78.
23. See Emese Lafferton, "The Magyar Moustache: The Faces of Hungarian State Formation, 1867–1918," *Studies in History and Philosophy of Biological and Medical Sciences* 38, no. 4 (December 2007): 706–32.
24. As, for example, in Lajos Méhely, *A fajtagadás hóbortja: Somogyi Józsefnek válaszul* (Budapest: Mérnökök ny., 1940).
25. On Méhely, see Gyurgyák, *A zsidókérdés Magyarországon*, 387–97.
26. Lajos Méhely, *Fajvédelem és fajnemesítés* (Budapest: Held János könvynomdája, 1936).
27. See Marius Turda, "The Biology of War: Eugenics in Hungary, 1914–1918," *Austrian History Yearbook* 40 (2009): 238–64.
28. István Milotay, "Gli ungheresi e il problema razziale," *La difesa della razza* 3, no. 20 (1940): 21.
29. Lajos Méhely, *Vér és faj* (Budapest: Bolyai Akadémia, 1940).
30. Méhely, *Vér és faj*, 5.
31. Méhely, *Vér és faj*, 45–46.
32. The experience of the Hungarian Soviet Republic (1919) had a traumatic impact on racial nationalists and antisemites like Méhely. See Paul Hanebrink, "Transnational Culture War: Christianity, Nation, and the Judeo-Bolshevik Myth in Hungary, 1890–1920," *Journal of Modern History* 80, no. 1 (March 2008): 55–80.
33. Quoted in Bernard Klein, "Hungarians and the Jewish Question in the Inter-war Period," *Jewish Social Studies* 28, no. 2 (1966): 92.
34. Quoted in Moshe Y. Herczl, *Christianity and the Holocaust of Hungarian Jewry* (New York: New York University Press, 1993), 87.
35. László Németh, "Faj és irodalom," in *Készülődés: A Tanú előtt* (Budapest: Magyar Élet kiadása, 1941), 32.
36. For a list of Méhely's publications, see Gyurgyák, *A zsidókérdés Magyarországon*, 690–92.

37. Malán did fieldwork with Méhely in Borsod and Veszprém Counties. See Mihály Malan, "Zur Augen-und Haarfarbe der Ungarn," *Verhandlungen der Deutschen Gesellschaft für Rassenforschung* 9 (1938): 99–105. See also János Gáspár, *Méhely Lajos és a tudományos fajvédelem Magyaroszágon* (Budapest: Stephaneum, 1931).
38. See Hans-Walter Schmuhl, *The Kaiser Wilhelm Institute for Anthropology, Human Heredity, and Eugenics, 1927–1945: Crossing Boundaries* (Dordrecht: Springer, 2008).
39. See, for example, Mihály Malán, "Magyar vér-oláh vér," *Magyar Szemle* 39 (1940): 187–92.
40. Malán, "A magyar fajkutatás problémái," 5.
41. Malán, "A magyar fajkutatás problémái," 5.
42. Marius Turda, "From Craniology to Serology: Racial Anthropology in Interwar Hungary and Romania," *Journal of the History of Behavioral Sciences* 43, no. 3 (2007): 361–77.
43. See Kunó Klebelsberg, *Neonacionalizmus* (Budapest: Athenaeum, 1928); Klebelsberg, *Ungarns weltgeschichtliche Sendung und seine Kultur* (Budapest: Pester Lloyd-Gesellschaft, 1930).
44. Lajos Méhely, "A magyar élettudomány problémái," *A Cél* 15, no. 12 (1925): 354–65.
45. Lajos Méhely, *Állítsunk fel magyar fajbiológiai intézetet* (Budapest: Held János, 1927), 12–13. See also Tamás Tóth, "Méhely Lajos és népbírósági pere," *Válóság* 54, no. 2 (2011): 63–74; Attila Kund, "Méhelÿ Lajos és a magyar fajbiológia kísérlete (1920–1931)," *Múltunk* (forthcoming 2013).
46. Méhely, *Állítsunk fel magyar fajbiológiai intézetet*, 13.
47. Méhely, *Állítsunk fel magyar fajbiológiai intézetet*, 14–15.
48. Méhely, *Állítsunk fel magyar fajbiológiai intézetet*, 15.
49. Méhely, *Állítsunk fel magyar fajbiológiai intézetet*, 17.
50. See Thomas Lorman, *Hungary, 1920–1925: István Bethlen and the Politics of Consolidation* (Boulder CO: East European Monographs, 2007).
51. See Enikő A. Sajti, *Délvidék 1941–1944: A magyar kormányok délszláv politikája* (Budapest: Kossuth, 1987); Gyula Juhász, *Magyarország külpolitikája, 1919–1945* (Budapest: Kossuth, 1969); Carlile A. Macartney, *October Fifteenth: A History of Modern Hungary, 1919–1945* (Edinburgh: Edinburgh University Press, 1961).
52. See Turda, "Entangled Traditions of Race."
53. See Holly Case, *Between States: The Transylvanian Question and the European Idea during World War II* (Stanford CA: Stanford University Press, 2009); László Kürti, *The Remote Borderland: Transylvania in the Hungarian Imagination* (Albany: State University of New York Press, 2001).
54. *A Magyar Nemzetbiológiai Intézet megalapítása: Az Intézet programmja* (Budapaest: Az EPOL kiadása, 1940).

55. Lajos Antal, *A magyar népesedés kérdései* (Budapest: EPOL, 1942).

56. Lajos Antal, *Biologismus als eine neue Lebensanschauung die ungarische Biopolitik* (Budapest: EPOL, 1940), 3–5.

57. By 1941 the Institute of National Biology had published seven books; a further nine were scheduled for publication in 1942 and 1943.

58. See Schmuhl, *Kaiser Wilhelm Institute*, and Alain Drouard, "Les trois ages de la Fondation française pour l'étude des problèmes humains," *Population* 38, no. 6 (1983): 1017–47. The French Foundation, for example, consisted of six departments, dealing with the biology of population, the biology of infancy and adolescence, biotypology, labor, economic productivity and rural economy, and biosociology.

59. János Gáspár, *A Magyar Nemzetbiológiai Intézet Öröklési, Fajbiológiai és Eugenikai Osztályának programmja* (Budapest: Egyetemyi, ny., 1940), 1–3.

60. László Benedek, *Állásfoglalás a házassági törvény kiegészítéséről és módosításáról szóló törvényjavaslat tárgyában* (Budapest: EPOL, 1941).

61. Dorottya Szikra, "A szociálpolitika másik arca: Fajvédelem és produktív szociálpolitika az 1940-es évek Magyarországán," *Századvég* 48, no. 2 (2008): 39–79.

62. Antal, *A magyar népesedés kérdései*, 4, 6.

63. See Gábor Doros, *Magyarság biológiai erőállománya* (Budapest: Stephaneum, 1944); Doros, *A magyarság életereje: A nemzettest biológiája, fajegészsége és eugéniája* (Budapest: Turul kiadás, 1944).

64. For other debates and cultural definitions of the nation, see Balázs Trencsényi, "'Imposed Authenticity': Approaching Eastern European National Characteriologies in the Inter-war Period," *Central Europe* 8, no. 1 (2010): 20–47; Trencsényi, *A nép lelke: Nemzetkarakterológiai viták Kelet-Európában* (Budapest: Argumentum Kiadó, 2011); Gergely Romsics, "*Magyar Szemle* and the Conservative Mobilization against *Völkisch* Ideology and German *Volksgeschichte* in 1930s Hungary," *Hungarian Studies* 24, no. 1 (2010): 81–97.

65. Lajos Antal, "Népesedésbiologia," *A Cél* 34, no. 5 (1944): 114–21; Gábor Doros, "Fajegészségügy," *A Cél* 34, no. 6 (1944): 142–46.

66. Quoted in Gábor Kádár and Zoltán Vági, *Self-Financing Genocide: The Gold Train, the Becher Case and the Wealth of Hungarian Jews* (Budapest: Central European University Press, 2004), 39.

67. Editorial, "A zsidókérdés," *A Cél* 35, no. 5 (1944): 65.

68. See, for example, Randolph L. Braham, *The Politics of Genocide: The Holocaust in Hungary* (New York: Rosenthal Institute for Holocaust Studies, 1994); Róbert Vértes, ed., *Magyarországi zsidótőrvények és rendeletek, 1938–1945* (Budapest: Polgár kiadó, 1997); László Karsai, *Holokauszt* (Budapest: Pannonica, 2001); Judit Molnár, ed., *The Holocaust in Hungary: A European Perspective* (Budapest: Balassi kiadó, 2005).

69. See [B. Z.], "Nemzetbiológiai Intézet," *A Cél* 30, no. 6 (1940): 15–6. While recognising the importance of the Institute of National Biology, the author regarded its activities as too broad, arguing instead for the creation of an institute dedicated exclusively to racial biology, namely the Institute of Racial Biology (Magyar Fajbiológiai Intézet).
70. See Alan E. Steinweis, *Studying the Jew: Scholarly Antisemitism in Nazi Germany* (Cambridge MA: Harvard University Press, 2006); Michael Wedekind, *Nationalsozialistische Besatzungs-und Annexionspolitiki in Norditalien 1943 bis 1945* (Munich: Oldenbourg, 2003); Patricia von Papen-Bodek, "The Hungarian Institute for Research into the Jewish Question and Its Participation in the Expropriation and Expulsion of Hungarian Jewry," in *Constructing Nationalities in East Central Europe*, ed. Pieter M. Judson and Marsha L. Rozenblit (New York: Berghahn, 2005), 223–42; Christian Gerlach and Götz Aly, *Das letzte Kapitel: Realpolitik, Ideologie und der Mord an der ungarischen Juden, 1944–1945* (Stuttgart: Deutsche Verlags-Anstalt, 2002).
71. Zoltán Bosnyák, "Állítsuk fel a Zsidókerdést Kutató Magyar Intézetet," *A Cél* 32, no. 1 (1942): 17–21.
72. As acknowledged in Zoltán Bosnyák, *A magyar fajvédelem úttörői* (Budapest: Stádium, 1942). Mehély also wrote the preface to Bosnyák's *Magyarország elzsidósodása* (Budapest: Held János, 1937).
73. The sterilization of the Roma was often discussed, both in scientific publications and in Parliament. See János Bársony and Ágnes Daróczi, eds., *Pharrajimos: The Fate of the Roma during the Holocaust* (New York: IDebate Press, 2008), 30–33. See also László Karsai, *Cigánykérdés Magyarországon, 1919–1945: Út a cigány holocausthoz* (Budapest: Cserépfalvi kiadó, 1992); Szilveszter E. Vizi, "Az 'Orsós-ügy' avagy a tudós felelőssége," *Magyar Tudomány* 39, no. 3 (1994): 326–34. The Nazi journal *Volk und Rasse* (Nation and race) reported in 1937 that Hungarian officials were contemplating the sterilization of the Gypsies (Roma). See "Ansiedlung von Zigeunern," *Volk und Rasse* 12, no. 5 (1937): 211.
74. See, for example, Zoltán Bosnyák, *Az idegen vér: A zsidókérdés fajpolitikai megvilágításában* (Budapest: Magyar Kultúrliga, 1938); Bosnyák, *A zsidókérdés* (Budapest: Stádium, 1941); Bosnyák, *Szembe Júdeával* (Budapest: Centrum, 1941); Bosnyák, *A harmadik zsidótörvény és a házasság* (Budapest: Stádium, 1941).
75. The institute's founding members were László Endre, Mihály Kolosváry-Borcsa, Lajos Zimmerman, Géza Lator, Ákos Doroghi Farkas, Ferenc Réthy-Haszlinger, and Zoltán Bosnyák. For a description of the institute's aims, see "A Zsidókérdéskutató Magyar Intézet szervezési szabályzata," in *Vádirat a nácizmus ellen: Dokumentumok a magyarszági zsidóüldözés történetéhez*, ed. Ilona Benoschofsky

and Elek Karsai (Budapest: A Magyar Izraeliták Országos Képviselete kiadása, 1958), 1:166–68.

76. Quoted in Papen-Bodek, "Hungarian Institute for Research," 231. Endre had already announced his support for the Final Solution in Hungary in his article "Das Judenabzeichen keine Stigmatisierung," published in *Pester Lloyd* 91, no. 74 (1 April 1944). The article is reproduced in Benoschofsky and Karsai, *Vádirat a nácizmus ellen*, 1:93–94.

77. Quoted in János Bársony and Ágnes Daróczi, "Roma holocaust-tények és tagadások," *Népszabadság*, 14 September 1998, 12.

78. For a detailed discussion, see, especially, David Cesarani, *Genocide and Rescue: The Holocaust in Hungary, 1944* (Oxford: Berg, 1997) and Gerlach and Aly, *Das letzte Kapitel*.

79. See Aristotle Kallis, *Genocide and Fascism: The Eliminationist Drive in Fascist Europe* (New York: Routledge, 2009).

80. See Benno Müller-Hill, *Murderous Science: Elimination by Scientific Selection of Jews, Gypsies and Others, Germany 1933–1945* (New York: Oxford University Press, 1988); Michael Burleigh and Wolfgang Wippermann, *The Racial State: Germany, 1933–1945* (Cambridge: Cambridge University Press, 1991).

81. Asher Cohen, "Continuity in the Change: Hungary, 19 March 1944," *Jewish Social Studies* 46, no. 2 (1984): 131–44. See also Kinga Frojimovics, "The Special Characteristics of the Holocaust in Hungary, 1938–1945," in *The Routledge History of the Holocaust*, ed. Jonathan C. Friedman (London: Routledge, 2011), 56–66, and, especially, Krisztián Ungváry, *A Horthy-Rendszer mérlege. Diszkrimináció, szociálpolitika és antiszemitizmus Magyarországon* (Budapest: Jelenkor Kiadó, 2012).

10

In the Shadow of Ethnic Nationalism

Racial Science in Romania

VLADIMIR SOLONARI

In *Eichmann in Jerusalem: A Report on the Banality of Evil,* Hannah Arendt called Romania "the most anti-Semitic country in pre-war Europe."[1] Recent scholarship has amply demonstrated the extent of antisemitism in Romanian culture, both popular and intellectual.[2] Antisemitism of the most virulent kind was propagated in Romania by such important figures as the national poet Mihail Eminescu, philosopher and professor of law at the University of Iaşi Vasile Conta, professor of political economy and finance at the University of Iaşi and member of the Romanian academy Alexandru C. Cuza, and physician and professor of medicine at the University of Bucharest Nicolae Paulescu, a scientist who made a significant contribution to the discovery of insulin. In their publications one can find all kinds of antisemitic stereotypes expressed in the most violent language.

Thus, by the time Romania became Nazi Germany's ally, it had a long history of racist discourse. The same, however, cannot be said about racial science and physical anthropological research, which were in their formative stage on the eve of the Second World War. Given that the word *racism* has different meanings in different contexts, it is not surprising that one scholar has emphasized the prominent role that racism—which he equated with antisemitism—occupied in the discourse of Romanian fascists, while another asserted that the most important Romanian eugenicist, Iuliu Moldovan, repudiated racism.[3] This apparent contradiction alerts scholars to the necessity of better defining the nature of Romanian racism and explaining its relationship with racial science within the confines of what Katherine Verdery has called "national ideology."[4]

Racist Antisemitism in Romania

Perhaps the best way to understand this issue is to look more closely at the writings of the most prolific Romanian antisemite, Alexandru C. Cuza. Educated in France and Germany, Cuza was a nineteenth-century positivist and materialist who believed in the ability of science to give definitive answers to all societal problems. His interest in the "Jewish problem" was provoked by the steady, and seemingly unstoppable, increase in the absolute size and share of the Jewish population in Moldovan cities, where they dominated the most lucrative crafts and professions. This process that had begun in the early nineteenth century resulted in the creation of "the hole in the middle," a situation in which ethnic Romanian elites controlled the state institutions and ethnic Romanian farmers toiled the fields, while the middle strata of society was rapidly losing its "national" character and becoming "Judaized."[5] Cuza and his fellow antisemites, all of whom came from Moldova, saw this process in apocalyptic terms.

In his 1899 magnum opus, *Despre poporație: Statistica, teoria, politica* (On population: Its statistics, theory, and politics), Cuza reinterpreted Thomas Malthus's theory of population in nationalistic terms and concluded that "in a given territory there can peacefully reside only one single ethnic nation." Otherwise, Cuza predicted, one of the two groups would be annihilated, expelled, or enslaved.[6] Consequently, Romanians had to rid themselves of the Jews, whom Cuza considered unassimilable, by any means. Simultaneously, other ethnic minorities had to be encouraged to assimilate.[7]

As Cuza himself stated, he undertook this and similar studies with the aim of alerting the Romanian public to the danger of "Jewish penetration" and pressuring the government to take a resolute stance against it.[8] That is to say, his antisemitic conviction came first, and in order to substantiate his preconceived conclusions he investigated various sciences in an attempt to find evidence of "the Jews' pernicious activity."[9] Not surprisingly, he later claimed that history, anthropology, theology, political science, and political economy all led to that same conclusion, thus forming the "science of anti-Judaism."[10]

Cuza developed his anthropological arguments most comprehensively in his 1908 book *Naționalitatea în arta* (Nationality in art). Heavily borrowing from Western antisemitic orientalists and racial ideologists, he argued that Jews belonged to an inferior race for two reasons. First, they were Semites, not Aryans. Second, even as Semites they were "impure," since they descended from the interbreeding of Semites with other races. That made them racial "bastards" and thus particularly dangerous.[11] Cuza appropriated these arguments from Houston Stewart Chamberlain's infamous *Die Grundlagen des Neunzehnten Jahrhunderts* (1899), which he frequently cited.[12] At the same time, he ignored, or rather had to ignore, those aspects of that same theory that were certain to undermine Romanians' own racial worth.

Indeed, Chamberlain treated racial bastardization as inevitably leading to biological degeneration. According to the established Romanian historiographical tradition, however, Romanians were themselves bastards—a mixed breed of pre-Romans and Roman colonists who had been settled in the newly conquered province of Dacia by the victorious emperor Trajan. Remarkably, while Cuza evoked the supposed cultural sterility of contemporary Greece as proof of the danger of interbreeding (modern Greeks supposedly descended from the interbreeding of their glorious ancestors with inferior, mostly Slavic and Turkic, peoples), he failed to notice that the same logic could be used to disparage his own nation.[13]

Thus Cuza's racism was effectively an outgrowth of his antisemitism. He selectively appropriated those elements of Western European racial theories that suited his antisemitism and ignored others that threatened to undermine Romanian racial worth. In this sense, his was an inconsistent racism. Intrinsically, Cuza remained a radical ethnic nationalist.[14] The same can be said of a number of other intellectual antisemitic ideologists, including Cuza's former student, disciple, and political rival Corneliu Zelea Codreanu. In 1923 Codreanu broke with Cuza's National Christian Party to form his own Legion of Archangel Michael, better known under its latter name, the Iron Guard. In contrast to Cuza, Codreanu was prone to religious mysticism and claimed that the idea of creating his own party had been inspired by the archangel himself. In his propaganda Codreanu skillfully combined readily recognizable religious symbols with denunciation of the pervasive corruption of the ruling elite and calls for martyrdom and moral regeneration. Like Cuza, Codreanu occasionally used the term *race*, but mostly as a synonym of *neam*, and certainly not in its more "scientific" sense. Violently antisemitic, he, too, was a radical ethnic nationalist obsessed with "the Jewish problem."[15]

The Dilemmas of Racial Anthropology in Interwar Romania

The first scholar to undertake extensive anthropological research on ethnic Romanians was the Swiss professor Eugène Pittard.[16] His conclusions as to the racial characteristics of ethnic Romanians were in several respects deeply unsettling to the conventional historiography of early Romania. According to this historiography, after the Roman withdrawal from the territory north of the Danube between 270 and 275 AD, the remnants of the Romanized population found refuge in the Carpathian Mountains. Having reinvented themselves as transhumant pastoralists, they survived through the end of the barbarian invasions. They then returned to the valleys of Transylvania, Wallachia, and Moldova, where they interbred with other foreign ethnic groups such as Slavs and Tatars.[17]

This widely shared account implied that ethnic Romanian mountaineers were better poised to preserve their pristine purity than Romanians from the valleys. Mountaineers were imagined to have been somehow nobler than denizens of the lowlands, and as such they were expected to be taller and more dolichocephalic (long-headed). Dolicocephaly was especially important from the vantage point of Nordic theory, the most influential racial theory in Europe at that time, according to which the cephalic index was central in establishing the biological worth of any given race: the longer the head and the lower the index, the more likely it was that a particular population belonged to the Nordic race. Pittard's anthropometric measurements, however, showed the exact opposite: mountain-dwelling Romanians were brachycephalic (broad-headed) and had shorter heads than Romanians from the valleys. Even worse, Romanians in general proved more brachycephalic than southern Slavs, Serbs, and Bulgarians.[18] Yet Pittard's findings were not politically motivated. He was no adherent of Nordic racial theory and he abhorred the idea of racial superiority, at least insofar as European races were concerned.[19] Pittard emphasized that none of the European nations consisted of a single race and that the achievements of

European civilization could not be ascribed solely to the Nordic race.[20] He also had a very favorable opinion of ethnic Romanians and generally considered them the most "cultured" of all Balkan peoples; he therefore saw no contradiction in both emphasizing their non-Nordic nature and praising their cultural creativity and potential for development.[21]

Writing in 1924, Pittard expressed bafflement as to why his "Romanian friends" were slow in researching racial characteristics of their prehistoric ancestors.[22] Pittard's comment proved to be prophetic: in the ensuing two decades physical anthropology remained underdeveloped in Romania. As Olga Necrasov asserted in her doctoral dissertation on anthropological research in northeastern Moldova at the University of Iaşi in February 1940, until that year no synthetic study of the racial composition of the Romanian population, whether of the country as a whole or of its individual provinces, had appeared. Those studies that had been published were "preliminary contributions concerning mostly geographic variation of different anthropometric characteristics."[23]

One possible explanation for the underdevelopment of physical anthropology in Romania might be structural. The humanities and social sciences were subject to severe institutional and financial constraints in interwar Romania. As Katherine Verdery has cogently argued, cultural production in interwar Romania was conditioned by an overabundance of intellectuals, a shortage of resources, and a virtual monopoly of state institutions on the disbursement of funds.[24] Romania emerged from the First World War having doubled its size in population and territory: this was at the expense of the country's erstwhile powerful neighbors, the former Austro-Hungarian and Russian Empires (with one additional province, southern Dobrudja, annexed from Bulgaria). However, the Soviet Union, Hungary, and Bulgaria laid revisionist claims on the lost provinces. Besides, Romania's administrative resources were overstretched. The rule of the Old Kingdom's bureaucracy over the new provinces provoked resentment, especially among nonethnic Romanians. At the same time, the overrepresentation of national minorities in the most lucrative professions, especially in the new provinces, was perceived by ethnic Romanians as an affront. Consequently, integral nationalism appeared as the simplest and surest way to legitimize and solidify this new country.[25] Many intellectuals eagerly participated in the production of nationalistic discourses, often advertising their own disciplines as capable of reflecting reality in the most vivid, persuasive, and "scientifically correct" way.

Given the nature of Pittard's findings, physical anthropology appeared to be a minefield. Why would a young researcher stake his future on it? Why would the government support this type of research, unless it was intended as a rebuttal to diversionary publications in neighboring countries seeking to undermine Romanians' self-esteem? Besides, as Marius Turda has convincingly argued, physical anthropology had to establish itself in competition with more traditional and deeply entrenched university disciplines such as history, ethnography, folkloric studies, musicology, philosophy, or even sociology.[26] All these disciplines had their own ways of producing a discourse on the nation, while some of the practitioners were very likely aware of physical anthropology's potential to question their underlying assumptions.

Unsurprisingly, therefore, those few researchers who had published on physical anthropology in the 1920s and 1930s often tried to make their findings appear both acceptable from the perspective of the dominant ideology of integral nationalism and politically useful as a means of boosting Romanians' claim to a higher status among the peoples of Southeastern Europe. This point can be illustrated by an article discussing the skulls of Romanians from Transylvania published in 1923 by Victor Papilian, professor of anatomy at the University of Cluj and a member of the Paris-based Society of Anthropology. Papilian claimed that his research (conducted on 230 men from the Cluj army garrison) had demonstrated that the cephalic index was lower among Romanians than it was among Hungarians. According to Papilian, earlier Hungarian researchers had incorrectly measured the longitude of the Hungarians' heads, thus moving the Romanian cephalic indices upward. Among competing methods of calculating the cephalic index, Papilian seemed to have selected the one that better suited his agenda. As a result, he could claim that the cephalic index of Transylvanian Romanians, though higher than that of Romanians from the Old Kingdom, was still lower than that of Hungarians.[27]

Hungarian figures nonetheless continued claiming the opposite. In 1928 the Hungarian anthropologist Jenö Davida published an article summarizing his measurements of a collection of skulls from the Institute of Anatomy at the formerly Hungarian University of Cluj (Hungarian: Kolozsvár). Having studied 141 skulls of ethnic Hungarians and Romanians in Transylvania, Davida—who sought to appear nonpartisan in his analysis—came to the conclusion that practically all criteria, including brain mass and cephalic index, were more favorable to Hungarians.[28] This viewpoint appears to have enjoyed wider support outside of Romania. As late as 1942, one of the most vocal proponents of the Nordic idea in Fascist Italy, Guido Landra, having summarized Papilian's findings for the Italian audience, noted that, "according to the commonly held view, the opposite was true."[29]

Another example of the use of new anthropological methods for political purposes was serological research conducted by Petre Râmneanțu on the Széklers, a Hungarian-speaking minority in the southeastern part of Transylvania. Since the Széklers resided in the center of Greater Romania, they constituted "a problem" from the Romanian nationalist point of view. Indeed, Hungarians did evoke the Székler presence as proof of their "right" to repossess Transylvania. Râmneanțu, a prominent member of the so-called Cluj School of eugenics, led by his mentor, Dr. Iuliu Moldovan, endeavored to find proof of the "Széklerization" of ethnic Romanian settlements, which supposedly had taken place in the past.[30]

Râmneanțu's rationale requires additional explanation. The use of serological research for anthropological studies in Romania had been introduced in 1924 by Dr. Sabin Manuilă, another disciple of Moldovan and a would-be prominent Romanian statistician.[31] Manuilă's research, conducted in collaboration with Gheorghe Popoviciu, was based on the study of patterns of hemagglutination and led him to some unsettling conclusions.[32] He found that the biological index of Romanians varied greatly from one region to another. Consequently, in order to calculate this index for the entire Romanian "race," the research had to include people of different ages and sexes in all parts of the country. On the other hand, having calculated this same index with regard to "eight races residing in Romania" (evidently ethnic groups), Manuilă and Popoviciu concluded that it was practically identical to the one of Balkan peoples such as the Bulgarians, Serbs, and Greeks.[33] Therefore, according to Manuilă, those peoples were kindred races to the Romanians.[34]

This proposition was self-evidently subversive, given that received nationalist wisdom emphasized differences between the Romanians and the peoples of the Balkan Peninsula, especially the Slavs. But so was the notion of biological heterogeneity among ethnic Romanians, since integral nationalism postulated homogeneity of the Romanian ethnic nation. Small wonder, then, that Manuilă, despite his initial enthusiasm for serological research, failed to pursue it any further, preferring instead to concentrate on demography and ethnic statistics. In early 1940 Olga Necrasov could cite—besides the works of Manuilă, Râmneanțu, and herself (on the Turkish-speaking Gagauz minority)—only four minor and geographically limited serological studies, all of whose authors eschewed broad conclusions.[35]

Undeterred, Râmneanţu ignored the inconvenient findings of Manuilă and Popoviciu and manipulated their methodology in order to substantiate a politically expedient thesis. By comparing the biological index of Széklers with that of ethnic Romanians from the Székler-settled area of Transylvania and ethnic Romanians from the Old Kingdom on the one hand and ethnic Hungarians from Hungary on the other, he concluded that that the Széklers' index was very similar to that of ethnic Romanians but quite different from that of Hungarians from Hungary. Thus, he concluded, Széklers were "in reality" Széklerized Romanians; consequently, unlike ethnic Hungarians from Hungary, they were not Asiatics but Europeans on par with ethnic Romanians. Furthermore, ethnic Romanians from the rest of Transylvania were biologically closer to ethnic Romanians from the rest of the country than they were to their kin in Hungary.[36]

Not all Romanian anthropologists, though, were as willing as Papilian and Râmneanţu to brazenly distort their methodologies, research agendas, and conclusions to suit the political climate at any given moment. Some of them did indeed internalize Nordic racial theory, even though they were aware of its negative implications for the evaluation of ethnic Romanians' biological worth. One such academic was ethnographer Ion Chelcea. In his 1935 article on the types of Romanian crania from Transylvania, he cited the assertions of German pro-Nazi anthropologists regarding the racial superiority of the Nordic race and concluded that "these assessments are generally known; they are also scientifically proven."[37]

Chelcea's article outlined his research on ethnic Romanian crania held in the Vienna Museum of Natural History. As an intern for one semester in 1934, Chelcea employed the racial taxonomy developed by the Austrian anthropologist Viktor Lebzelter, the director of the museum.[38] Chelcea's findings confirmed that most Romanians were brachycephalic, even though he deliberately raised the threshold for dolichocephaly to 80.000–99.000 to ensure a greater number of dolichocephals than would have been the case if the more widely accepted threshold index of 75.900 had been applied. Racially, they belonged to Romano-Mediterranean, Nordic, Kurgan, Dinaric, Dacian, and Avaro-Turanic races, he argued. The presence of Nordic racial types was due to the invasions of Germanic tribes during the period of great migrations. The Dacian racial type was identified with ethnic Romanian inhabitants in the Western Carpathian Mountains

(Munţii Apuseni), known as Moţi, and had the following characteristics: short stature, brachycephaly, and dark hair. This latter finding was particularly important since, according to the dominant historiography, it was in the thick of the Carpathian forests that the descendants of intermarried Dacians and Roman settlers had hidden themselves during the period of the great migrations. Thus Moţi were imagined as most closely related to the Dacians.[39]

Lebzelter himself conducted research on four thousand soldiers from the Royal Romanian Army in 1932. He interpreted his findings in correlation with the dominant interpretations of early Romanian history and linguistics. It was Lebzelter who identified the Dacian racial type. In addition, he identified the Romano-Mediterranean racial type, particularly among the inhabitants of the lowlands of Wallachia and the Danube River plains. Dark haired, of short stature, and brachycephalic, they were descendants of Roman settlers, according to Lebzelter. Finally, among the ethnic Romanian inhabitants of Transylvania and the mountainous regions of Muntenia he found many "Slavic types" related to the Slavs in Yugoslavia. Thus, Lebzelter concluded, "racial analysis of the Romanian people showed a manifest correlation between the history of the people, of the language, and of the race."[40]

Ironically, at the time when Lebzelter made this bold statement Romanian historiography was increasingly moving away from emphasizing ethnic Romanians' Latin roots and toward celebrating their supposed "autochthonism" instead. *Dacianism*, as Lucian Boia had branded this tendency, was purely speculative. In spite of this it gained traction not only with the Romanian right but also among the most respectable historians and archeologists, keen on ascribing special nobility to Dacians as well as to Gets and Thracians, from whom Dacians supposedly descended.[41] Dacianism can best be understood as a way of staking the Romanian claim to a special status among the peoples of East Central Europe.[42] Lebzelter and Chelcea's "Dacian type," however, was ill-suited for this purpose since it was far removed from the iconic Nordic race—tall, dolichocephalic, and blond.

Nevertheless, Lebzelter and Chelcea's conceptualizations perfectly dovetailed with earlier interpretations of anthropological data on ethnic Romanians, as formulated by Pittard, for example. Interwar Romanian anthropological research tended to support their findings: a mostly brachycephalic ethnic Romanian population of short stature in the mountains and increasingly taller and more long-headed people as one descended into the plains. Given the potentially embarrassing nature of their findings, it comes as no surprise that Romanian scholars as a rule refrained from interpreting them within the framework of the prevalent discourse on Romanian ethnogenesis.[43] By the same token, they abstained from evoking Nordic racial theory or ranking various races according to their respective biological worth.

Still, Romanian anthropologists never questioned the heuristic value of the category of *race* and assumed its immutability. They continued to do so despite Franz Boas's research, which demonstrated as early as 1912 that skull forms changed from generation to generation among European immigrants to the United States, suggesting, in Boas's own words, "a plasticity (as opposed to permanence) of types."[44] One of Romania's leading eugenicists, Gheorghe Banu, discussed these findings in his 1939 monograph *L'hygiène de la race.*[45] Banu referred to Eugen Fischer's 1923 *Spezielle Anthropologie: Rassenlehre,* in which this influential German anthropologist had discussed Boas's research.[46] Neither Fischer's nor Banu's discussion made any notable impact on Romanian anthropologists, however.

Iordache Făcăoaru and the Ranking of European Peoples

Among Romanian anthropologists Iordache Făcăoaru alone attempted to systematically address the problem of the purportedly low biological worth of the Romanian nation. Born in 1897 in southern Moldova, Făcăoaru studied philosophy at the University of Bucharest and later worked as a high school teacher. Between 1929 and 1931 he studied anthropology, pedagogy, psychology, heredity, and racial hygiene at the University of Munich. His mentor in the latter discipline was Fritz Lenz, the influential anthropologist and eugenicist who in 1937 became a member of the Nazi Party and head of department at the Kaiser Wilhelm Institute for Human Heredity Sciences and Eugenics in Berlin.[47] In 1932 Făcăoaru became an assistant researcher in anthropology at the Institute of Hygiene and Social Hygiene at the University of Cluj, headed by Iuliu Moldovan.[48] From that time on he became one of the two most important and prolific members of the Moldovan School of eugenics (along with Petru Râmneanţu). He participated in a number of the so-called monographic campaigns initiated by the students of Dimitrie Gusti, professor of sociology at the University of Bucharest. The aim of those campaigns was to meticulously describe all aspects of the life of a particular ethnic Romanian village.[49] As a member of Gusti's teams, Făcăoaru conducted anthropological measurements; as someone with strong right-wing convictions, at one point he joined Codreanu's Legionary Party. During the short-lived National Legionary regime (September 1940–January 1941) he served as director of the Department of Higher Education, Cults, and Arts in the Ministry of National Education, a position that he used to (unsuccessfully) promote eugenic legislation on marriage control.[50] He was also a member of the government-created commission to purge universities of faculty considered hostile to the regime.[51]

Făcăoaru viewed anthropology as an applied science whose mission was to guide the policies of the state toward the biological improvement of a *neam*. He was also concerned about the low opinion that foreign anthropologists tended to have of the Romanians' biological worth. He attributed this "grave stigmatization of a *neam*" to the lack of reliable information about the racial composition of Romanians. This situation had to be urgently addressed: "We can remain indifferent to many things. There is, though, one thing toward which we cannot remain indifferent, namely how our place in the spiritual hierarchy of nations is presented."[52] He devoted his efforts to changing this perception.[53]

In a series of articles in *Buletin eugenic şi biopolitic* and *Sociologie românească* from 1936 to 1938, Făcăoaru expanded on his findings about the racial composition of various ethnic Romanian villages that he studied within the framework of Gusti's monographic campaigns. In a 1940 article he dismissed Lebzelter's interpretation as methodologically unsound and confusing. In particular, he rejected identification of the Dacian type with short, brachycephalic, dark-haired Romanians from the Moţi region. Following Romanian archeologist Vasile Pârvan, he suggested that those Romanians were actually "tall, vigorous, blond, and blue-eyed." Furthermore, the existence of Nordic elements within the Romanian nation had to be explained not by the invasions of Germanic tribes in the first centuries AD, as Lebzelter had suggested, but by the prehistoric substratum of the autochthonous population. Făcăoaru calculated the racial composition of ethnic Romanian villagers and ethnic Romanian university students; indices for university students were calculated both for the country as a whole and for individual provinces. Following the usual practice at the time, Făcăoaru made separate calculations for males and females. His formula for male villagers, for example, was as follows: M31A21N12O10D10E8X6Da1.8Mo0.2–Br79+Bl21, where *M* stands for Mediterranean race, *A* for Alpine race, *N* for Nordic race, *O* for Orientalid race, *D* for Dinaric race, *E* for Esteuropoid race, *X* for Antlantid race,

Da for Dalic race, *Mo* for Mongoloid race, *Br* for Brachycephals, *Br* for dark haired, and *Bl* for blond.[54]

Calculating the racial composition of the Romanian nation was, however, just the first phase of an ambitious project. In 1939 in *Zeitschrift für Rassenkunde* he published an essay titled "Beitrag zum Studium der wirtschaftlicher und sozial Bewährung der Rassen" (Contribution to the study of the economic and social verification of races), in which he aimed to gauge the intellectual endowment of various European races based on their success, or lack thereof, in the economic and social spheres.[55] His selection included 1,290 ethnic Romanian peasants from Transylvania and Banat and 866 students from the University of Cluj (464 of them ethnic Romanians, 244 ethnic Hungarians, 63 ethnic Germans, and 95 Jews). He divided peasants into three categories based on the wealth and income of their households: rich, middle, and poor. Students were divided into five categories based on the income and status of their fathers, ranging from very poor to very rich. Subsequently, Făcăoaru identified the racial origin and identity of each ranked individual. For further analysis he selected only those whom he identified as belonging to one particular race or a mixture of two races; all those whose immediate forebears were descended from more than two races were excluded. This made his selection unrepresentative, since, to give just one example, there was just one individual in the group of monoracial ethnic Romanian students placed in the "very poor" income category. Făcăoaru acknowledged as much and tried to resolve the issue by adding two groups, mono- and biracial, together, even though this raised a number of methodological issues that he was unable to solve, as he himself conceded.[56] Indeed, if purity of race was to be praised above anything else—as German racial theorists, for whom this essay was intended, believed—than biracial individuals had to be evaluated separately from monoracial ones.

Subsequently, Făcăoaru calculated the percentage of each particular race in each of those categories. He considered as highly endowed those races that were overrepresented in the categories "rich" and "very rich" and as poorly endowed those who were overrepresented in the categories "poor" and "very poor." In summarizing his findings, he concluded that European races could be divided into three groups: the Atlantid, Nordic, and Dalic races were, he argued, highly endowed; the Orientalid, Mediterranean, Dinaric, and Preasiatic races averagely endowed: and the Esteuropoid, Alpine, and Mongoloid races poorly endowed.[57]

Făcăoaru's career suffered a setback after the January 1941 Legionary rebellion against General Ion Antonescu's government. As an active member of the Iron Guard, he might have been for a time under investigation. However, shortly afterward, probably in 1942, he resumed his anthropological research, this time as an employee of the Romanian civil administration in Transnistria (a region in southwestern Ukraine occupied by Romania) under Governor Gheorghe Alexianu.[58] In Transnistria he conducted anthropological research in ethnic Moldovan villages.[59] In late 1943 he summarized his years-long research in an article in which he developed an original methodology for calculating "bioracial value" and applied it to the main European nations and ethnic Romanians from various parts of Romania.

His starting point was an assumption that a race's biological value remained the same irrespective of which European people its specimen belonged to. In other words, he believed "that the place of every race in the racial hierarchy of each particular *neam* remained the same."[60] This is why he insisted that indices of the biological value of various European races that he calculated on Romanian material were valid for a specimen belonging to any European nation (which he regarded as ethnic nations, or *neamuri*). He proceeded to calculate the "bioracial values" of the major European nations. Furthermore, he incorporated into his analysis blood-composition rankings of various European peoples (more "European" versus more "Asiatic") as well as rankings of their intellectual capacities using IQ tests conducted on American troops during the First World War.[61] Having applied this methodology to his own findings as well as data collected by other Romanian anthropologists (though not Lebzelter and his Romanian followers, such as Chelcea), he constructed the following hierarchy of European and North American nations: (1) English, (2) Schwabians [*sic*], (3) Swedes, (4) Dutch, (5) North Americans, (6) Danes, (7) Germans, (8) Belgians, (9) Serbs, (10) Norwegians, (11) Scots, (12) Yugoslavs, (13) French, (14) Romanians, (15) Bulgarians, (16) Polish Jews, (17) Austrians, (18) Greeks, (19) Czechs, (20) Italians, (21) Turks, (22) Poles, (23) Russians, (24) Hungarians, and (25) Gypsies.[62]

It is notable that Russians (with whom Romania was at war) and Hungarians (with whom tensions regarding the Transylvanian issue were at fever pitch) were at the bottom of the list. The same applied, not surprisingly, to the Gypsies: Făcăoaru harbored an implacable hostility toward them, while a part of Romania's Gypsy population had just been subjected to deportation by Antonescu's regime.[63] In this sense, Făcăoaru's ranking neatly fit into the pattern of Romanian prejudices and political expedience. In other respects, however, his hierarchy was nothing short of scandalous. To begin with, while Romanians were not at the bottom of the list—unlike in the writings of those Western anthropologists whom Făcăoaru had vehemently attacked earlier in his career—they still occupied its lower part, just below the only slightly better endowed French. Worse still, Germans, officially allied with the Romanians in a war against the English and Americans, ranked substantially lower than both of the latter. No less provocative from the political point of view should have been the fact that Italians, who only a few months earlier had been considered the second most important member of the Axis bloc, were ranked twentieth. *Horribile dictu*, "Polish Jews" were ranked higher, though just a notch, than Austrians!

The second part of Făcăoaru's article was devoted to the ranking of the ethnic Romanian population. He divided ethnic Romanians according to province: western (Bukovina, Banat, Transylvania, and Crişana-Maramureş), eastern (Moldova, Bessarabia, and Transnistria); and southern (Oltenia, Muntenia, and Dobrudja). The western provinces exhibited the highest share of overendowed races, followed by the eastern and southern. The same was true of the blood indices of their populations: more European in western provinces, with the greater share of "Asiatic" blood in eastern and still more in southern provinces. Since Făcăoaru had no comparative data available as to the economic and social standing or mental capacities of the ethnic Romanian populations of these provinces, he resorted to impressionistic statements. His trained eye, he assured his readers, was sharp enough to grasp the reality even in the absence of reliable statistical data: "In comparison with the peasant of other Romanian provinces, the peasant from Transylvania is more equitable, more disciplined, more well-to-do, more level-headed, more serious, more orderly, has a more adequate judgment of people and things, is more sound in his work, is more humane and less conceited." It was on the basis of such observations that Făcăoaru claimed that "the social structure of a village is a reflection of the physical [read: biological] structure of its residents."[64]

Although Făcăoaru was upbeat about the heuristic and pragmatic value of his theory, his mentor Moldovan was not. Expressing his concern about the potentially offensive and divisive nature of Făcăoaru's disquisitions, he inserted an editorial note in the same issue of the journal, making it clear that this article was published "with the aim of acquainting the readers with various points of view, even though the editorial board does not agree with [its author's] conclusions as to the ranking of ethnies."[65]

Ethnic Nationalism Biologized, or *Neam* as "Race"

As a matter of fact, Moldovan had been in doubt for quite some time about the political expedience of racial research in Romania. In 1934, in an address to the general assembly of the cultural association of Romanians from Transylvania, Astra, he stated that "the racist doctrine presents one's own race as the most distinguished in the world, in this case the blond, tall, dolichocephalic Nordic race, which allegedly embodies, in the most ideal proportions, the most select qualities of human nature . . . [thus] distancing itself disparagingly from other races and inferior peoples."[66] However, by the late 1930s and early 1940s Moldovan's fascination with the Nazi regime, and specifically its ruthless application of eugenicist doctrine, grew.[67] It was during this period that, together with some other members of his school of eugenics, he embarked on the project of redefining the traditional Romanian notion of *neam* in explicitly biological terms. The aim was to endow it with the qualities of solidity and immutability characteristic of the notion of "race," at the same time normalizing the primeval synthesis of various races and *ethnies* that, according to the Romanian national mythology, led to their ethnic nation's *ethnogenesis*.

For two years, from November 1936 to September 1938, Moldovan's journal *Buletin eugenic şi biopolitic* ran a series of articles by one of his disciples, the Iron Guard member Ovidiu Comşia, which were clearly intended to lay out the new doctrine of the school. Comşia's ponderous prose and convoluted logic can be summarized as follows. He started with an attack on what he called the "racist dogma," that is, the concept of the superiority of the Nordic race. He emphasized that "no race, no ethnic collectivity seems to be disposed to accept the superiority of another race or collectivity." It was ethnicity, or *neam*, not race, which was the ultimate reality, even though this concept was more complex and difficult to formulate. In an attempt to define the biological substance of *neam*, Comşia praised the benefits of ethnic interbreeding that had occurred a long time back and that had been followed by a protracted period of biological "stabilization": this was the case with Romanians, who supposedly had survived a thousand years in isolation in the Carpathian Mountains after the withdrawal of the Roman legions. Comşia even defined *neam* as "a stabilized mix."[68] Such stabilization occurred in an organic unity with a particular space. Comşia argued that the interbreeding that had taken place at a later stage (in this context he called it "hybridization") was dangerous because it involved elements from outside the ethnicity's "biological space" and led to "social disequilibrium." He further claimed that the decline of ancient civilizations was a result of "hybridization" and that modern progress was increasing the danger of the "biological decline" of a *neam*. Such a negative development could be prevented by concerted action by the government, which should be charged with halting excessive interbreeding.[69]

Moldovan followed the same line of thought in a series of essays that he penned between 1938 and 1943. While in 1926 he had emphatically denied that Romanians were "direct descendants of a single ancestral family," by 1942 he insisted that "today there is not a single family that did not, somewhere in the past, share common ancestors with every other Romanian family." On this occasion he summarized his thinking on the problem of race and ethnicity as follows: "When others talk of race, we speak of *neam* or, more precisely, of ethnobiological substance [*fond*]. *Because all manifestations of life, including spiritual ones, are biological* [emphasis in the original]."[70]

Petru Râmneanţu expressed the same idea in a 1943 article appropriately titled "Înrudire de sânge," (Blood relationship): "Our *neam*, perhaps more than anyone else, is a community of blood, which progressively becomes more and more homogeneous."[71] In a public lecture, *Sânge şi glie* (Blood and soil), that he delivered in November 1943 at the School of Medicine at the University of Cluj, he emphasized, following Comşia, the significance of environmental factors for the formation of a particular *neam*.[72] Thus, no matter how much intermixing happened in the past, the *neam* in the present had to guard its purity.

Comşia's conceptualization of the notion of *neam* represented the position of the majority of the Moldovan School of eugenics. As Marius Turda has convincingly argued, the school offered "a biological model of identity," which was, it might be added, in complete accord with traditional national mythology.[73] Without bothering too much about the hierarchy of races and ethnic nations, Moldovan's associates indulged, in the *Buletin eugenic şi biopolitic* and various pamphlets in the late 1930s and early 1940s, in nationalistic fantasies of national "purification," ethnic segregation, and the violent removal of minorities from Romania's territory—in the first place those whom they branded as "biologically inferior," such as Gypsies, Turks, Tatars, and Hungarians.[74]

Such a conceptualization was shared in Romanian nationalistic circles outside of the narrow confines of the eugenicist community. A curious attempt to reconcile the celebration of the biological purity of *neam* with Christian Orthodox religion was undertaken by Liviu Stan, at the time professor of theology at the University of Sibiu.[75] Stan, like some other Christian Orthodox clerics, was ideologically close to the Iron Guard. In early 1941 in the Legionary newspaper *Cuvântul* he published an article calling, in the most violent language, for the segregation, "total isolation," and eventual expulsion of Gypsies, who supposedly posed the gravest danger to the purity of Romanian blood, race, and spirit.[76] In 1942 he published a brochure in which he explored the relationship between racism and Christianity.[77]

Stan believed that racism and, in particular, the exultation of one's own race, or "myth of one's own blood," could play a positive spiritual role since it could lead to "a new ascension from the low depth in which a substantial part of humanity lingers," that is, from atheistic materialism and hedonism, which he identified with capitalism and modernity. At the same time, Stan warned that racism could never replace Christianity and that its role should not be exaggerated. He inveighed heavily against the German theorists of racism such as Chamberlain and Alfred Rosenberg, who insisted that "true" religion was a reflection of the "racial soul"; that Christianity, as a Semitic religion, corrupted the soul of the Nordic race, and especially that of Germans; and that it should be replaced by a resuscitated Aryan religion. All of these opinions were deeply flawed, Stan asserted, since they neglected the fundamental truth that religion was a transcendental reality revealed by God to all humanity, not to a particular race. Race could exercise some influence over religious practices, but only outwardly. It could not affect its spiritual core, which had nothing to do with the soul of individual races. The bulk of Stan's book was devoted to a resolute denunciation of the anti-Christian teachings of German racists. In spite of this, the book ended with an apology for traditional Romanian ethnic nationalism and xenophobia, which Stan called "Romanian racism." Supposedly, Romanian racism was legitimate while the German variant was not, since the former did not intend to demote (Orthodox) Christianity from its proper place as the highest form of spirituality. Notably, while Stan was close to Comşia, Moldovan, and Râmneanţu in his emphasis on environmental factors as decisively contributing to the formation of a race, he did not explicitly refer to their publications.[78]

Mystical Racism

Some Romanian racists tried to conceptualize race in mystical terms. In so doing they followed the example of Chamberlain and Rosenberg and the Italian racist theorist Julius Evola.[79] In 1936 the Romanian poet and Iron Guard sympathizer Ion Foti published a book appropriately titled *Concepția eroică a rasei* (The heroic concept of race). In the first part of the book he outlined his vision of Romanian national identity in racial terms. According to Foti, the Romanian race had played the dominant role in the history of Eastern Europe as a bulwark of civilization and Aryanism against all kinds of barbarians—Turks, Hungarians, Russians, and Jews. The rest of the book was effectively both a summary and an apology for Rosenberg's *Myth of the Twentieth Century*. In just one respect did Foti express disagreement with his idol, namely his animus toward Christianity. Unlike Rosenberg, Foti was a devout Christian and believed that Orthodox Christianity was perfectly compatible with the Romanian racial soul, even if, in the case of some other Orthodox peoples such as Russians, Orthodoxy led to the development of unhealthy passivity. The lingering danger of a similar spiritual corruption among Romanians could be countered by the development of race consciousness.[80]

Foti's book was badly written and overly dependent on Rosenberg's treatise. Alexandru Randa provided a somewhat more astute narrative on Romanian racial history. A son of a colonel in the Austro-Hungarian General Staff and a Romanian aristocrat of Greek origin, he was raised at the family estate in Bukovina.[81] A doctor of law and philosophy at the University of Cernăuți, this right-wing publicist and Legionary sympathizer published in 1939 a veritable panegyric of Italian Fascism and German Nazism called *Europa eroică* (The heroic Europe).[82] Randa's racial theory was further developed in a 1941 pamphlet, *Rasism românesc* (Romanian racism).[83] In the aftermath of the January 1941 Legionary rebellion, however, Randa was effectively removed from the Romanian intellectual scene. Along with other important Legionaries who had been prosecuted by the Antonescu government, he was smuggled out of the country by the Germans. Together with other Legionaries he spent the rest of the war in special quarters in the Schutzstaffel (SS) concentration camps of Buchenwald and Sachsenhausen-Oranienburg, where they were kept as an alternative government in waiting in case Romania changed sides in the war. After the ousting of the Antonescu government in an August 1944 coup they were released, and they created a pro-Nazi puppet Romanian government in exile.[84]

Rasism românesc opened with a statement disclosing its intended political usage and affiliation to Rosenberg's discursive strategy: "The creation of the Romanian racial consciousness is a primordial condition for the affirmation of Romanianism [Românism] in the international arena. This Romanian racism will be based on the Aryan myth."[85] Randa built his doctrine on the nationalistic fantasies of the Romanian archeologists Vasile Pârvan and Ioan Andrieşescu and the historian Nicolae Densuşianu, who extended Romanian history far beyond the Roman conquest and the Dacian and Getic periods into the Thracian and Pelasgian epochs (the second millennium BC), about which practically no reliable information is available.[86] Randa went a step further by postulating that the Romanian territory served, from as early as 2000 BC, as an "avant-garde, as a gate of exit, and as a point of departure of Aryanism into Asia," that is, as a "cradle of the world." It was from the Carpathian Mountains that "the mass of Italics" invaded the Apennine Peninsula; it was also from here that the Aryans reached Palestine, where they founded Jerusalem. The "bearers of civilization" of ancient China and Japan supposedly originated from the "Transylvanian stronghold of Aryanism," too. Thus, he continued, "Today's Romanianism constitutes a rear guard of the great creative Aryan mass that descended four thousand years ago into Greece, in Persia, and in India."[87] Needless to say, Randa had no evidence whatsoever to substantiate his claims, save for the cautious suppositions of various historians as to the geography of migrations of ancient peoples. Among such suppositions he selected only those that suited his agenda and read them as established facts.

The rest of the book reinterpreted the past two thousand years of Romanian and Balkan history from the racial standpoint. Following Chamberlain's view of the Roman Empire as a "chaos of peoples," Randa asserted that the conquest of Dacia by the Roman emperor Trajan in the early second century AD led to the settlement of the new province by heterogeneous elements from all over the empire, thus weakening its "racial substance." Nonetheless, the race miraculously survived, so that in the late imperial period "Germans and Dacians . . . were allies against Mediterranean internationalism." During the period of barbaric migrations, "Romanian people"—so called in spite of the fact that the name *Romanian* had been invented much later—stood "in a place of honor in the great European front against the Orient."[88]

This assumption allowed Randa to discern in the actions of Roman emperors and generals who fought against "Asiatic" invaders in the east the spirit of racial Dacians (or Gets, Thracians, or Illyrians, seen as racially identical), even if they simultaneously fought against "Aryan" Germans in the west. The same was true for Byzantine emperors and generals as well as their Bulgarian and Serbian foes: they were racial Dacians. While all north Balkan peoples had identical "racial substance," Slavization ultimately destroyed it so that "with the decay of the Byzantine Empire, the name of Romans transferred to the elite nation called Romanian, which was the only one that managed to destroy invading tribes [i.e., Slavs, Tatars, Turks, and Hungarians] and to preserve the spiritual treasures of antiquity." At the same time, paradoxically, the Ottoman Empire's elites were populated by elements of Thracian, Illyrian, or Dacian stock.[89] It was those same racial elites who established their rule over the Romanian principalities of Wallachia and Moldova.

According to Randa, in the medieval period Romanians were under the undue influence of an alien Slavic culture, which found its expression in the use of the Old Slavonic language in the church and state administration, as well as in the Cyrillic alphabet. In the late eighteenth and nineteenth centuries this influence was replaced by that of "Latinist" intellectuals who overemphasized Roman roots and imported too many French cultural forms. It was only in the twentieth century that Romanians came to fully appreciate their Thracian-Dacian racial essence and thus to rediscover their racial soul, he concluded.[90]

Purely intuitive and mystical, Randa's conceptualization lies beyond the ambit of scholarly discourse, as do the discourses of his idols Chamberlain, Rosenberg, and Evola. Nevertheless, it may be useful to note that Randa not only based his inferences on far-fetched interpretations of historical sources (almost all of them secondary) but also remained oblivious to the views of the majority of physical anthropologists, who considered pre-Roman residents of Romania to be of non-Nordic origin. At the same time, however, Randa's fantasies were in sync with Făcăoaru's readings of anthropological data. It is conceivable, therefore, that had the war ended differently their discourses might have provided the Romanian regime with a somewhat more coherent racial ideology.

The Romanian wartime government remained largely disinterested in racial theories. Ion Antonescu was a self-confessed integral nationalist who saw himself as a defender of the Romanian *neam*. He believed that, although this *neam* was great and noble, it lingered in poverty and backwardness due to the incessant malignant activity of ethnic foreigners (*străini*) who had penetrated the ethnic Romanian nation. His antisemitism and Roma-phobia were especially virulent. However, such sentiments were commonplace among the Romanian radical right, and Antonescu's views did not need to be informed by formal or "scientific" racial doctrines. When he used the term *race* (*rasă*, or *rassă*), he meant it as a synonym of *neam*. Although his distant relative and second in command between 1941 and 1944, Mihai Antonescu, did belatedly develop an interest in German racial thinking and seemed to have been awed by it, this interest was transitory.[91] All in all, Antonescu's administration did very little to promote racial research and conceptualization.

Ion Antonescu's main adviser on issues pertaining to "population policy," Sabin Manuilă, was a militant nationalist who passionately hated the Roma. Between 1940 and 1942 his nationalism hardened further and he, too, adopted some elements of biological racial discourse. In an article he wrote on 6 August 1940, he suggested that the promotion of a scientifically determined "population policy" had to become the main aim of the state, "as suggested many years back by the father of our racial science Mr. Professor Iuliu Moldovan."[92] Ironically, as already discussed, Moldovan consistently opposed the importation of racial doctrine as potentially detrimental to Romanian national self-esteem. His strategy, like that of his associates, was to imbue the traditional notion of *neam* with biological meaning, thus making it a "quasi-race." While lacking even the semblance of intellectual consistency and being deeply parochial, this strategy had the advantage of fitting better into the structure of Romanian nationalistic prejudices and perceived national interests. "Biologically" hardened ethnic nationalism thus remained the dominant discourse in Nazi-allied Romania. Racial theories lingered in its long shadow.

Notes

1. Hannah Arendt, *Eichmann in Jerusalem: A Report on the Banality of Evil* (New York: Viking Press, 1963), 172.
2. On Romanian popular antisemitism, see Andrei Oişteanu, *Inventing the Jew: Antisemitic Stereotypes in Romanian and other Central East-European Cultures* (Lincoln: University of Nebraska Press, 2009). On intellectual antisemitism, see

Leon Volovici, *Nationalist Ideology and Antisemitism: The Case of Romanian Intellectuals in the 1930s* (Oxford: Pergamon Press, 1991); Radu Ioanid, *The Sword of the Archangel: Fascist Ideology in Romania* (Boulder CO: East European Monographs, 1990); International Commission on the Holocaust in Romania, *Final Report* (Iaşi, Romania: Polirom, 2005), 21–31; Marius Turda, "Fantasies of Degeneration: Some Remarks on Racial Anti-Semitism in Interwar Romania," *Studia hebraica* 3 (2003): 336–48.

3. Cf. Ioanid, *Sword of the Archangel*, 116–31; and Maria Bucur, *Eugenics and Modernization in Interwar Romania* (Pittsburgh: University of Pittsburgh Press, 2002), 36, 38–39, 99, 108–9, 212, 215.
4. See Katherine Verdery, *National Ideology under Socialism: Identity and Cultural Politics in Ceauşescu's Romania* (Berkeley: University of California Press, 1995).
5. I borrow the term "hole in the middle" from Terry Martin, *The Affirmative Action Empire: Nations and Nationalism in the Soviet Union, 1923–1939* (Ithaca NY: Cornell University Press, 2001), 179.
6. Alexandru C. Cuza, *Despre poporaţie: Statistica, teoria, politica ei; Studiu economic politic* (Bucharest: Imprimeriile "Independenţa," 1929), 524–25. Here I translate the Romanian words *neam* and *popor*, which Cuza used interchangeably, as "ethnic nation/group." The first meaning of the term *neam* is "kin." Somewhat similar to the German *das Volk*, it strongly implies the vision of an ethnic nation as a large extended family. On Romanian synonyms for "ethnic nation," see Katherine Verdery, *What Was Socialism, and What Comes Next?* (Princeton NJ: Princeton University Press, 1996), 248.
7. Cuza, *Despre poporaţie*, 597.
8. Alexandru C. Cuza, introduction to *Studii economic-politice (1890–1930)* (Bucharest: Editura Casei Şcoalelor, 1930), lxxiv–lxxvii.
9. See the list of Cuza's publications in Paul Shapiro, "Faith, Murder, Resurrection: The Iron Guard and the Romanian Orthodox Church," in *Antisemitism, Christian Ambivalence, and the Holocaust*, ed. Kevin E. Spicer (Bloomington: Indiana University Press, 2007), 139.
10. As quoted in Guido Landra, *La problema della razza in Romania* (Bucharest: Instituto italo-romeno di studi demografice e razziali, 1942), 175–76. Landra failed to indicate which of Cuza's texts he was quoting.
11. Alexandru C. Cuza, *Naţionalitatea în arta: Expunere a doctrinei naţionaliste; Principii, fapte, doctrine; Ediţia a doua, cu anexe* (Bucharest: "Minerva," Institut de arte grafice şi editura, 1915), 114.
12. For more on Chamberlain's influence on Cuza and other Romanian nationalists and racists, see Marius Turda, "Conservative Palingenesis and Cultural Modern-

ism in Early Twentieth-Century Romania," *Totalitarian Movements and Political Religions* 9, no. 4 (2008): 437–53.

13. Cuza, *Naţionalitatea în arta*, 125.
14. For an informative discussion of the interrelationship and differences among racism, ethnocentrism, and antisemitism, see John P. Jackson Jr. and Nadine M. Weidman, *Race, Racism, and Science: Social Impact and Interaction* (New Brunswick NJ: Rutgers University Press, 2004), xiv–xv.
15. On Codreanu and his party, see, e.g., Eugen Weber, "Romania," in *The European Right: A Historical Profile*, ed. Hans Rogger and Eugen Weber (London: Weidenfeld & Nicholson, 1965), 501–74; Nicholas M. Nagy-Talavera, *The Green Shirts and the Others: A History of Fascism in Hungary and Rumania* (Stanford CA: Hoover Institution Press and Stanford University, 1970), 246–344; Ioanid, *Sword of the Archangel*; Francisco Veiga, *La mística del ultranacionalismo: Historia de la Guardia de Hierro, Rumania, 1919–1941* (Bellaterra, Spain: Publicaciones de la Universidad Autónoma de Barcelona, 1989); Armin Heinen, *Die Legion "Erzengel Michael" in Rumänien: Soziale Bewegung und politische Organisation; Ein Beitrag zum Problem des internationalen Faschismus* (Munich: Oldenbourg, 1986); Constantin Iordache, *Charisma, Politics and Violence: The Legion of the "Archangel Michael" in Inter-war Romania* (Trondheim, Norway: Trondheim Studies on East European Cultures and Societies, 2004); George L. Mosse, *Towards the Final Solution: A History of European Racism* (New York: Howard Fertig, 1978), 198.
16. The history of Romanian anthropology has been extensively studied within a wider European and Western context by Marius Turda. See esp. "The Nation as Object: Race, Blood, and Biopolitics in Interwar Romania," *Slavic Review* 66, no. 3 (2007): 413–42; *Eugenism şi antropologie rasială în România* (Bucharest: Cuvântul, 2008), esp. 91–144; "Rasse, Eugenik und Nationalismus in Rumänien während der 1940er Jahre," in *Holocaust an der Peripherie: Judenpolitik und Judenmord in Rumänien und Transnistrien 1940–1944*, ed. Wolfgang Benz and Brigitte Mihok (Berlin: Metropol, 2009), 161–71; "Controlling the National Body: Ideas of Racial Purification in Romania, 1918–1944," in *Health, Hygiene and Eugenics in Southeastern Europe to 1945*, ed. Christian Promitzer et al. (Budapest: CEU Press, 2011), 325–50.

    In the late nineteenth–early twentieth century Pittard conducted several anthropological expeditions in the Balkans, including Romania. See Eugène Pittard, *Les peuples des Balkans: Recherchers antropologiques dans la péninsule des Balkans spécialement dans la Dobroudja avec 149 figures, graphiques et cartes dont 91 illustrations d'après des photographies prises par l'auteur* (Genéve at Lyon: Georg & Co, 1920), 61.
17. For an exposition and critique of the "ethnogenesis" of Romanian historiogra-

phy, see Lucian Boia, *History and Myth in the Romanian Conscience* (Budapest: Central European University, 2001), 83–111.

18. Pittard, *Les peuples des Balkans*, 95–96, 619. Pittard wondered whether one could surmise that the bearers of the so-called Kurgan culture of the southern Russian steppe who were dolichocephalic could be the ones who brought dolichocephaly to the Balkans, while "primitive Romanians" were brachycephalic.
19. Eugène Pittard, *Race and History: An Ethnological Introduction to History* (1926; repr., New York: Kegan Paul, 2003), 48.
20. Pittard, *Race and History*, 49–139, 155–73.
21. Eugène Pittard, *Peoples des Balkans: Esquisses antropologiques avec 4 cartes et quelques figures* (Paris: Attinger frères, Editeurs Neuchatel, 1918), 58; Pittard, *Race and History*, 283.
22. Pittard, *Race and History*, 283.
23. Olga Necrasov, *Recherches anthropologiques dans le nord-est de la Roumanie* (Iaşi, Romania: "Presa bună," 1940), 8.
24. Verdery, *National Ideology under Socialism*, 27–71.
25. Irina Livezeanu, *Cultural Politics in Greater Romania: Regionalism, Nation Building and Ethnic Struggle, 1918–1930* (Ithaca NY: Cornell University Press, 1995).
26. Marius Turda, "Craniology and Racial Identity in Interwar Transylvania," *Anuarul Institutului de Istorie "G. Bariţ" din Cluj-Napoca* 14 (2006): 124–26.
27. Victor Papilian, "Nouvelles recherches anthropologiques sur la tête des roumains de Transylvanie," *Revue anthropologiques* 33, no. 9–10 (1923): 335–41.
28. Jenö Davida, "Beiträge zur Kraionogie der Magyaren und der siebenbürgeschen Wallachen," *Anatomischer Anzeiger* 66, no. 1–3 (1928): 30–42.
29. Landra, *La problema della razza*, 30. In 1941–42 Landra was head of the Italo-Romanian Institute for Demographic and Racial Studies in Bucharest, a shadowy organization created for the purpose of providing him with employment after his debacle as the head of the Italian Racial Office, from which he was fired in February 1939, only six months after it began its operations. See Aaron Gillette, *Racial Theories in Fascist Italy* (London: Routledge, 2002), 85, 130, 178.
30. Pierre Râmneanţu, "Origine ethnique des Séklers de Transylvanie," *Revue de Transylvanie* 2, no. 1 (1935): 45–59. See also Petru Râmneanţu, *Cercetări asupra originei etnice a populaţiei din Sud-Estul Transilvaniei pe baza compoziiţiei serologice a sângelui* (Cluj, Romania: Institutul de arte grafice "Ardealul," 1935). On Râmneanţu and his views, see Bucur, *Eugenics and Modernization*, esp. 36–37; Turda, "Nation as Object."
31. For more information on Manuilă and his views, see Vladimir Solonari, *Purifying the Nation: Population Exchange and Ethnic Cleansing in Nazi-Allied Romania*

(Washington DC: Woodrow Wilson Center Press; Baltimore: Johns Hopkins University Press, 2010), 75–80, 88–94.

32. On the methodology of serological research, see Turda, "From Craniology to Serology: Racial Anthropology in Inter-war Hungary and Romania," *Journal of the History of Behavioral Sciences* 43, no. 4 (Fall 2007): 369–70.
33. Sabin Manuilă and G[heorghe] Popoviciu, "Recherches sur les races roumaine et hongroise en Roumanie par l'isohémagglutination," *Comptes rendus des séances de la Société de biologie et des ses filiales* 90, no. 7 (1924): 542–43; Manuilă, "Recherches séro-anthologiques sur les races en Roumanie par la method de l'isohémagglutination," *Comptes rendus des séances de la Société de biologie et des ses filiales* 90, no. 14 (1924): 1071–73.
34. Manuilă, "Recherches séro-anthologiques sur les races," 1072.
35. Necrasov, *Recherches anthropologiques*, 10.
36. Râmneanțu, "Origine ethnique des Séklers de Transylvanie," 57–59.
37. Ion Chelcea, "Tipuri de cranii românești din Ardeal: Cercetare antropologică," *Academia Română: Memoriile secțiunii științifice* ser. 3, vol. 10 (1934/35): 21. Chelcea referred to Egon Freiherr von Eickstedt's *Rassenkunde und Rassengeschichte der Menschen* and Bruno Kurt Schultz's *Erbkunde, Rassenkunde, Rassenpflege*.
38. Chelcea, "Tipuri de cranii românești din Ardeal," 1.
39. Chelcea, "Tipuri de cranii românești din Ardeal," 20–22.
40. Viktor Lebzelter, "La repartition des types raciaux romano-méditerranéens en Roumanie," *L'anthropologie* 45, no. 1–2 (1935): 65–69, quotation at 69.
41. Boia, *History and Myth*, 95–100.
42. Revealingly, Pittard was among the first to suggest that an emphasis on the pre-Roman racial roots of ethnic Romanians might better serve Romanians' own national interests than the then more widely known Latinist discourse. See Pittard, *Les peuples des Balkans*, 48.
43. See C.[?] Ureche et al., L'indice céphalique chez les roumains et chez les autres nationalités de Transylvanie," *L'anthropologie* 45, no. 3–4 (1934): 334–36; Fr. I. Rainer, *Enquêtes anthropologiques dans trois villages roumains des Carpathes* (Bucharest: Monitorul Oficial și Imprimeriile Statului, 1937); Necrasov, *Recherches anthropologiques*, esp. 156–67.
44. Jackson and Weidman, *Race, Racism, and Science*, 133.
45. Gheorghe Banu, *L'hygiene de la race: Étude de la biologie héreditaire et de la normalization de la race* (Bucharest: Imprimeria națională, 1939), 9–16, 201–2. See also Bucur, *Eugenics and Modernization*, 42–43.
46. Eugen Fischer, *Spezielle Anthropologie: Rassenlehre* (Leipzig: B. G. Teubner, 1923).
47. Michael Wedekind, "Wissenschaftsmilieus und Ethnopolitik im Rumänien der 1930/40-er Jahre," in *Herausforderung Bevölkerung: Zu Entwicklungen des mod-*

ernen Denkes über die Bevölkerung vor, im und nach dem "Dritten Reich," ed. Josef Ehmer et al. (Wiesbaden: VS Verlag für Sozialwissenschaft, 2007), 243–44. On Lenz, see Jonathan Marks, *Human Biodiversity: Genes, Race, and History* (New Brunswick NJ: Transaction, 2009), 88.

48. Wedekind, "Wissenschaftsmilieus und Ethnopolitik," 245.
49. On Gusti and its monographic campaigns, see Solonari, *Purifying the Nation,* 80–88.
50. Turda, "Nation as Object," 439.
51. Wedekind, "Wissenschaftsmilieus und Ethnopolitik," 245.
52. Iordache Făcăoaru, "Socialantropologia ca ştiinţă pragramatistă," in *Antropologia în stat ca ştiinţa şi obiect de învăţământ* (Cluj, n.d.), 41–54. Among the Western anthropologists whose negative opinion of Romanians he referred to were Madison Grant (*The Passing of the Great Race,* 1925), Augustus H. Keane (*Man Past and Present,* 1920), and William Ripley (*The Races of Europe,* 1900).
53. In addition, in 1937 he published an article contending that, from the racial point of view, Széklers were closer to the Romanians than they were to Hungarians. See Iordache Făcăoaru, "Copoziţia rasialăla la Români, Secui şi Unguri," *Buletin eugenic şi biopolitic* 8, no. 4–5 (1937): 124–42. This was exactly the point Râmneanţu was making on the basis of serological research.
54. Iordache Făcăoaru, "Despre structura rasială a populaţiei rurale," *Revista de igienă social* 10, no. 1 (1940): 86, 88, 95.
55. Iordache Făcăoaru, "Beitrag zum Studium der wirtschaftlicher und sozial Bewährung der Rassen," *Zeitschrift für Rassenkunde* 9, no. 1 (1939): 1, 26–39.
56. Făcăoaru, "Beitrag zum Studium," 31–32.
57. Făcăoaru, "Beitrag zum Studium," 34.
58. Făcăoaru carried out most of his research between 1933 and 1940. See his article "Valoarea biorasială a naţiunilor ş provinciilor româneşti," *Buletin eugenic şi biopolitic* 14, no. 9–10 (1943): 278–79. In Transnistria, apparently, he was a protégé of Traian Herseni, a notable Romanian sociologist of the Gusti school, an erstwhile member of the Iron Guard, a right-wing publicist, and director of culture in the Transnistria governorship. See Herseni's deposition in his postwar investigation file, U.S. Holocaust Memorial Museum, Washington DC, RG-25.004M, reels 98 and 118.
59. See the only extant part of his 1943 research paper in U.S. Holocaust Memorial Museum, Washington DC, RG 31-004M, reel 2.
60. Făcăoaru, "Valoarea biorasială," 279.
61. On the uses of "A" and "B" indices and the meaning attributed to them as well as the history of serology in Romania, see Turda, "From Craniology to Serology," 369–73. On the IQ testing of American troops during the First World War and

their racist reading by U.S. eugenicists, see Jackson and Weidman, *Race, Racism and Science*, 115–17.

62. Jackson and Weidman, *Race, Racism and Science*, 280, 287.
63. On Romanian-Hungarian tensions over Transylvania during the Second World War, see Holly Case, *Between States: The Transylvanian Question and the European Idea during World War II* (Stanford CA: Stanford University Press, 2009). On Făcăoaru's hostility toward Gypsies, see Solonari, *Purifying the Nation*, 70–72, 269–90.
64. Solonari, *Purifying the Nation*, 304–5.
65. "Nota redacţiei," *Buletin eugenic şi biopolitic* 14, no. 9–10 (1943): 352.
66. Quoted in Pamfil Matei, *"Asociaţiunea transilvaneană pentru literatura română şi cultura poporului român" (ASTRA) şi rolul ei în cultura naţională (1861–1950)* (Cluj-Napoca, Romania: Editura Dacia, 1986), 104.
67. Iuliu Moldovan, *Introducere în etnobiologie şi biopolitică* (Sibiu, Romania: Editura subsecţiei eugenice şi biopolitice a Astrei, 1944), 169–70.
68. Ovidiu Comşia, "Spaţiul biologic (partea I)," *Buletin eugenic şi biopolitic* 7, no. 11–12 (1936): 322–23, 328.
69. Comşia, "Neamul regenerat," *Buletin eugenic şi biopolitic* 8, no. 10–12 (1937): 304–16.
70. Moldovan, *Introducere*, 14, 21.
71. Petru Râmneanţu, "Înrudire de sânge," *Buletin eugenic şi biopolitic* 14, no. 7–8 (1943): 236.
72. Petru Râmneanţu, *Sânge şi glie: Extras din "Buletin eugenic şi biopolitic," Vol. XIV, Nr. 11–12, 1943* (Sibiu: "Cartea Românească din Cluj," 1943).
73. Turda, "Nation as Object," 435.
74. Solonari, *Purifying the Nation*, 69–73.
75. See the entry on Liviu Stan in Mircea Pacurariu, ed., *Dictionarul teologilor români* (Bucharest: Editura Enciclopedica, 2002).
76. Liviu Stan, "Rasism faţa de ţigani," *Cuvântul*, 18 January 1941.
77. Liviu Stan, *Rasâ şi religiune* (Sibiu, 1942).
78. Stan, *Rasâ şi religiune*, 7, 18–19, 35, 45, 124–29, 141–45.
79. On Chamberlain's racism, see George L. Mosse, "Introduction to the 1968 edition," in *The Foundations of the Nineteenth Century*, by Houston Stewart Chamberlain (New York: Howard Fertig, 1968), 1:v–xx. On Rosenberg, see James B. Whiskers, "Introduction to the English Edition," in *The Myth of the Twentieth Century: An Evaluation of the Spiritual-Intellectual Confrontations of Our Age*, by Alfred Rosenberg (Torrance CA: Noontide Press, 1982), xxix–xivii.
80. Ion Foti, *Concepţia eroică a rasei* (Bucharest, 1936), 23–27, 44–47.
81. On Randa, see Anna-Dorothee von den Brincken, "Nekrolog: Alexander von Randa," *Historische Zeitschrift* 222, no. 2 (April 1976): 516–17.

82. Alexandru Randa, *Europa eroică* (Bucharest: Institutul de Arte Grafice "Universul," 1939).
83. Alexandru Randa, *Rasism românesc* (Bucharest: "Bucovina" I.E. Torouţu, 1941). Besides Chamberlain and Rosenberg, Randa was inspired by Julius Evola's racial fantasies. On Evola, see Gillette, *Racial Theories in Fascist Italy*, 154–75.
84. Brincken, "Nekrolog," 516; Heinen, *Die Legion "Erzengel Michael,"* 459–61.
85. Randa, *Rasism românesc*, 2.
86. For an informed critique of this ultranationalistic discourse, see Boia, *History and Myth*, 96–100.
87. Randa, *Rasism românesc*, 3–6.
88. Randa, *Rasism românesc*, 8.
89. Randa, *Rasism românesc*, 29–62.
90. Randa, *Rasism românesc*, 6.
91. Solonari, *Purifying the Nation*, 136–41, 148, 268, 271–72, 377–78, 405.
92. See the Sabin Manuilă Collection in the National Archives of Romania, Bucharest, XII/195/1940.

11

Building Hitler's "New Europe"

Ethnography and Racial Research in Nazi-Occupied Estonia

ANTON WEISS-WENDT

Racial discourse was commonplace in wartime Europe. What makes Estonia stand apart from the rest of the Nazi-occupied countries of East Central Europe is that many Estonian academics and scientists not only talked racial science but also acted it out, without subscribing to Nazi ideology. In retrospect, Estonians proved simultaneously the object and the subject of Nazi racial grand designs. This chapter argues that the local discourse concerning the biological health of the Estonian nation was far more attuned to the views of German, and later Nazi, racial experts than has previously been assumed. The relatively lax occupation regime introduced by the Nazis in Estonia and the idea of Finno-Ugrian ethnographic order influenced a substantial number of Estonian scientists and scholars to both intellectually and practically contribute to the Nazis' radical reshaping of Europe. By advancing racial research and participating in population transfers, prominent members of the Estonian scientific and academic elite unwittingly contributed to the building of Hitler's "New Europe."

Existential Fear: The Nation in Peril

Population-related issues had preoccupied the Estonian literati since the time Estonians emerged as a nation in the second half of the nineteenth century.

The impending danger of Germanization and/or Russification, coinciding with increased migration within the Russian Empire, raised the question of the sustainability of the Estonian nation. In 1870 one of the leading Estonian intellectuals, Jakob Hurt, formulated what he called a "categorical imperative of Estonian nationalism." According to Hurt, the Estonians, although small in numbers, could attain greatness spiritually and culturally.[1] Estonian national identity rested on three pillars: language, culture, and a myth of common descent. By emphasizing the uniqueness of the Estonian language and its Finno-Ugric base, Estonian intellectuals helped to legitimize the existence of the new nation. A cultural component thus came to play a central role in Estonian nationalist ideology. The idea of the Estonian nation as a linguistic and cultural community became dominant even among those influenced by social Darwinism. One of the founders of the Estonian state, Jaan Tõnisson, for example, interpreted the struggle for existence mainly in cultural terms.[2] The Estonians had neither a high cultural tradition nor the memory of political independence and therefore developed only a weak sense of history. Therefore the latter had to be manufactured by such nineteenth-century ideologists as Carl Robert Jakobson.[3]

Some Estonian intellectuals, however, argued that the sheer size of the population determined the spiritual health of a nation, reflecting widely held popular opinion that the territory of Estonia could sustain a much larger population. In the wake of the program of Estonianization in the late 1930s, leading Estonian politicians had accommodated this *idée fixe* into their programmatic speeches. Thus Jaan Tõnisson believed in the possibility of increasing the size of the Estonian population from 1 million to 4 million. Likewise, President Konstantin Päts in his New Year address to the nation in December 1937 urged the Estonians to make natural growth a priority. The paternalistic, pronatalist regime of Päts projected the long-term effect of a low birthrate on the labor force and the country's capability to defend itself against potential aggressors.[4]

The human losses inflicted by the First World War and subsequent civil war contributed to the growth in popularity of Darwinism and Mendelism in Estonia. Although military victories had boosted collective morale, the war was viewed as harmful: healthy young men had been killed, while diseased or handicapped individuals had returned home and propagated. From the Estonian nationalist perspective, the country's low birthrate, indeed, one of the lowest in Europe, threatened the nation with extinction. The fact that the Russian minority constituted the only population group with a high birthrate in Estonia exacerbated these fears. According to the neurologist Voldemar Üprus, in three hundred years the Estonian people would be vastly outnumbered by the Russians. At the other end of the spectrum, some Estonian politicians had calculated the possible impact of the application of the Nazi concept of *Lebensraum* to their country. Considering the voracious appetite of the leaders of the Third Reich, they prophesied, it would not take long before the Germans set their eyes on the sparingly populated territory of Estonia.[5]

Positive eugenics first made inroads in Estonia in the form of an antialcoholism movement. The emphasis on the protection of mothers and children added credibility to eugenic programs. The establishment of the Estonian Eugenics Society Healthy Breeding program (Eesti Eugeenikaselts Tõutervis) in 1924 institutionalized ideas that had long been part of the mainstream discourse within educated circles. In 1927 and 1935 the society held the Convention of National Upbringing, which discussed such issues as race, fertility, and society, thus introducing a biological perspective on the nationalist debate.[6] Otherwise the boundary between positive and negative eugenics was rather thin. In times of crisis, promotion of better genetic stock quickly degenerated into restrictive measures against deficient hereditary "carriers." According to some estimates, the proportion of inferior racial elements in Estonia ran as high as one-third of the total population.[7] Forced sterilization appeared to be the most efficient way of fighting the hereditary ills of society. In that respect, proponents of eugenic intervention in Estonia tended to follow trends found in the Scandinavian countries, which, in turn, closely observed developments in Germany and the United States.

Most of the twenty-six known sterilization laws in Europe and North America were adopted in the 1930s. The issue of sterilization received most attention in Protestant countries, notably in Scandinavia. Denmark introduced preventive legislation in 1929, followed by Norway in 1934 and Sweden in 1935. The sterilization law that had the most immediate impact on Estonia was that passed by the Finnish parliament in 1935. In addition to historical, cultural, and linguistic ties that bound the two countries together, Estonia and Finland had developed a similar set of basic principles that guided them throughout the interwar period. Adherence to peasant values and defense of Western civilization against the Communist danger were probably the two most important components of both the Finnish and Estonian worldview. As regards racial hygiene, Finland and Estonia differed in one particular respect. In Finland eugenics enjoyed enthusiastic support from the Swedish minority, whereas in Estonia ethnic minorities did not participate in the public debate at any level of discussion. One of the possible explanations is that with the loss of the status of a ruling class, the Swedes found themselves increasingly threatened by the Finnish-speaking majority.[8]

In Estonia the issue of sterilization emerged for the first time at the Estonian Physicians Convention in 1924. However, it was only under the weight of the economic depression in the 1930s, and in particular as a reaction to the severe pressure placed on the country's welfare system, that eugenics laws were formally introduced. Adopted in 1936, the Law on Forced Sterilization and Abortion targeted mainly the mentally ill, whose supposed excessive sexual potency had been seen as dangerous for society at large. Also mentioned in the law were epileptics, incorrigible alcoholics, and sexual murderers; sexual murder was to be combated by means of forced castration. At the drafting stage, Estonian legislators specifically drew, among others, on the Nazi law on sterilization.[9] During the next few years a total of forty-one people (all but four women) were sterilized. The respective numbers for Scandinavian countries were notably higher, ranking from several hundred (in Finland) to several thousand (in Sweden).

Proponents of eugenics were motivated by the desire to push back the boundaries of scientific knowledge rather than by ideological or political concerns. Less well known is the impact eugenics exerted on the growth of authoritarian tendencies in interwar Europe. Intellectuals such as Juhan Vilms, an Estonian physician and politician, saw the root cause of demographic problems in Estonia in dismantling a traditional, patriarchal society. As a remedy, Vilms advocated the creation of small businesses and individual agricultural holdings, a system of bartering, and the institutionalization of family values. The corporate organization of society would facilitate the transfer of power from corrupt politicians into the hands of experts at both the legislative and executive levels. And since people did not display equal mental capacity, Vilms argued, they could not possess equal rights. Thus, according to Vilms, if a democratic country such as Estonia were to introduce science-based social laws that favored gifted and talented people over substandard individuals, it would provide a bulwark against what he considered to be dangerous socialist ideas about equality.[10]

Both positive and negative eugenics drew additional popular support by opposing all things Communist. Indeed, despite some common trends, for example, crime prevention, the Soviets notoriously rejected eugenics as a "fascist science."[11] In the course of the discussion on population policies in Estonia in the 1930s two major threats were identified: the Russians and the Germans. But it was a different type of danger that each of these two countries posed to Estonia. The Russians had raw power, whereas the Germans were in possession of a potentially dangerous ideology. It is hard to say to what extent leading Estonian politicians listened to the advice of the self-styled experts on eugenics when they first dismantled democracy in Estonia in 1934 and then partially substituted a free-market economy with a corporate system. One way or another, the objective was achieved.

The response of the intellectual elite to growing pressure from Nazi Germany can be encapsulated in one word: self-reliance. The "German danger" could supposedly be averted if, and only if, the Estonian people came together as one. Political activism was substituted for the metaphysical idea of spirit: trust in the nation's strength and giving life a new meaning. Leading Estonian intellectuals had promoted autarky and cultural isolationism as a means of protecting the country against possible aggression. Jüri Uluots, a law professor at Tartu University (TU), elaborated on "the Estonian ideal of culture" in a paper that he presented before the Association of Friends of Estonian Culture several months after Hitler had been appointed chancellor. According to Uluots, Estonians had to rid themselves of any foreign, that is, German, influences in order to develop their own unique ideal of culture. Uluots predicted a bright future for the Finno-Ugrian peoples, who had all the prerequisites required to become a great nation. Overall, Uluots occupied a middle ground between individualism and authoritarianism. He championed "social solidarity," implying that individual liberty could be restricted in the interests of society.[12] Oskar Loorits, a well-known Estonian linguist, echoed Uluots's sentiment. Claims that the Germans had introduced Christianity and European culture to the Estonians were nothing more than a shameless attempt to justify their seven-hundred-year-long domination. In fact, Loorits contended, the Estonians were an old *Kulturvolk* (people of culture) that had awoken from a century-long coma. All spiritual and material means should be mobilized to realize the spiritual and physical regeneration of the Estonian nation.[13] This reasoning represented an attempt to substitute the Nazi millenarian utopia with still another, national, utopia. At the very least, the literati tried to convince their audience that the ideas of spirit and Finno-Ugrian unity could stand up to the challenge of the Nazi *völkisch* ideal.

Estonia was hardly an exception when it came to the promotion of natalism and eugenics. Much of interwar Europe resorted to the natural sciences in search of a solution to demographic problems. However, what for other countries was only theory turned to reality for Estonia, Latvia, and Lithuania. To discuss acute demographic problems, supporters of eugenics in Estonia planned to come together at the Third Nationalist Upbringing Convention in 1940. Not only did the Soviet occupation of summer 1940 frustrate the eugenic agenda (sterilization, for example, was outlawed as contradictory to Soviet morale) but it also seemingly heralded the demographic disaster that adherents of the eugenics movement had been warning about for so long. As they had predicted, the Russian masses would sweep over Estonia to destroy the entire people. The subsequent mass deportation of Estonian civilians by the Soviet occupation authorities convinced even the most skeptical and least nationalistic among Estonians that the Russians intended to eradicate the Estonians as a nation.

Nazi racial experts shared the view of those Estonian scholars who predicted a gloomy future for the Estonian nation. In early 1939 the Reich Security Main Office (Reichssicherheitshauptamt, or RSHA) released a hundred-page document titled "Die Bevölkerungsverhältnisse in Baltikum" (Population-related issues in the Baltic). By having interpreted the available statistics along ideological lines, the planners at Reichsführer-SS (RFSS) Heinrich Himmler's Office disregarded the reality of interethnic relations in Estonia. The sole purpose of the document was to justify military intervention in the Baltic States, which were portrayed as incapable of defending themselves against the Communist danger. The report criticized the Baltic governments for their inability to address population issues. According to the report, the Estonian nation was a dying nation. It explained that due to a low birthrate the Estonians were incapable of withstanding the drive from the East. Of all ethnic groups living in Estonia, only the Russian minority had a high birthrate. The process of urbanization prompted not only internal migration but also the growth of a mass proletariat. Pauperized Russian settlers from the densely populated eastern provinces were taking over the "exclusively Estonian" western provinces, whose population, in its turn, moved to the cities. Consequently, the health of the Estonian nation was undermined through mixed marriages. The internal Russian expansion underway in western Estonia created a justification for Soviet intervention, the report continued. In an attempt to cut Germany off from the sea, Poland was allegedly cooperating with Russia by sending seasonal workers to the Baltic.

Estonian authorities were doing nothing to confront this problem, the report complained, while Estonians—intrinsically materialistic—remained exposed to Communist ideas.[14] Needless to say, the RSHA report teemed with contradiction. The Nazis acknowledged that the number of Polish workers in Estonia was negligible and that in ten years fewer than ten thousand Russians had settled in the country's western provinces. A further influx of the Slavic population was impossible due to the sealed border between Estonia and the Soviet Union. There were also no indications that the number of *Mischlinge* (people of mixed origin) had significantly increased over the preceding decade. For one thing, the ongoing process of Estonianization had reduced the numbers.

Racial Anthropology in Interwar Estonia

Nothing similar to Nazi racial science had developed in interwar Estonia. The concept of racial superiority was alien to Estonians, partly because the Estonians themselves did not rank very high in the Nazi racial hierarchy. Nazi biological concepts of race had little appeal among either intellectual or ordinary Estonians. Nonetheless, they did view themselves as a distinctive race, outside the common Nazi usage of this word. Altogether, the concept of *race* was not synonymous with racial biology and had been used interchangeably with the concept of ethnicity in the 1920s. References to *race* could be found in books and newspapers and in governmental and diplomatic records; Americans used it as often as the Nazis did.[15] Thus the American consul in charge in Reval (Tallinn) explained that Estonians were different from the Russians in race, language, and religion. American citizens of Estonian descent shared this view by stating that, ethnologically, the Estonians were a separate race that belonged to the Finno-Ugrian group.[16] The anticolonial image of the United States did not prevent its diplomats from mixing racial and psychological characteristics. The U.S. consul in Riga, for example, described Estonians as "the best human stock of all, the keenest, the most broadminded, and the most tractable."[17] In short, the word *race* lacked precise meaning. Therefore it makes no sense to look for underlying connotations when German diplomats talked about "the Estonian race" or Americans about "the German racial minority in Estonia."

However, there was another discipline related to racial science that did benefit from Estonian researchers' input, namely physical anthropology. The significance of physical anthropometrics and craniometry in interwar Estonia can be understood only in the context of the nineteenth century, when the first pseudoscientific assessments of the Estonians as a race were made. The interest of the newly proclaimed republic in the study of the physical and mental characteristics of the Estonian people stemmed from a natural desire to refute racial and cultural stereotypes imposed by German colonists. Until the mid-nineteenth century, civic society in predominantly Protestant Latvia and Estonia consisted exclusively of the Baltic German minority, the Deutsch. Estonian peasants constituted the lower strata, the Undeutsch. By the nineteenth century the Estonian word for "Germans," *sakslased,* (and its shortened version for "master," *saks*) came to describe any educated, white-collar, German-speaking individual, regardless of nationality. The local population distinguished itself by referring to newcomers as "Balts" or "Baltic barons." As late as 1860 most Estonians had not yet developed "the consciousness of belonging to a lasting political entity," viewing themselves as the *Landesvolk,* the "people of the country."[18] The lack of historical memory of a separate cultural or political entity created a sense of inferiority among the native population. Ethnic self-identification was further hampered by the relatively small numbers of Estonians, who were among the smallest of national groups in Eastern Europe. The question of ethnic viability loomed large as another roughly hundred thousand Estonians settled between 1860 and 1897 in other parts of the Russian Empire.[19]

The Estonians' problematic self-image coincided with, or was further reinforced by, the racial stereotyping of European, mainly German, scholars. In the nineteenth century scholars believed that peoples who belonged to the same language family had common roots. Ingrained in comparative linguistics, one dominant theory held that Finno-Ugrians had common ancestors. Some linguists found commonalities between the Finno-Ugric and Turkic languages, which led to a search for physiological similarities between Finns and Estonians and the Mongols. This hypothesis reinforced the view of those European anthropologists and philologists who considered Finno-Ugrians to belong to the Mongoloid race. By the end of the nineteenth century most Finns did indeed believe that they were Mongols or Turanians. Finno-Ugric nationalities, known in Russian as Chud', had been treated in the Tsarist Empire as inferior human beings. Russian anthropologists and politicians presented Finno-Ugric minorities as primitive peasants with deficient racial characteristics. For the most part, Russian scholars, too, considered Finno-Ugrians Mongoloid.[20]

Based on the false assumption that linguistic and biological traits were correlated, the Estonians were classified as the "yellow" race. According to Arthur de Gobineau, the yellow race had only a "culture-bearing" as opposed to a "culture-building" function, as in the case of the Aryans. In a bizarre twist of logic, the linguistic factor came to dictate physical features. Karl Ernst von Baer, a famous nineteenth-century natural scientist who had taught at TU, described the Undeutsch as "phlegmatic," "insensate," and "blockish." Even after the feudal dependence of the peasants was abolished, Estonians remained for the most part hostile to the Deutsch. This hostility helped to project an image of the Estonians as a rather reserved and unfriendly people, while their substandard material conditions and poor hygiene revived the notion of Estonians' "ugliness." Not surprisingly, the 1905 peasant unrest in Estonia targeted mainly German landlords: the subsequent heightened tensions between the Baltic German and Estonian groups made the former look for an answer in Gobineau's theory of rebellion of "less valuable" races against the master race.[21]

Nevertheless, by the end of the nineteenth century the Estonians' racial image had improved. By that time the international (or rather, German) scientific community started observing the light skin color of the Estonians and Finns and this "won" back these two peoples for Europe. Thus one German anthropologist wrote in 1878 that due to a substantial improvement of social conditions in the course of the nineteenth century the Estonians' physical outlook as well had changed for the better. Another German scholar concluded his 1901 study of the skulls and craniums of Estonian men as follows: "In general Estonians are well equipped by nature both physically and psychologically. They share all the traits of the white race and under certain circumstances can join the best part of humanity in pursuit of their goals."[22] Despite the seemingly positive assessment of the physical and mental capacities of the Estonian people, the underlying prejudices are obvious.

The combination of nineteenth-century racial theories, positive eugenics, imperial thinking, and social determinism informed collective German perceptions of Estonians during the interwar period. General Rüdiger von der Goltz, whose troops were defeated by the Estonian forces in June 1919, expressed the anger felt by many dispossessed Baltic Germans when he proposed full Germanization of Estonians and Latvians—those "barbarians with a West-European waste of manners."[23] Until 1945 German encyclopedias such as *Brockhaus* or *Meyer* continually described Finns as Mongolians. Friedrich Max Müller, who introduced to the world the theory of Aryan languages, declared Finno-Ugric languages "nomadic."[24] In 1939 German scholars were still arguing that because the Estonians were related to both Finns and Mongols they belonged predominantly to the East Baltic racial type. Nazi racial experts attributed vestiges of the Nordic race in Estonia to contacts with ethnic Germans and, to a lesser degree, Swedes.[25]

To fight the Mongoloid theory on its own terms, in 1924 the Finnish Academy of Science initiated an extensive anthropometrical study. Two years later Rolf Nordenstreng produced a new racial type—the East Baltic type—which he used to bury the Mongoloid theory. The East Baltic race was placed among other "blond" European races; the Finns and Estonians thus became "Europeans."[26] By losing "eastern" traits, Finns and Estonians attempted to reclaim their place in Europe as well as to distance themselves culturally and politically from Russia. Simultaneously, Finns and Estonians promoted Finno-Ugric solidarity as a way of strengthening their national identity.

Finns led the way with the establishment of the so-called Finno-Ugrian Society in 1883. Over the next thirty years the Finno-Ugrian Society organized and sponsored numerous ethnographic and linguistic expeditions to Siberia. Unable to exercise political pressure, the society attempted to influence the central Russian authorities indirectly, by means of fieldwork among the Finno-Ugric peoples of Russia. Estonian scholars followed suit in the 1920s, making a significant contribution to the field of Finno-Ugric studies in the process. Both Finland and Estonia played a major role in launching the Congress of Finno-Ugric Peoples, which took place for the first time in Helsinki in 1921. One area of engagement was, for example, collecting money on behalf of the Livonians (a tiny Finno-Ugric minority residing in the coastal area northwest of Riga). Otherwise, along with Hungary, Finland and Estonia often stressed the history of foreign oppression they had experienced in the recent past. In joint declarations adopted at the Finno-Ugric congresses, the three largest Finno-Ugric nations presented themselves as the outpost of Western civilization against the "Eastern enemies of culture."[27] Beyond that, the notion of Finno-Ugrian kinship has little substance: the only common feature that sixteen Finno-Ugric peoples in Eurasia share are about five hundred words with common roots.

Estonia's most prominent anthropologist was Dr. Juhan Aul, who dedicated his life and career to discovering "ideal" physical and racial characteristics of the Estonians. A professional biologist, Aul proclaimed as his goal the establishment of an Estonian anthropological institute, which would build a database for the study of the biological and social health of the Estonian people. According to Aul, physical anthropology was part of genetics and had wider application for society, including in the fields of pedagogy and criminology. As one of his scientific objectives, Aul intended to demonstrate a correlation between the mental and physical abilities of humans. Aul proved himself a strong believer in biological and environmental determinism. Thus Aul concluded that children in urban centers, particularly those belonging to the upper classes, tended to become physically less and mentally more capable. He proposed to apply the result of his studies to the internal colonization of Estonia, suggesting that settlers should be chosen from among the fittest of the population. Hereditary mental capabilities were another issue that had preoccupied Aul for years. His interest in zoology and genetics propelled him to become the author of the first Estonian textbook on genetics (1926). In his textbook Aul argued that physical and mental characteristics, alongside cultural and economic characteristics, determined national history. Although Aul considered the possibility of negative selection, he disapproved of the violent methods of intervention proposed by some of his colleagues from the Estonian Eugenics Association. Instead, Aul preferred education and medical checkups to abortion and sterilization.[28] This ambivalence might have reflected his complicated origins. Born into a Russian Orthodox family, until 1931 he bore a German name—Johan Klein.

The transition from physical anthropology to racial anthropology was not inevitable. Juhan Aul scorned racial pseudoscience when he learned about the unprecedented support that the Nazis gave to it. Aul discarded the notions of "Aryan race" and "German blood" that the Nazis had used to extol the German nation at the expense of other nationalities, including the Finno-Ugrians. What disturbed Aul most was the thesis that Germans, as the sui generis northern race, deserved credit for major cultural achievements. From an anthropological point of view, Aul argued, Swedes, English, and Scots actually scored higher than the Germans. Aul concluded that the Nazis used political anthropology to reinvent the hackneyed story of the spiritual and cultural superiority of the Germans.[29] By so doing he established a nexus between anthropological studies and the preservation of national culture. In 1936 Aul successfully applied to a foundation called Estonian Cultural Capital (Eesti Kultuurkapital) for financial support. Aul justified his anthropological project by referring to a struggle for survival that the Estonian people allegedly had been waging. Although this struggle had been carried out mainly in the field of culture, Aul wrote, anthropology played an important role too. Those who wished ill to Estonia distorted the facts about the anthropological makeup of the Estonians. That could have been avoided through the publication of original anthropological studies like his, Aul concluded.[30]

Thanks to Aul's anthropological studies, Estonia became one of the best-researched countries in the world. Aul achieved remarkable results by collecting anthropometrical measurements of fifty thousand Estonians, that is, 5 percent of the total population. In May 1939 the Estonian Ministry of Education approved the curriculum of a newly established Eugenics Institute at TU. Aul's former supervisor, Hans Madissoon, assumed the position of head of the institute. The Eugenics Institute's curriculum reflected the ideas that Aul and his fellow eugenicists had propagated. The institute was to study the racial particularities of the Estonians and other Finno-Ugrian peoples, as well as social and biological factors that contributed to the low birthrate in Estonia. Medical and theology students, in addition to those obtaining a teaching diploma, were obliged to take courses administered through the Eugenics Institute. Among other things, students learned about the negative impact of interbreeding on birthrates; children as state value; prevention of procreation by substandard individuals by means of isolation, sterilization, eugenic abortion, and castration; qualitative and quantitative population policies; appreciation of racial values in society; and eugenic legislation. To implement such an extensive research and educational program, the Eugenics Institute requested subscriptions to a number of professional journals, including the *Eugenics Review*, *Archiv für Rassen-und Gesellschaftsbiologie*, and *Volk und Rasse*.[31]

As the titles of the journals suggest, the bulk of the anthropological scholarship came from Nazi Germany, which Aul visited on a research trip in 1939. As one of his colleagues in the Anthropological Institute at Breslau University confirmed, "Anthropology makes great progress in Estonia" (die Anthropologie in Estland kräftig marschiert).[32] Aul had come a long way, from criticizing the Nazis in 1933 to inadvertently incorporating some elements of Nazi racial theory into his research by 1939 to fully cooperating with them in the period between 1941 and 1944. To better understand this transformation, one may consider how German occupation authorities handled one other academic discipline, ethnography.

Ethnography: Hijacking the Nationalist Agenda

Ethnology (ethnography), known in German as *Völkerkunde*, was another academic discipline that enjoyed the support of the Nazis. Although ethnology had developed independently from racial science and had no direct connection to the latter, the merger of the two disciplines in the Third Reich was perhaps inevitable. However, this process happened only gradually. In fact, most ethnologists in Nazi Germany were not members of the Nazi Party and rejected racial ideology, despite the fact that both Alfred Rosenberg and Heinrich Himmler had promoted the study of ethnology.[33] Given the nature of integral nationalism, ethnography had played an important role in shaping national identity in East Central Europe. Estonians, for their part, had prided themselves on the extensive collections of the Museum of Estonian Literature and the Estonian National Museum, both in Tartu. The Nazi authorities, too, came to appreciate it. An official at the Reich Ministry for the Occupied Soviet Territories (Reichsministerium für die besetzten Ostgebiete, or RMO) found just one problem with the exhibit at the Estonian National Museum: the mannequins tended to have East Baltic physiognomic features. This mistake could be easily corrected, he suggested, by changing the mannequins.[34] In the summer of 1942 a group of German ethnologists visited the Baltic States on behalf of Operations Staff Rosenberg (Einsatzstab Reichsleiter Rosenberg, or ERR). They returned with a pledge of cooperation from the leading Estonian scholars in the field. In spite of his image as an Estonian nationalist, the director of the Estonian National Museum, Dr. Eerik Laid, was prepared to collaborate with the Germans academically. His colleague Helmi Kurrik had been well known in Germany since at least 1937, when she had participated in a workshop of the Nordische Gesellschaft (Nordic Society). Helmi Üprus, the curator of the folk art collections, was also willing to cooperate, as was the head of the Estonian Folkloric Archives, Dr. Oskar Loorits. A well-established scholar, Loorits was eager to participate in the solution of larger scientific problems. Of all the scholars in the Baltic States, Loorits praised German ethnology the most.[35]

But what exactly did the Nazis want to achieve by engaging Estonian ethnologists and folklorists? The Nazis grasped the importance of ethnology for the Estonian political establishment, which in the 1930s had sometimes used it for propaganda purposes. By extolling the past, the government successfully diverted attention from the acute problems confronting the Estonian state. The use of ethnology also served to strengthen Estonian national identity. These were the motivations behind the state-of-the-art ethnographic museum and the folkloric archives, with their excellent collections and highly qualified personnel. In spite of the profound development of Estonian *Völkerkunde* over preceding decades, much less attention had been paid to organization and leadership than ideology. As a result, the existing power vacuum enabled the German authorities to appropriate it. Under the condition of occupation, one of the ERR officials wrote, folklore was an asset, the importance of which should not be overlooked. It was in German interests to encourage the academic intelligentsia involved in the research of the Estonian past. Without German support, however, those public figures might join the anti-German opposition, employing the slogans of Estonian folk culture. Keeping in mind the objectives of the program of Germanization, any attempts at suppressing indigenous culture and language risked the danger not only of provoking resistance and breeding anti-Nazi opposition but also strengthening the ranks of a "culture-proletariat." Therefore, it appeared much more prudent to acknowledge the importance of ethnographic and folkloristic research while co-opting it for the benefit of the occupying power. Since most scholars in the Baltic were ready to collaborate with their German colleagues, it was just a matter of seizing the momentum.[36]

In practical terms, the ERR intended that all the Baltic scholars and academic institutions involved in ethnographic and folkloristic research would be organized into a single network supervised by the Reich Commissariat Ostland (Reichskommissariat Ostland, or RKO) Office in Riga. The first joint project, called "German Heritage in the Ostland," was designed to forge a sense of cultural community based on blood (*das Gefühl einer blutsbedingten Kulturgemeinschaft*). This was to be accomplished by demonstrating the common Indo-Germanic heritage of German and Baltic folk cultures. In order to justify the necessity for German leadership over indigenous cultures, the project was supposed to emphasize the overall German impact on Eastern Europe throughout its history. To keep the morale of Baltic scholars high, their German colleagues proposed to promote bilingualism in academic publications, including journals. However, any research that had broader implications was to be published in German, something which was designed to make local Estonian scholars feel part of an international academic community. For the ERR, the first step in that direction involved soliciting prospective publications for German journals such as *Europäischen Quellen zur Volkskunde.*[37] By ostensibly accommodating Estonian nationalist sentiments, German occupation authorities prepared the ground for a more extensive deployment of local scientists and academics for the purpose of a radical restructuring of the ethnographic map of Europe.

Racial Studies at Tartu University, 1941–44

Under watchful German eyes, Estonian scholars actively pursued folkloristic and anthropological studies intended to enable the Estonian people to obtain their rightful place in Hitler's New Europe. While the Soviets did not sabotage the population studies initiated in interwar Estonia, the Nazis actively supported them. The initiative originated at the Health Institute at TU, which in June 1942 launched a study of natural population growth. The study focused on the social and physiological causes of low natural growth rates in Estonia.[38] On 26 March 1943 the president of TU, Edgar Kant, announced the establishment of a chair in anthropology and racial studies. By unanimous vote, the professorship went to Juhan Aul, then a member of the faculty at the Zoology Institute. On 30 September 1943 Kant filed a request with the Estonian Ministry of Education to establish the Institute of Anthropology and Racial Science. Two weeks later permission was granted.[39]

It is difficult to tell now whether it was Aul who sought out the German Security Police in Estonia or if they reached out to him.[40] Either way, in March 1942 the Estonian Security Police School in Tallinn asked Aul to deliver a series of lectures on anthropology and racial science. Aul felt enthusiastic about the assignment yet denied that he advocated racial inequality.[41] Thus in the spring of that year Aul was lecturing to future police officers. This did not go unnoticed by the RKO Office in Riga, Latvia, which invited Aul for consultations. The university paid Aul to visit an RKO adviser on racial issues. The subject of the conversation in Riga on 29 August was Aul's anthropological research, which apparently satisfied the Nazi racial expert. Upon his return to Estonia, Aul was assigned by President Kant to teach a four-hour course, consisting of lectures and tutorials, on anthropology and racial science to TU students. Following the establishment of the chair in anthropology and racial science, Aul's teaching load increased to six hours. Aul taught Introduction to Racial Science and other, related courses to students in both sciences and humanities.[42] According to RMO officials, the more Estonian students who took courses in anthropology at TU, the better; some of them might later continue their studies in the Reich.[43] Simultaneously, Aul continued with his research. In the spring and summer of 1942 Aul and his aide Leeni Sõrmus spent three months organizing an anthropological photo collection as well as processing anthropological data on both Estonian Russians and ethnic Estonians from the country's two southernmost provinces.[44]

The Nazis wanted to make Estonians and Latvians believe that they were racially related to Germans. The stronger the belief, the easier it would be to Germanize these two peoples in the future. Local scholars could strengthen the credibility of the blood-relation theory by lending their support to the claim.[45] This is why the Nazis were so encouraging of the kind of research that Aul had been doing in Estonia. From a political point of view, Aul was totally reliable, they believed.[46] In their racial assessment of the Estonian people, the German Security Police drew heavily on Aul's anthropometrical database. The conclusion read as follows: judging by their physical characteristics, Estonians were closer to Swedes and Norwegians than to Latvians or Russians.[47] Aul's findings made their way as far as Himmler's desk in Berlin. Aul marked height, head-index, and skin, hair, and eye color for each of the twenty thousand people he had studied. Aul passed on information regarding the racial composition of Estonians, of whom 24.8 percent belonged to the Nordic race and 29.2 percent to the East Baltic race. It must have been particularly pleasing for Himmler and his colleagues to read the following excerpt from Aul's study of Estonian Russians: "In racial terms, the Pechory region looks particularly bad. The region has been known as a transit area for various eastern peoples and is currently settled predominantly by Russians and the ethnically kin Setus. For the most part, one finds in the Pechory region an admixture of primitive East European and East Baltic races, interspersed with Asian characteristics. There are numerous signs of physical degeneration, which, however, point to heavy contamination with syphilis rather than racial interbreeding."[48]

On 15 April 1943 a group of distinguished guests, including the commissar general of Estonia (Generalkommissar Estland) Karl Litzmann, the head of the German Security Police in Estonia (Kommandeur der Sicherheitspolizei und des SD Estland) Dr. Martin Sandberger, and SS and police commander (SS-und Polizeiführer Estland) Johann Hinrich Möller, attended an event at TU, called Science in the Service of the Military. The event drew considerable attention in academic circles. Instead of the anticipated seventy students close to four hundred attended.[49] Committing to deliver one of the two academic lectures, Aul chose to title his talk "Anthropological and Racial Studies in Estonia." In his presentation, Aul focused on the issue that had troubled many Estonians for decades, namely the place of Estonians between the East and the West. Drawing on his anthropometrical research, Aul classified Estonians as belonging to the European racial type and proclaimed the Mongoloid theory dead. The Estonians were tall people with light hair color, like the Swedes (elsewhere Aul argued that Swedes and Estonians were the best representatives of the Nordic race).[50] However, it was not the Swedes, Aul argued, who had introduced the Nordic racial type in Estonia. In fact, the Nordic element in contemporary Estonia could be traced as far back as the Stone Age. According to Aul, Nordic racial characteristics accounted for 35 percent of Estonian inhabitants, while East Baltic characteristics accounted for 34 percent. The Nordic racial type was prevalent in the western part of the country, while the Baltic type was more common in the areas bordering Latvia in the south and Russia in the east. Even then, the East Baltic racial type, as evident in Estonia, should be considered autochthonous to the region. It should be clear to every intelligent Estonian, Aul explained to his audience, that the East Baltic racial type was very different from that to which the Russians belonged. Aul stressed that in order to get an accurate anthropological overview of the Estonian people one had to study the neighboring peoples. To some degree, he said, such research had already been conducted on Latvians, Estonian Swedes, and Russians. At the end of his lecture, Aul thanked the German authorities for establishing the chair in anthropology and racial science at TU, because anthropology "is very important for both understanding our national history and resolving population-political questions." Aul concluded his presentation with a quotation from the German anthropologist Hans Weinert: "Now we finally know that dealing with racial differences in humans is nothing like a meaningless hobby such as a herbarium or butterfly collection. We need the whole of anthropology as a politically important national science." By introducing the "West Estonian type of Nordic race" Aul effectively contributed to racial science, as devised and promoted by Nazi ideologists.[51] Georg Leibbrandt—one of the people who participated in drafting Master Plan Ost—cited Aul, according to which the Estonians and the Germans were two racially related peoples (*rassisch nahestehendes, artverwandtes Volk*).[52]

TU was involved with several book projects promoting the exclusivity of the Estonian nation. In the summer of 1942 head of the Estonian civil administration Dr. Hjalmar Mäe began discussing with Edgar Kant, Juhan Aul, Hans Madissoon, and Jüri Uluots the publication of two prospective books that would introduce Estonia to German audiences. The books were supposed to be published in German by the Institute for Border and Eastern Studies (Institut für Grenz-und Ostlandstudien) in Berlin. A thousand copies of the first book—more accurately, a thirty-page pamphlet—were to be made available for limited distribution only, while the second book would have a larger circulation. Mäe presented a tentative outline of both books when he addressed the TU faculty. Mäe advised the authors to demonstrate that the Germans and the Estonians were two different nations, but with similar racial and psychological traits. He hinted that German colonization of Estonia was a bad idea by emphasizing that Germans in general and German farmers in particular would be unable to adjust to the harsh climatic conditions of Estonia. The group of authors was further supposed to prove that contacts with the Germans and Swedes had had positive effects on Estonia, while similar contacts with the Russians had caused only misery. Potential readers should be led to deduce that the Estonians had survived due to their superior features, which justified their claims to the region between the Baltic Sea and Karelia. To make it sound more convincing, the university professors were also to provide an overview of the origins of the Estonian people, including a discussion of the Mongoloid theory versus the Nordic racial theory. A reader should be able to draw the conclusion that Estonian independence came about as a result of a historical development.[53]

Edgar Kant and TU had also planned to publish a popular book on Estonian ethnography. This time it was to be an Estonian-language edition printed by an Estonian publisher. According to an outline prepared in January 1943 by Juhan Libe, a chapter on anthropology was to demonstrate that the Estonians had begun settling in the Baltic region in prehistoric times. Archaeological findings were to be used to study the racial composition of the early settlers as well as to disprove the Mongoloid myth. By emphasizing the links between the Finno-Ugric and Indo-European languages, linguistic analysis was supposed to further demonstrate why the Mongoloid theory was false. A thorough perusal of vocabulary should have led to the conclusion that during Tacitus's times the area settled by the Estonians, or other Finno-Ugrians, stretched as far south as Prussia. The book used references to folklore to validate the connection between Estonia and northern Europe, particularly during Viking times, as well as the connection between Estonia and the Finno-Ugric peoples of Siberia. For this reason, the authors writing the historical overview were advised to cover both Estonia and the area of potential Estonian settlement in Ingermanland, a territory east of the Estonian border.[54]

A further book project, meanwhile, dealt extensively with the issue of race, framing the subject explicitly within the Nazi racial discourse and attuned to the demands of Nazi race theory. According to the book outline, the following topics were to be covered: interbreeding between German males and Estonian females, the development of racial and political consciousness among the Estonian population, the growing preference for the Nordic race over the East Baltic race, and the awareness of the racial value of the Estonian nation. A handwritten note in the margins added that, according to data from the Folklore Archives in Tartu, Estonians had developed a negative attitude toward other peoples, such as Jews, Gypsies, Russians, and Latvians.[55]

Estonian elites considered the notion of kinship among the Finno-Ugric peoples of vital importance to the Estonian nation and therefore framed it as an issue of national survival. Fear of extinction made many Estonian intellectuals seek out sources of perceived threat and ways to avert it. This preoccupation is evident from research topics proposed by and for TU. The Estonian Security Police and the Estonian Relief Agency (Eesti Rahva Ühisabi, or ERÜ) were responsible for suggesting essay topics for university students; the best essays were to receive awards. The Security Police came up with three topics that, by and large, underpinned the principles of its conduct: "Nazi perspectives on the notion of proper behavior," "police tasks necessary for providing state security," and "differences in criminal legislation between Greater Germany and Estonia, and their influence on public order." The police authorities were particularly eager to teach the theory of crime prevention to Estonian youth. The ERÜ encouraged university students to ponder the "legal status of mother and child in Estonia and Germany," "causes of mortality among small children," "establishing youth courts in Estonia," "consequences of Bolshevik terror on population growth in Estonia,"; and "Ingermanland: settlement patterns and population." With regard to the last two topics, the director of the ERÜ, Otto Leesment, provided the following explanation. The Estonian people had lost thousands of their strongest members to Bolshevik terror. Since that had undoubtedly had an impact on national population growth, research should suggest ways to sustain the Estonian population. In the context of Hitler's "New Europe," Estonians were expected to have special relations with Ingermanland and its population. A regional population study was also important because of the substantial number of ethnic Estonians and racially kindred peoples living in Ingermanland. All of them were destined to become part of the Estonian nation.[56] Juhan Aul at the Zoological Museum conducted research on the local Russian population under the title "Anthropology of the Russians Living on the Estonian Soil: Anthropological Particularities of the Estonians and Their

Racial Origin." On a scale of A through D, his study was graded B, which meant research deemed important for the reconstruction of the country.[57] Aul had promised to finish his manuscript by the end of 1944. Also mentioned was a nearly completed study, "Contributions to the Study of the Anthropology of Latvians."[58]

The Eugenics Institute experienced a renaissance under the Nazis, thanks to the institute's hyperactive former director Hans Madissoon. The Soviet authorities had tolerated the institute, which had operated under a different director and a more general name, the General Biology Laboratory.[59] The Nazis transferred the equipment and the remaining employees of the defunct laboratory to the Madissoon-led research unit. But the revival was limited. Although under the same academic roof, the Eugenics Institute fared worse than the Institute for Anthropology and Racial Science. Apparently it was less of a priority for the Nazis. Stricken with financial problems, the Eugenics Institute had to constantly lobby the Estonian Self-Government for additional funds. The monies, occasionally allocated to Madissoon and his institute, usually came from the "cultural reserves" of the Estonian Self-Government. As late as March 1944 the Eugenics Institute received 3,000 Reichsmarks for conducting population studies. At that time, four salaried employees were conducting research on behalf of the institute in the Central Archives in Tartu.[60] Madissoon and the Eugenics Institute were working on a study titled "The Question of *Völkisch* Development of the Estonians and Causes for the Decline of Birthrates in Estonia," yet as of April 1944 Madissoon had managed to produce only a few chapters of that study.[61]

Population Transfers in Northwestern Russia, 1941–43

The way the Nazis applied racial policy in Estonia enabled the Estonians to be at the same time the object and the subject of study. While the Estonians had been classified in accordance with Nazi racial standards, there was no need for Estonians to protest because they were ranked higher than Russians, Latvians, Lithuanians, or even Finns. Head of German Security Police in Estonia Sandberger observed a rather weak racial consciousness among the Estonian population. The liberal intelligentsia rallied against Nazi racial theory, which they interpreted from a narrow, nationalistic perspective. In particular, they were concerned that the theory of race extolled the Germans as a ruling class at the expense of the Estonians, whose Finno-Ugrian origins inadvertently placed them alongside the Mongols. However, Sandberger argued, the situation was changing for the better. Sandberger observed the prerequisites for the fostering of racial thinking among the Estonians.[62] The key factor in that "racial awakening" was the existence of Finno-Ugric minorities in the occupied Russian territories.

The Nazis capitalized on the Estonian quest for the Finno-Ugric ideal rather than pushing them in that direction. Estonian president Konstantin Päts, shortly before his arrest by the Soviets on 30 July 1940, produced a document commonly referred to as Päts' political testament. Among other things, Päts suggested the relocation of the Estonian and Finnish borders eastward so that they could include the Finno-Ugric minorities. Päts also proposed replacing Estonia's Russian population with racially similar peoples from Russia proper. According to Päts, the Finns should have carried out a similar resettlement program in Karelia.[63] The 1939 Soviet aggression against Finland made many Estonians feel guilty for not having helped a brother nation in need. The Estonians could now claim to repair that moral failure by assuming the symbolic role of caretaker of the Finno-Ugric nations in the German-occupied territory of the Soviet Union. Considering the close ties between Finland and Estonia—even if the former had usually played the role of leader and the latter of follower—the Estonians were now psychologically relieved of a possible inferiority complex. In practical terms, however, the Estonians could do little to help their linguistic kin in Ingermanland. The only thing Estonians could effectively do was to extend to the Finno-Ugric peoples of Russia the notion of a common culture and spirit. Before the war this attempt was realized mainly through ethnological studies, while in the context of Nazi occupation it came to be conveyed through anthropological and racial studies.

In early August 1941 acting president of TU Kant approached the German military commandant, asking if the German authorities would be interested in advancing national linguistic studies by granting a group of TU scholars access to a Soviet POW camp. In the camp the researchers aimed to study Finno-Ugric languages as spoken by inmates. For the benefit of scholarship, Kant suggested, it would be desirable if the Germans henceforth concentrated in Tartu those POWs who spoke Finno-Ugric languages.[64] In cooperation between TU and Sonderkommando 1a of the German Security Police (SK 1a), the Geography Institute prepared a series of maps outlining the racial (eleven maps) and ethnic (twelve maps) composition of Estonia and Ingermanland.[65]

The Nazi "New Europe" was to be created through population transfer. The resettlement of Baltic Germans in the fall of 1939 (so-called *Umsiedlung*) marked the first phase of the ethnic restructuring of the continent. In the northeastern corner of Europe, Estonia and Finland became actively involved in executing the racial blueprints of the Nazis. This involvement distinguished Estonia from the rest of occupied Eastern Europe, which had, for the most part, remained a passive recipient of the population policies emanating from Berlin. What Estonian and Finnish ethnologists and linguists had until now only dreamed of—a greater Finno-Ugrian entity—was becoming reality. Although it did not result in any significant increase of political representation, the demographic remodeling of the areas located between the Baltic Sea in the west and Lakes Ilmen and Ladoga in the east echoed the interwar rhetoric regarding a pan-Finno-Ugrian union of sorts. Once again, the Nazis attempted to build a bridge to the hearts and minds of Estonians. The repatriation of ethnic Estonians from former Soviet territories had long been on the agenda of Estonian politicians and intellectuals. The German Security Police supported this endeavor as long as it concerned racially fit Estonians.[66] According to Head of the German Security Police in Estonia Sandberger, incorporating small ethnic groups from Ingermanland into the Estonian majority would advance the concept of the Nazi "New Europe." When asking whether there was "more of a valuable Estonian and Finnish ethnic stock left in the Soviet Union," Sandberger argued that the Estonians would cherish the idea that they had saved even a tiny bit of precious human material. Indeed, by winning it back to the Estonian nation they would have prevented the imminent decline of kindred peoples.[67] Head of the Estonian Self-Government Mäe went one step further, proposing to exchange Estonia's Russian population for ethnic Estonians and other Finno-Ugrians living in Russia.[68]

In mid-October 1941 the Eighteenth German Army started registering ethnic Germans, Estonians, Finns, Latvians, and Lithuanians residing on the territory under its control, that is, Ingermanland. SK 1a, with its provisional headquarters in Gatchina, some fifteen miles southwest from Leningrad, was placed in charge of the operation. On 11 December the German Security Police in Tallinn dispatched a special registration commission (*Erfassungsabteilung*) to Ingermanland. Until March 1942 the commission, consisting of eighteen Estonian officers and a number of local auxiliaries, registered a total of eighty-one thousand members of ethnic minorities. As it went along, the commission identified Communists, partisans, and NKVD officials as well as potential conscripts for the police battalions. The resettlement began shortly after the registration process was completed. Almost all the three thousand ethnic Germans who had been identified were immediately transferred to Germany proper. Other minorities were given an option to voluntarily resettle in Estonia and Latvia. By the end of 1942 some twenty-five thousand people, including seven thousand Finns, had moved to Estonia. The general evacuation of the Estonian population from Ingermanland commenced in November of that year. In 1943 the Finnish government arranged for the transfer of sixty-three thousand ethnic Finns via Estonia to Finland. To administer this transfer, the occupation authorities set up a quarantine camp not far from Tallinn, near Klooga. A "thank-you" note, congratulating it on the successful population transfer, was sent to the Eighteenth German Army, the Security Police, and the Estonian Self-Government.[69]

Between 17 August and 15 September 1942 a group of TU professors and graduate students conducted a population study near the town of Kotly in Ingermanland. Among the most noted participants in the research expedition were the ethnologist Gustav Ränk and the linguist Paul Ariste. Another member of the expedition was an Estonian agronomist in the service of the Security Police in Tallinn. The main objective of the study was to map villages with a predominantly Finno-Ugrian population. The Wehrmacht, which sponsored the research, was particularly interested in reviving Finno-Ugrian settlements in that area.[70]

Capitalizing on the success of the preceding year's expedition, the Estonian Security Police organized another field trip, which took place between 5 and 25 August 1943. The makeup and objectives of the second expedition were different from the first. The collecting of folkloristic data was abandoned, and of the three ethnologists who had conducted fieldwork in 1942 only one remained. Instead, the anthropologist Juhan Aul joined the group. A party of five departed for Kotly, just east of the Estonian border, from where they conducted research trips to Ust-Luga, Karakol'e, and surrounding villages.[71] The German Security Police in Estonia and the Wehrmacht endorsed the expedition. The certificate that SK 1a issued to Aul and his colleagues obliged the local authorities to provide the scholars from TU with all assistance and protection necessary to their work. According to the certificate, Aul was supposed to carry out racial, population-political, and folkloristic research in western Ingermanland.[72] For three weeks, Aul took anthropometrical and craniometrical measurements of all local Votes,

Izhorians, and Russians between nineteen and forty-five years of age. Each of the 353 individuals was subjected to twenty-six different measurements, and a quarter of them were photographed. The main objective of this data collection was to map the racial features of the Finno-Ugrians and Russians.[73] According to Aul, the procedure went smoothly thanks to the cooperation of the local village administration, which ordered all residents to appear at a certain time in a designated place. Aul continued taking measurements the next year in Estonia. However, the Soviet air raid of March 1944 brought Aul's ambitious project to a halt.[74]

The anthropometrical databank that Juhan Aul accumulated over a number of years is comparable only to that of Roger Ritter and his Berlin-based research institute for racial hygiene and biology, which had gathered data on some twenty thousand German Sinti and Roma.[75] None of the other parts of the RKO could boast of such a high-quality racial research program as Estonia's.[76] Although Professor Jēkabs Prīmanis in Riga had also conducted an anthropometrical study, which he published as a book in 1937, the Nazis did not elevate him to the same status as Aul in Tartu. The Estonians found themselves at the forefront of a gargantuan campaign designed to permanently change the ethnic map of Europe. While German anthropologists were conducting racial examinations of candidates, including Estonians, for the Reich Labor Service (Reichsarbeitsdienst), Aul and his Estonian colleagues were studying other Finno-Ugrians.[77] What until then only the most daring ethnologists and politicians could dream about now became a reality: the Estonians were helping their linguistic kin, including ethnic Finns, in Russia proper; the Estonian Security Police manned outposts in the northwestern Russian borderland. Although Estonia was an occupied country, in areas such as ethnography, the military, or the judiciary, Estonians tended to behave more like Nazi allies.

Estonians did not waste time speculating about "Greater Estonia," which their status as an occupied country could not afford them anyway. Instead, Estonians capitalized on the territorial gains that Nazi propaganda had promised them. Even then, Estonians approached the matter seriously by conducting scientific research on Finno-Ugric peoples living in the former Soviet territories. By doing so, Estonian scholars continued pursuing the trends first introduced in independent Estonia. The only substantial difference was the introduction of racial studies—a new dimension that the Estonian researchers added to their academic inquiry. Racial studies may be considered a natural outgrowth of the anthropological and eugenic studies that had blossomed throughout the interwar period, but it was also a concession in the direction of the Germans. However, the Estonian initiative was not a mere facsimile but rather a genuine attempt to come to terms with new demographic and political realities. For example, Estonians projected a very different future for Leningrad than did the Nazis. A "Russian fist" that separated the two Finno-Ugric nations, Leningrad was meant to become a center of education for the "Estonian-Finnish world."[78] The Estonian 185th Police Battalion was supposed to assume its duties in Leningrad. In anticipation of the takeover of the city, Estonian policemen were learning Russian and watching Leningrad-themed films.[79] Sometimes the very term *Finno-Ugric peoples* was rendered as "peoples related to Estonians" (*eestisugu rahvad*).[80]

What many Estonians really wanted to achieve under the condition of Nazi occupation was to proudly assert themselves as a nation. When Estonian scholars went to Russia to study kindred peoples—from ethnographic and linguistic but also racial perspectives—they sought to reaffirm their own self-worth rather than to prove the inferiority of other ethnic groups. The old, discredited Mongoloid theory was still very much alive. The German Security Police were able to ascertain the hidden agenda of Professor Juhan Aul when he talked about racial science in Estonia: while speaking about the virtues of the Aryan race and the high percentage of Nordic blood in the Estonian nation, Aul was actually trying to provide a scientific rationale for Estonia's independence.[81] Aul considered the Finno-Ugric minorities in Russia to be part of the Nordic and East Baltic racial group, while simultaneously denying any "eastern" traits to Estonians.[82] Remarkably, the Nazis were eager to use Aul's services, regardless of the fact that he had never preached racial antisemitism. However, as in the case with positive versus negative eugenics, once the Estonian academic elite resolved to improve Estonians' image in the Nazis' eyes, they inevitably slid into negative stereotyping, not necessarily racial stereotyping but certainly cultural and political. Otherwise, at no point, before or during the war, did Estonians subscribe to Nazi racial ideology. Whatever Estonians said or did, their ultimate goal was to rehabilitate the Estonian state, which had collapsed like a house of cards in 1940. In order to be successful at fighting what they perceived as a Russian Communism that had robbed Estonia of its independence, many Estonian intellectuals, academics, and scientists sought to demonstrate the superiority of Estonian culture over that of the "Asiatics."

Estonians did not feel threatened specifically by Jews or Russians but, given their relatively small population size, lived in fear of annihilation. Soviet deportations, executions, and even mobilization seemed to corroborate the views of those scholars who argued in the 1920s and 1930s that the Estonian nation was on the brink of biological extinction. The prewar scare of a low birthrate that had allegedly undermined the Estonian nation was amplified dramatically with the onset of Soviet terror. Nazi propaganda reinforced those fears, claiming that German troops had prevented the imminent destruction of the Estonian people. Thus in September 1941 one Nazi expert asked rhetorically, "Are the Estonians a dying nation, like the Livonians? They do not want to die, they are afraid to die!"[83] Yet a few months later, in January 1942, Himmler proposed to transfer the orphaned children of Estonian and Latvian deportees to the Reich for Germanization.[84]

By the time of the Nazi occupation, Estonians were still wrestling with negative racial stereotypes that had gained currency in the nineteenth century—ironically, thanks to German scholars. Now, however, the German authorities actively supported the views of those anthropologists and politicians who placed Estonians at the top of a racial pyramid that encompassed all peoples of the rump Soviet Union. Although many Estonians loathed Germans, they never comprehensively questioned the tenets of Nazi ideology. To win the overall support of the Estonian population, the Germans granted them a range of important psychological concessions. Thus they constantly stated that the Estonians were a highly developed, racially healthy people of high culture, encouraged the pseudoscientific research that was supposed to demonstrate the exceptional qualities of the Estonian people, and allowed the Estonians to take part in a pan-European program of population transfer, while promising future territorial acquisitions. The Estonians wanted to regain both the biological and allegorical vitality of their nation in order to be able to govern themselves again. The widely held belief was that the more Estonians contributed to the German war effort, the greater the chance they had of winning political autonomy.[85] As the head of the Estonian Security Police, Ain Ervin Mere, once reported to a subordinate, it was essential for the Estonians to make a contribution to the New Order in Europe in case Germany won the war.[86] Eventually, however, the quest for independence came to include a massive dislocation of the population both within and outside Estonia's borders.

Notes

Abbreviations Used in the Notes

BAB	Bundesarchiv-Berlin (German Federal Archives, Berlin)
EAA	Eesti Ajalooarhiiv (Estonian Historical Archives, Tartu)
ERA	Eesti Riigiarhiiv (Estonian State Archives, Tallinn)
ERM	Eesti rahva muuseum (Estonian National [Ethnographic] Museum, Tartu)
NARA	U.S. National Archives and Records Administration, Washington DC
TU HRKO	Tartu Ülikooli Haruldaste Raamatute, Käsikirjade Osakond (Tartu University Rare Books and Manuscripts Division, Tartu, Estonia)
USHMM	U.S. Holocaust Memorial Museum, Washington DC

1. Cited in Ken Kalling, "Kolm miljonit eestlast—see oleks juba midagi! Esseelaadne sissejuhatus iibetemaatikasse," *Annales Litterarum Societatis Esthonicae, 2000–2001* (Tartu, Estonia: Õpetatud Eesti Selts, 2003), 183–84.
2. Jaan Tõnisson served several terms as prime minister, head of state, and foreign minister, respectively. Besides the long-serving president Konstantin Päts, Tonnisson was the second most recognizable politician in independent Estonia.
3. Toivo Raun, "Nineteenth- and Early Twentieth-Century Estonian Nationalism Revisited," *Nations and Nationalism* 9, no. 1 (2003): 140–41.
4. Kalling, "Kolm miljonit eestlast," 184–85. See also Andres Kasekamp, *The Radical Right in Interwar Estonia* (New York: St. Martin's Press, 2000), 123. In 1922 Estonia's population stood at 1,107,059; by 1939 the population had increased by a mere 1.3 percent. The systematic campaign of Estonianization was introduced by President Päts in the wake of the bloodless coup d'état in 1934 with the purpose of diverting public attention from current political and economic problems. The staple of the campaign was Estonianization of names.
5. Kalling, "Kolm miljonit eestlast," 186–89. Kalling characterizes the discussion of population-related issues in Estonia in the 1920s and 1930s as "natural growth hysteria."
6. Ken Kalling, "The Self-Perception of a Small Nation: The Reception of Eugenics in Interwar Estonia," in *"Blood and Homeland": Eugenics and Racial Nationalism in Central and Southeast Europe, 1900–1940*, ed. Marius Turda and Paul Weindling (Budapest: Central European University Press, 2007), 254–55, 258.
7. Ken Kalling, "Näitlik juhtum eesti meditsiiniloost: Dr. Juhan Vilms ja eugeenika," *Eesti Arst* 3 (March 2002): 182–83.

8. On the particularities of the eugenics debate in Finland, see Marjatta Hietala, "From Race Hygiene to Sterilization: The Eugenics Movement in Finland," in *Eugenics and the Welfare State: Sterilization Policy in Denmark, Sweden, Norway, and Finland*, ed. Gunnar Broberg and Nils Roll-Hansen (East Lansing: Michigan State University Press, 1996), 195–258.
9. *Deutsche Justitz*, 28 February 1936, referring to *Völkische Beobachter*, BAB, R-22/2501. The text of the Estonian Law on Sterilization was printed in *Riigi Teataja*, 4 December 1936.
10. Kalling, "Näitlik juhtum eesti meditsiiniloost," 184.
11. David Hoffmann, "Soviet History in Its International Context: New Approaches, New Perspectives," paper delivered at the Thirty-fifth National Convention of the American Association for the Advancement of Slavic Studies, Toronto, 21 November 2003.
12. U.S. legation in Estonia to the secretary of state, 27 August 1938, NARA, T-1170/11.
13. German Consulate in Tallinn, political report, 25 April 1933, NARA, T-120/3507.
14. RSHA, report "Die Bevölkerungsverhältnisse in Baltikum," [1939], BAB, R-58/1063.
15. See, e.g., W. F. Reddaway, *Problems of the Baltic* (Cambridge: Cambridge University Press, 1940), 14, 22, 28.
16. American consul in charge in Reval to the secretary of state, 2 March 1921; Delegation of American Citizens of Estonian Decent, petition to U.S. president Warren Harding, 10 March 1922, NARA, M-1170/4.
17. U.S. consul in Riga, "Racial Characteristics of the Estonians," 21 May 1928, NARA, M-1170/5.
18. Gershon Shafir, *Immigrants and Nationalists: Ethnic Conflict and Accommodation in Catalonia, the Basque Country, Latvia, and Estonia* (Albany: State University of New York Press, 1995), 132–33; Andres Niitepõld, "Eine Analyse des heutigen Standes der Beziehungen der baltischen Völkergruppen zueinander und untereinander," *Jahrbuch des baltischen Deutschtums* (1983): 151.
19. Toivo Raun, "The Latvian and Estonian National Movements, 1860–1914," *Slavonic and East European Review* 64, no. 1 (January 1986): 67–71.
20. Aira Kemiläinen, *Finns in the Shadow of the "Aryans"* (Helsinki: SHS, 1998), 40, 65–70, 187, 214, 218–19.
21. Ken Kalling, "Body as a Measuring Device," in *The Human Being at the Intersection of Science, Religion and Medicine*, ed. Anne Kull and Tartu Ülikool (Tartu, Estonia: Tartu University Press, 2001), 64–67. For an on overview of nineteenth-century racial theories, see Michael Burleigh and Wolfgang Wippermann, *The Racial State: Germany 1933–1945* (Cambridge: Cambridge University Press, 1991), 23–44.
22. Quoted in Kalling, "Body as a Measuring Device," 67, 69.

23. German Foreign Office, Reich Minister's Office, Records on Estonia, 16 December 1924, NARA, T-120/1464.
24. Kemiläinen, *Finns in the Shadow of the "Aryans,"* 68–71.
25. Roderich Ungern-Sternberg, *Die Bevölkerungsverhältnisse in Estland, Lettland, Litauen und Polen: Eine demographisch-statistische Studie* (Berlin: Verlagsbuchhandlung von Richard Schoetz, 1939), 3–4, 8–9.
26. Kemiläinen, *Finns in the Shadow of the "Aryans,"* 72–73, 85, 176–87, 214–17.
27. Sirkka Saarinen, "The Myth of a Finno-Ugrian Community in Practice," *Nationalities Papers* 29, no. 1 (2001): 41–45, 49, 50. Saarinen does not elaborate on who specifically were meant by "the Eastern enemies of culture."
28. Ken Kalling, "Professor Juhan Aul and Eugenics," *Papers on Anthropology* 7 (Tartu, Estonia: Tartu University Press, 1997): 175–77.
29. Juhan Aul, "Poliitilisest antropoloogiast," *Erk* 7 (1933): 190–92.
30. Juhan Aul's grant application to the Estonian Cultural Capital, 19 October 1936; Ministry of Education to Aul, 14 June 1938, TU HRKO, Juhan Aul Collection (unprocessed at the time of use).
31. Correspondence between Estonian Ministry of Education and Medical School at Tartu University, February–June 1939, EAA, 2100/4/214.
32. J. Schwiderzky to Aul, 27 August 1939, TU HRKO, Juhan Aul Collection.
33. Hans Fischer, *Völkerkunde im Nationalsozialismus: Aspekte der Anpassung, Affinität und Behauptung einer wissenschaftlichen Disziplin* (Berlin: Dietrich Reimer Verlag, 1990), 27, 31, 39, 51–53, 96–97, 141.
34. RMO office, Firgau, activity report from 25 October through 30 November 1943, 3 December 1943, NARA, T-454/38.
35. ERR, Strobel, report, "Volkskundliche Lage im Ostland," 10 June 1942, BAB, NS-30/152; ERR, Strobel, report, "Über den volkskundlichen Einsatz im Osten vom 22 April bis 18 Mai 1942," BAB, NS-30/79.
36. ERR, Strobel, report, "Volkskundliche Lage im Ostland," 10 June 1942, BAB, NS-30/152.
37. ERR, Strobel, report, "Volkskundliche Lage im Ostland," 10 June 1942, BAB, NS-30/152.
38. Health Institute at TU to State Central Bureau of Statistics, 18 June 1942, EAA, 2100/13/36. TU fully resumed operations in January 1942. Until the end of the war, TU was the only such institution in the German-occupied Baltic States. The College of Technology in Riga and the Veterinary College in Vilnius were the other two institutions of higher learning in the Baltic.
39. Dean of Mathematics Department to President of TU Kant, EAA, 2100/2/47; Director of Education Massakas to President of TU Kant, 13 October 1943, EAA,

2100/13/10. Edgar Kant was one of the founders of the so-called Club of Estonian Nationalists. Following the dissolution of the Estonian right-wing Vabs organization in 1934, Kant provided them with a much-needed political tribune by publishing articles in *Vaba Sõna.*

40. Following the German conquest of Estonia in the late summer and early fall 1941, SK 1a was restructured as the Office of the German Security Police in Estonia. The office had a supervisory function over the much larger, Estonian branch of the police, known as the Estonian Security Police. For more details on the Security Police in Estonia, see my book, *Murder without Hatred: Estonians and the Holocaust* (Syracuse NY: Syracuse University Press, 2009), 84–94, 108–22.
41. Juhan Aul, unpublished memoirs (1972), 94, TU HRKO, Juhan Aul Collection.
42. Correspondence between President of TU Kant, GK Estonia Litzmann, Head of the Estonian Security Police School Johannes Unt, and Juhan Aul, March–October 1942, EAA, 2100/2/47; President of TU Kant, memo to the deans of schools, 25 October 1943, EAA, 2100/15/60.
43. RMO Office, Leibbrandt, report, "Gedanken zur Frage der Eindeutschbarkeit der Völker des Ostlandes," 7 November 1942; RMO Office, Wetzel, report, "Richtlinien zur Frage der Eindeutschbarkeit der Esten, letten u. Litauer," 17 March 1943, BAB, R-6/160.
44. Aul to President of TU Kant, 1 July 1942, EAA, 2100/2/47.
45. RMO office, Dept. I, to Leibbrandt, 3 May 1943, NARA, T-454/99. On Nazi plans for Estonia and Estonians, see Weiss-Wendt, *Murder without Hatred*, 71–74, 78–83.
46. RMO Office, Firgau, activity report from 25 October through 30 November 1943, 3 December 1943, NARA, T-454/38.
47. Situational report of the German Security Police (Ereignismeldung UdSSR) no. 135, 19 November 1941, BAB, R-58/219.
48. RFSS Himmler to RSHA, 2 November 1942, ERA, R-4365/1/10.
49. Head of German Security Police in Estonia Sandberger, situational report "Estonia B," no. 34, 3 May 1943, USHMM, RG-15.007M/9.
50. Nikolai Kann to Juhan Aul, 7 September 1943, TU HRKO, Juhan Aul Collection.
51. Juhan Aul, report, "Über die anthropologische Forschungsarbeit in Estland und deren Ergebnisse," 15 April 1943, EAA, 2100/15/80. Most ideas that Aul presented in 1943 came from his dissertation, which he defended in 1938. See "Dissertation Committee Report on Ph.D. Candidate Aul's dissertation—'Lääne-Eesti maakondade eestlaste antropoloogilisi tunnuseid ja tõuline kuuluvus,'" EAA, 2100/1/735.
52. RMO Office, Leibbrandt, memo, 23 June 1943, BAB, R-6/159.
53. Head of Estonian Self-Government Mäe to President of TU Kant, 25 August 1942, EAA, 2100/15/58.

54. Juhan Libe, "Üld-rahvaslik (populaarne) väljaanne Eesti rahvusteaduste alalt," January 1943, EAA, 2100/15/58.
55. Thematic outline of a book, no date, EAA, 2100/15/58.
56. Head of Estonian Security Police Mere to Estonian Self-Government, 12 September 1942; Head of ERÜ Leesment to Estonian Self-Government, 11 September 1942; Head of ERÜ Leesment to Director of Education Massakas, 11 November 1943, EAA, 2100/13/13. The ERÜ was set up in imitation of the Nazi Welfare Organization.
57. Report, "Forschungsarbeiten an der Universität Dorpat," March 1943, EAA, 2100/13/75.
58. Report, "Der Stand der für Landesaufbau wichtiger Forschungsarbeiten (B-Themen) nach den Fakultäten," 1 April 1944, EAA, 2100/13/77.
59. Remarkably, neither Madissoon nor Aul were among the 10,200 people deported by Soviet authorities from Estonia in June 1941. The mass deportation of June 14–15 mainly affected the political and military elite, members of the police and various paramilitary groups, businessmen, and well-to-do peasants. Apparently the Soviet regime did not view eugenicists and racial anthropologists as a potential threat.
60. Correspondence between Head of the Eugenics Institute Madissoon, President of TU Kant, and the Estonian Self-Government, November 1941–March 1944, EAA, 2100/4/214.
61. Report, "Der Stand der für Landesaufbau wichtiger Forschungsarbeiten (B-Themen) nach den Fakultäten," 1 April 1944, EAA, 2100/13/77.
62. Head of German Security Police in Estonia Sandberger, annual report from July 1941 through 30 June 1942, ERA, R-819/1/12.
63. Kalle Lõuna, *Petserimaa integreerimine Eesti Vabariiki 1920–1940* (Tallinn, Estonia: Eesti Entsüklopeediakirjastus, 2003), 114. Although some authors consider Päts's testament a forgery, the most recent research has proved its authenticity. Head of German Security Police in Estonia Sandberger, annual report from July 1941 through 30 June 1942, ERA, R-819/1/12.
64. Acting president of TU Kant to German HQ in Tartu, 7 August 1941, EAA, 2100/15/64. During the Second World War Finns collected ethnographic and linguistic material from POWs of Finno-Ugric descent.
65. President of TU Kant to field headquarters, 20 October 1941, EAA, 2100/15/64.
66. Commander of Einsatzgruppe A of the German Security Police Walter Stahlecker, consolidated report from 16 October 1941 to 31 January 1942, USHMM, RG-11.001M/R-14.
67. Head of German Security Police in Estonia Sandberger, "Registrierung der Esten auf der anderen Seite der Grenze," 12 December 1941, ERA, R-819/1/13.

68. Oskar Angelus, *Tuhande valitseja maa: Mälestusi Saksa okupatsiooni ajast 1941–1944* (1956; repr., Tallinn, Estonia: Olion, 1995), 222–23.
69. Tiit Noormest, "Eestlaste ja teiste rahvusvähemuste ümberasustamine Loode-Venemaalt Saksa okupatsiooni ajal 1942–1943," *Tuna* 4, no. 2 (2001): 39–41.
70. Kotly military commandant to TU, 16 November 1942, EAA, 2100/15/59.
71. Gustav Ränk to head of Estonian Security Police Julius Ennok, "Ingeri ekspeditsiooni aruanne," 19 September 1943, ERM, Gustav Ränk Collection.
72. Head of German Security Police in Estonia Sandberger, certificate issued to members of the expedition, 4 August 1943, ERM, Gustav Ränk Collection; commander of EK 1, certificate issued to Juhan Aul, 6 August 1943, TU HRKO, Juhan Aul Collection. Similar certificate issued to Gustav Ränk, ERM, Gustav Ränk Collection.
73. Gustav Ränk to Head of Estonian Security Police Julius Ennok, "Ingeri ekspeditsiooni aruanne," 19 September 1943, ERM, Gustav Ränk Collection.
74. Aul's memoirs, pp. 95–96, TU HRKO, Juhan Aul Collection.
75. "Juhan Aul," in *Eesti teaduse biograafiline leksikoon* (Tallinn, Estonia: Eesti Entsüklopeediakirjastus, 2000), 109. On Robert Ritter, see Guenter Lewy, *The Nazi Persecution of the Gypsies* (Oxford: Oxford University Press, 2000), 43–49.
76. RMO Office, Leibbrandt, report, "Gedanken zur Frage der Eindeutschbarkeit der Völker des Ostlandes," 7 November 1942, BAB, R-6/160. Research undertaken in Latvia was at best sketchy, no research was carried out in Lithuania, and the anthropological studies that the Soviets undertook in Belorussia were of limited value for the Nazis.
77. RKO Office, memo, 20 August 1942, Latvian State Historical Archives, Riga, R-70/5/74.
78. ERR, Dr. Wilhelm Brachmann, report on Estonia, 22 September 1941, BAB, NS-30/80.
79. R. Meriste, "Noorimana Julg.-grupp 185-ndas," in *Eesti riik ja rahvas II: Maailmasõjas* (Stockholm: EMP, 1959), 7:102.
80. RMO Office, Maurach, report on exhibition "Vital Struggle of the Estonian People," [1943], NARA, T-454/21.
81. Head of German Security Police in Estonia Sandberger, situational report "Estonia B," no. 34, 3 May 1943, USHMM, RG-15.007M/9.
82. RMO Office, Firgau, memo re: resettlement of Ingermanland Finns, 11 October 1944, NARA, T-454/106.
83. ERR, Dr. Wilhelm Brachmann, report on Estonia, 22 September 1941, BAB, NS-30/80.
84. Reich Commission for the Strengthening of Germandom, Staff Main Office, to

RFSS Himmler, 13 May 1942; Higher SS and Police Commander North Russia Friedrich Jeckeln to Himmler's staff, 13 June 1942, BAB, NS-19/2216. By early 1942 there must have been 1,167 orphans in Estonia and another 3,950 in Latvia.

85. Tartu District Commissar Meenen, memo, December 1943, BAB, R-90/3.

86. Interrogation of Jaan Tamm, Tallinn, 24 November 1960, USHMM, RG-06.026/12.

12

In Pursuit of Biological Purity

Eugenics and Racial Paradigms in Nazi-Occupied Latvia, 1941–1945

BJÖRN M. FELDER

In December 1941, six months after Nazi Germany invaded the Soviet-occupied Baltic States, Latvian anthropologist Lūcija Jeruma-Krastiņa wrote an article in the largest Latvian newspaper, *Tēvija*, under the heading "Par latviešu rasisko būtību" (On the racial fundamentals of the Latvians): "Our forefathers were fighting as Nordic people did in the past. . . . It is the Nordic blood that had for generations upheld the fighting spirit of our forefathers."[1] In her article Jeruma-Krastiņa, a leading Latvian anthropologist and the first Latvian woman to receive a PhD, not only directly asked her fellow countrymen to support the Nazi war effort against the Soviet Union but also pointed out the "Nordic" characteristics of the Latvians. Although Jeruma-Krastiņa did not use the term *Aryan*, anyone who read her article must have understood that she racially categorized the Latvians as "Nordic," thus placing them on the same level as the Germans. This particular piece of popular science dealt with racial anthropology differently from her prewar writings: they had emphasized a balance between brachycephalic (broad-headed) and dolichocephalic (long-headed) types and an admixture of "Nordic" and "East Baltic" traits among the Latvians.[2]

Like many other scholars in Europe, Jeruma-Krastiņa had subscribed to the paradigm of racial anthropology since her graduate studies at the University of Latvia in Riga. However, even though she had supported the theories of the German racial anthropologist Hans F. K. Günther since the late 1920s, the abovementioned article signified a qualitative change.[3] Printed in a series of essays on race, Jeruma-Krastiņa's article was not mere propaganda; rather, it placed the author within the realm of Nazi racial anthropology.

Neurologist Teodors Upners, by contrast, did use the term *Aryan race* vis-à-vis the Latvians. Upners served as the head of the eugenic division of the Working Group for the Advancement of National Vitality (Tautas dzīvā spēka veicināšanas darba kopa, or TDSVDK), established in December 1941 with the purpose of promoting eugenics in Latvia. As a part of the Latvian eugenic project, this working group belonged to the social department within the so-called Latvian Self-Administration (Selbstverwaltung) under German occupation. In his speech to scientists and members of the Latvian health administration, Upners expanded on the "purity of the Aryan people," that is, the Latvians. He feared "degeneration" caused by mixing with "inferior races."[4] In order to prompt the vitality of the Latvian race and simultaneously improve its quality, Upners evoked eugenic measures as exemplified by racial hygiene in Nazi Germany. Upners's argumentation was a radical departure from the prewar discussion on eugenics in Latvia. No Latvian scholar had ever called for compulsory sterilization, as Upners did, nor had any local eugenicist ever before suggested the adoption of eugenic measures like those introduced earlier in Nazi Germany.

Following a discussion of the paradigmatic shift in racial and eugenic discourse in Latvia in the wake of the Nazi invasion of 1941, this chapter analyzes the motivation of those Latvian physicians, scientists, and politicians who popularized and sometimes implemented new racial and eugenic concepts. It examines the extent to which the racial paradigm shift occurred due to the opportunism of Latvian protagonists and whether it was a genuine strategy to realize a specific agenda. While the history of eugenics has been a subject of intensive research in Western Europe, Germany in particular, only recently has it provoked significant academic interest among scholars of East Central Europe.[5] Similarly, as regards the Baltic States, historical research in this field is relatively new. In Lithuania, Arunas Germanivicius has just launched a research project examining Lithuanian psychiatry and eugenic debates during the interwar period.[6] Ken Kalling worked on the same subject in Estonia while paying attention to racial science as well.[7] By contrast, the chapter by Anton Weiss-Wendt in this volume is the first scholarly treatment of racial science in Nazi-occupied Estonia. In the case of Latvia, Andrew Ezergailis has contended that there is little evidence for the existence of biological nationalism, racial thinking, or even an indigenous eugenic movement.[8] Likewise, he has contested the claim that a eugenic curriculum was ever instituted at the University of Latvia following the Nazi occupation.[9] A brief overview of eugenics in prewar Latvia questions these assumptions.[10] In the ensuing debate, a fellow historian, Vita Zelče, while acknowledging that a eugenics movement existed in Latvia, nonetheless insisted that the eugenic project in Latvia was essentially benign—aimed mainly at increasing national birthrates and assisting poor Latvian families. In any case, she argued that writing a comprehensive history of Latvian eugenics was altogether impossible due to the fact that so much essential documentation had been destroyed in the aftermath of the occupation.[11]

In spite of the fact that eugenics is closely connected to race, studies examining the Nazi mass murder of Jews in Latvia rarely have dwelled on national debates, forms of nationalism, or racial science.[12] In general, the public and scientists alike used the term *race* in the sense of a genetic group, especially in the context of eugenics and biopolitics. In addition, anthropologists used the classical system of classification in which race was closely connected with moral values.[13] No wonder that anthropologists like Juhan Aul in Estonia and Jēkabs Prīmanis in Latvia proved the foremost proponents of eugenics. Both were leading scientists in their field and also founding fathers of national eugenic projects.[14] Whenever historians have addressed racism and racial science in the Baltic region, they have invariably focused on the Nazi racial program for the occupied East.[15] At the same time, the available literature on the history of racism and racial anthropology rarely mentions the Baltic States, if at all.[16] In fact, a preoccupation with issues of race and racial identity was not restricted to Nazi sympathizers and supporters but was common among European scientists, publicists, and politicians across the ideological spectrum. Using the case study of Latvia, this chapter aims to demonstrate that racial thinking and a eugenic agenda were prevalent among the European elite at least until the end of the Second World War. In fact, while different interpretations of the substance, aims, and practical implementation of racial science and eugenics certainly existed among scientific elites, their basic principles remained largely unchallenged.

Racial Identities and Eugenics in Independent Latvia, 1920–40

The high point of the Latvian national movement was 1918, when the independent Latvian state came into existence. For Latvians, ethnicity and national identity were as important as they were to other "late nations" of East Central Europe. The newly established republic was essentially a nation-state, even if Latvia, along with Estonia, had some of the most liberal legislation regarding minority rights in Europe.[17] By joining the community of European nation-states, Latvia also participated in contemporary debates about nation and race. Certainly, the Latvians longed for a racial identity that most other nations, such as the Germans, English, or French, had already achieved.[18] However, race was more than just a common cause. By situating themselves on a racial scale, the Latvians tried to legitimize their state. This was effectively the case with all East Central European states that came into existence in the wake of the First World War. Most of the "late nations" were described by Western anthropologists as Slavic and therefore categorized as "Eastern," meaning inferior to the western, "Nordic" races. The prominence of racial thinking, racism, and racial self-perception was part of the Zeitgeist and the conduit for self-legitimization. Throughout East Central Europe scholars and publicists were writing about race and the racial composition of their own people as well as of others. As a rule, racial studies featured a mixture of science and politics; scholarly texts often had an explicitly political subtext. Writing in the late nineteenth century, Bulgarian anthropologists distanced their nation from Russia and Serbia by stating that, although the Bulgarians had adopted a Slavic language, racially they were not Slavs. In a similar vein, Ukrainian anthropologists described their nation as "Nordic" but identified the Russians as an "Eastern" people, thus drawing a racial front line between "Nordic/Aryan" and "Eastern." Some Polish scholars believed that the Poles were "Nordic" but that Germans (because of alleged racial interbreeding in Germany), ironically, were not.[19] Described by Western scholars as "Mongols," Estonians were eager to change this image with the help of science.[20]

Latvian scientists were also interested in creating an appropriate racial identity for the Latvians. Professor Gaston Backman, a Swedish racial anthropologist who became the first head of anatomy at the newly established University of Latvia in 1920, introduced this discipline to Latvia. Backman, who was describing Latvians as the defenders of the Nordic race against the inferior Slavic race as early as the

First World War years, conducted anthropometrical research with the purpose of racially categorizing the entire Latvian nation.[21] The Latvian physician and anthropologist Jēkabs Prīmanis succeeded Backman in 1928. Prīmanis had begun his studies at the Military Academy of Medicine in St. Petersburg. He received a PhD from the University of Latvia where he also defended his *Habilitation* (professorial thesis) in 1927.[22] As a student of Backman, Prīmanis continued his mentor's racial research with the purpose of creating a racial identity for the Latvians. Prīmanis's research focused on the racial origins of the Latvians, engrained in the latest anthropological theories. His assistants measured the heads of young women, army recruits, and high school and university students, comparing their skulls with those found in medieval and ancient tombs. Remarkably, Prīmanis utilized *dainas*—folklorist songs and poems—to describe Latvians' physical characteristics.[23] Thus he resorted to literature, craniology, and paleoanthropology in his search for the "pure" Latvian race. By so doing he created a biohistory of the Latvians that seconded the nationalist narrative. Prīmanis's research culminated in his 1937 book *Latviešu abtropoloģiskais raksturojums* (The Latvian anthropological character), which was more of a political treatise than an anthropological study.[24] In his book he described the Latvians as "Nordic" people who had had "East Baltic" ancestors, meaning Estonians and Livonians. Although Prīmanis did not use the term *Aryan,* he did confirm the "racial" value of the Latvian nation. He argued that "East Baltic" heredity was only to a small degree influenced by Slavic tribes, who were often seen by European anthropologists as inferior. His assistant, Jeruma-Krastiņa, had taken anthropological measurements of nearly one thousand Latvian women for her doctoral thesis.[25] In 1936 her thesis appeared in the comprehensive compendium *Jaunais nācionālisms* (New nationalism) under the title "Latviešu rasiskās īpašības" (The racial character of the Latvians).[26] In her study Jeruma-Krastiņa fell back on the theories of Rudolf Martin, one of the founding fathers of German racial anthropology, who invented the methodology and measuring tools for the emerging science of anthropometry.[27] Jeruma-Krastiņa described the Latvians as an admixture of Nordic and East Baltic racial traits with the former dominating. Her claim that the Latvians were a "primordial Nordic race" was taken from a textbook by Günther.[28] Jeruma-Krastiņa's PhD thesis received academic awards from the Kulturas fonds and the Kr. Barons fonds, demonstrating that her anthropological theories were in the intellectual and political mainstream of thinking about Latvian racial identity.[29]

Racial Nationalism and the Dictatorship of Kārlis Ulmanis from 1934

Domestic and international politics had a decisive impact on debates about race and eugenics in Latvia. On 15 May 1934 the Latvian prime minister, Kārlis Ulmanis, declared a state of emergency and usurped power in the country. By so doing Ulmanis followed a trend in East Central Europe of replacing democratic forms of government with more authoritarian ones. Thus in December 1926 Antanas Smetona seized power in Lithuania, while Konstantin Päts did the same in Estonia in March 1934. Like his counterparts in Lithuania and Estonia, Ulmanis had roots in democratic politics: he played an important role in creating an independent Latvian state in 1918, served several terms as prime minister, and was the leader of the largest political party, the Latvian Farmers' Union.[30] While Ulmanis banned every political party, including his own, after the coup, he was keen on establishing national consensus. Thus he asked politicians from all the parties to support his regime. Simultaneously, he struck against the Social Democrats (the Communist Party was banned in the early 1920s) and the Fascist Pērkonkrusts movement. Yet some of his socioeconomic policies, for example, a program of nationalization providing social support to workers, gained Ulmanis considerable popularity in Latvia. His regime employed authoritarian and totalitarian approaches to economic organization, ideology, and rituals (e.g., a state corporate system, the designation *vadonis* (leader), and the Fascist salute) but avoided using extreme forms of violence. Indeed, under Ulmanis Latvia did not become a party state like Nazi Germany, nor did it have an equivalent of the Sturmabteilung (SA) militia. In terms of ideology, the Ulmanis regime was given to a form of *völkisch*, biological nationalism. This type of ideology was assiduously promoted by the national press, which effectively censored not only leftist views but also extreme nationalist and antisemitic views. As far as negative eugenics was concerned, the Latvian national project starting from 1938 included programs of voluntary sterilization and abortion. While Prīmanis and his assistants and students strictly adhered to scientific definitions, methodology, and categories developed in the field of racial anthropology, the late 1930s witnessed an upsurge of popular interest in racial nationalism and eugenics. Articles on either subject published after 1934 usually passed censorship.

Although Ulmanis himself did not contribute any fundamental writings or articulate a comprehensive program, several intellectuals close to him did try to establish a *völkisch* ideology for the new state under the label of "new nationalism." One such intellectual was the director of the Latvian War Museum, Ernests Brastiņš. Brastiņš was searching for a true "Latvian" religion as the foundation for national identity. He discovered it in a reconstructed Latvian paganism, which he used as the basis for the creation of a new church, Dievturība.[31] He also wrote a position paper on the *völkisch* state—the mélange of classic *völkisch* literature on mythology, culture, and nations. The raison d'être of the nation, the future Latvian state would be situated between authoritarian and totalitarian forms of rule, he believed. Brastiņš saw the nation as a biological entity, or a "race," and when describing the Latvians, he used the term *Aryan*.[32] Consequently, he promised the future Latvian nation would ultimately become a distinct biological group, the product of systematic "breeding."[33] Another literati and supporter of Ulmanis was Jānis Lapiņš, journalist and editor of the journal *Sējējs*. In the foreword to the programmatic volume *Jaunais nācionālisms* (New nationalism), which included contributions on racial psychology, racial anthropology, religion, and other subjects, Lapiņš called for "racial purity" of the nation.[34] The notion that race was an essential characteristic of a nation enjoyed a consensus among the Latvian elite, prompting a debate on the biological traits of the national body. In the late 1930s articles on "race" appeared not only in intellectual journals but also in the popular media, ranging from daily newspapers to the journal of the Latvian railway. These populist articles did not so much advance racial anthropological theories as blend science into a common understanding of race, for example, as a blood-related community embedded in folk myths.[35]

It was the racial anthropologist Prīmanis who was to become one of the founding fathers of the Latvian eugenic project, a project implemented with the purpose of reducing the percentage of the "inferior" population in Latvia.[36] Together with Pauls Stradiņš, a surgeon who successfully survived all political systems to become the leading figure in the medical profession in postwar Soviet Latvia, Prīmanis helped to realize the eugenic program that had been officially introduced in 1937.[37] The Latvian eugenic project consisted of three parts: positive eugenics in the form of propaganda and eugenic education; negative eugenics, exemplified by voluntary abortion and sterilization; and eugenic research. As early as 1933 medical doctors were allowed to carry out abortion on eugenic grounds.[38] Although the 1937 law provided for the creation of several eugenic commissions, only one such commission was established. Founded in January 1938, the commission held its sessions in the Ministry of National Welfare in Riga.[39] The eugenic commission comprised two medical experts, the psychiatrists Dr. Verners Kraulis and Dr. Kristaps Rudzītis; the prosecutor Voldemārs Salmiņš; and the chair, Oskars Alks, who otherwise served as the head of the ministry's health department.[40] The commission held biweekly meetings at the ministry building in Riga.[41] Acting for the "benefit of the natural selection of the people," the commission made decisions regarding abortion and sterilization in accordance with the preamble of the eugenic law.[42] In contrast to Scandinavian countries and some U.S. states, sterilization was permitted only after patients or their relatives consented to the operation. All in all, some 648 abortions and 60 sterilizations were performed in Latvia in 1938 and 1939.[43]

Eugenic propaganda and research fell within the purview of the Institute for Research on National Vitality (Tautas dzīvā spēka pētīšanas instituts, or TDSPI), established in 1938 under Prīmanis and Stradiņš. The TDSPI was divided into separate departments for anthropological, eugenic, and medical-demographic research.[44] Scientists at the institute collected and analyzed medical, demographic, anthropological, and genetic data on the Latvian population and drafted future population policies. The TDSPI came into existence through the efforts of the Society for the Promotion of Health (Veselības veicināšanas biedrība), a union of several medical and charitable organizations that comprised leading Latvian physicians and members of the Latvian elite. Headed by Stradiņš and Prīmanis, the Society for the Promotion of Health featured several prominent members, such as the former Latvian president Gustavs Zemgals.[45] Latvian eugenicists saw the Latvian nation as a biological entity, calling for the "improvement of the race" by means of reducing the number of "inferior" elements and increasing that of "superior" ones. Latvian hereditary hygiene (*iedzimtības higiēna*), as it was referred to, differed from Swedish "welfare" eugenics that sought to improve public health, drawing mainly on the German model of racial hygiene.[46] As regards reform eugenics, which was meant to eliminate hereditary diseases without promoting a racial utopia, it does not fit in either model, since quite a few young Latvian eugenicists studied or conducted research in Germany before and after 1933.[47] Those scholars had a certain racial-biological mindset, though certainly not as extreme as that of Nazi racial hygienists. Verners Kraulis, a neurologist and a member of the eugenic commission who became one of the leading exponents of eugenics, had ensured that in spite of negative birthrates and official prenatal propaganda, quality would take precedence over quantity when dealing with "inferior families."[48] The interwar eugenic project indicates that Latvian scientists had a broader biological-racial agenda than just "curing" society of hereditary diseases; similar to Nazi Germany, two-thirds of all cases of sterilization in Latvia involved the so-called feebleminded. Feeblemindedness was defined in both social and biological terms, encompassing the mentally disabled and socially underdeveloped as well as the uneducated. Indeed, the ultimate goal was the "biological improvement" of the Latvian nation, predicated on eliminating "inferior individuals."

The Soviet occupation that began in June 1940 fundamentally changed Latvian society. The entire state structure was dismantled, leading figures were removed, and society and the economy as a whole were restructured after the Soviet model. President Ulmanis was arrested and so were senior members of his government, except for those who had managed to escape abroad. Following the rigged parliamentary elections of July 1940, the Soviet authorities installed a Communist regime under the old Latvian Bolshevik Jānis Kalnbērziņš. The Soviet terror that followed culminated in the mass deportation of 14 June 1941. Among approximately 30,000 persons in Latvia who fell victim to Soviet terror, some 15,400 were deported in the summer of 1941. Since the Soviet Security Police specifically targeted the Latvian elite, many Latvians saw the terror as aiming to destroy their nation.[49]

As far as public health policies were concerned, the changes proved less drastic than they might have been otherwise, though health insurance provisions and civil servants' salary rates did change.[50] While the former minister of national welfare, Jānis Volonts—replaced with the Jewish physician Dr. Michael Joffe—and the head of the health department, Oskars Alks, were arrested, the staff of health institutions and hospitals not only survived but actually fared well. (Volonts was deported to a prison camp in Kirov province west of the Ural Mountains, where he was executed on 13 March 1943.)[51] Even the activity of the eugenic commission under Kraulis continued uninhibited. In spite of his proclivity for racial and biological thinking, he tried hard to advance his career under the Soviets. Although Kraulis no longer taught eugenics at the University of Latvia, in July 1940 he became head of the Association of Latvian Doctors (Latvijas Ārstu biedrība).[52] The eugenic commission could no longer address eugenics per se, due to the fact that in the Soviet Union of the late 1930s eugenics, genetics, and all biological auxiliary sciences were forbidden, replaced with neo-Lamarckian theories. The commission effectively became an institution for abortion, the usual method of contraception in the Soviet Union before 1937. The patients treated after June 1940 were mainly wives of Communist Party and military officials, predominantly Russians. One-third of the patients held Soviet citizenship.[53]

Latvian "Racial Hygiene": Paradigmatic Changes during the Nazi occupation, 1941–45

The terror that had defined the year of Soviet rule shaped the positive expectations toward the invading Germans. The war against the Soviet Union was at the apex of Hitler's political and racial agenda. The *Lebensraum* to be secured in the East was expected to become Germany's racial colony, whose indigenous population would serve as slave laborers or would be annihilated.[54] By June 1941 the Master Plan East was still in the preparation stage. One essential element of the Nazi racial utopia, however, was already underway—the mass killing of Jews. The mass murder of the Jews began with the German invasion of the Soviet Union, long before the "Final Solution" was decided at the Wannsee Conference.[55] In Latvia, which was incorporated into the Reich Commissariat Ostland, the Nazis camouflaged the killings as "spontaneous" pogroms by the local population. When this strategy failed, they co-opted local policemen to carry out mass executions. However, as participation in mass murder proved mentally taxing for many of the policemen, the German Security Police and the Security Service (Sicherheitspolizei und Sicherheitsdienst) established a special killing squad under Viktors Arājs, known as the Arājs Commando.[56] As a mobile killing unit subordinated to the Einsatzgruppe A, the Arājs Commando murdered not only Jewish civilians but, beginning in fall 1941, also Jewish patients in Latvian psychiatric hospitals. Beginning in January 1942 the German Security Police also began executing non-Jewish mental patients. All in all, the Nazi euthanasia campaign claimed twenty-four hundred lives in Latvia.[57]

The German-sponsored Latvian Self-Administration had no real power and took orders from the Reich Commissariat Ostland and its subsidiary offices. A retired Latvian general, Oskars Dankers, assumed the position of director general of the interior, supervising other "directors" among the surviving ministers and high-ranking civil servants of the former Latvian Republic. The Nazis tried to secure the support of Latvian politicians for the war effort with the promise of restoration of statehood in the near future. In reality, however, they wanted to transform Latvia into a Nazi colony. In the attempt to gain independence by means of providing military support, the Latvian Self-Administration eventually approved a general mobilization ordered by Hitler in 1943, leading to the formation of two Latvian Waffen-SS divisions.[58]

Especially enthusiastic about the German "liberation" in the summer of 1941 were members of the right-wing Pērkonkrusts (Thunder Cross), which had been banned since 1934. The Pērkonkrusts presented themselves as "Latvian National Socialists" and therefore as the only possible partners of the Nazis. Many ordinary Latvians were likewise convinced that the Pērkonkrusts under Gustavs Celmiņš and Ādolfs Šilde would play an important role from then onward. In July 1941 the Pērkonkrusts gained control not only over the Latvian press, including the major newspaper *Tēvija*, but also over the local administration and Latvian police.[59] Celmiņš and Šilde actively cooperated with the German Security Police and the Security Service, with which they had probably enjoyed contact since the mid-1930s.[60] In July 1941 they established their headquarters at the Latvian Society Building—a de facto center of political activity in Riga and Latvia. In July and August 1941 the Pērkonkrusts organized a series of self-promoting public lectures.[61] Among other subjects, the lectures tackled "Freemasons," "Jews in art and cultural life," "the Jewish Question," "National Socialism," "Fascism," and "Marxism and Liberalism."[62]

One of the lecturers was Edmunds Puksis, a leading member of the Pērkonkrusts since its foundation in 1932, who spoke on "Race, People, and State from the Perspective of National Socialism." In his other speech, called "Race," Puksis enumerated the five major European races and advanced a "scientific" view, according to which "race" was not the same as a people or nation (*tauta*) but represented the physical and mental characteristics of a certain group. In his discourse Puksis used the words *Aryan* and *Nordic* interchangeably. For him, the Latvians were a "Nordic" people with an "Eastern Baltic admixture." In contrast to the racial anthropologist Jeruma-Krastiņa, Puksis went beyond pure science: the main objective of his first two lectures was introducing and promoting Nazi racial theory. Thus he mentioned the struggle for racial survival and the superiority of the Aryans as the only race capable of creating civilization. The enemies of the Aryan race, according to Puksis, were the Jews, who "had never contributed nor will ever contribute to any civilization." When he called

Jews "the culture-destroying people" (kūlturās ārdoša tauta), he was actually quoting Hitler's *Mein Kampf* and his three-race theory (masters, slaves, and parasites). Puksis effectively approved of Nazi ideology, concluding that "a single person is only a transient being, but nation and race are eternal. . . . A nation and a race has the sacred task of creating culture."[63] Another important issue for Puksis was "racial purity," supposedly the foundation for any nation. In one of his lectures, following an overview of the history of eugenics in Nazi Germany, Puksis suggested using "racial hygiene" to secure racial purity. In retrospect, by proclaiming their support for Nazi racial ideology, the Pērkonkrusts attempted to present themselves as the only reliable Latvian group that the Nazis could entrust to lead a future Latvian state. By underlining the supposed "Nordic" or "Aryan" makeup of the Latvians, they posed as racially equal partners who should also be treated as such. For better or worse, the Pērkonkrusts did not survive for very long. In mid-August 1941 the Pērkonkrusts movement was banned by the Wehrmacht command. Even though some of its members ended up working for the German Security Police and the Security Service, ironically, the Pērkonkrusts eventually became part of the anti-German resistance.[64]

As the Pērkonkrusts controlled the local press, and for some time also *Tēvija*, they were able to promote their extremist views, including antisemitism. Even after the Pērkonkrusts had been banned, leading members of the organization such as Puksis and Celmiņš continued holding positions in the Latvian Self-Administration. Puksis, for example, continued as the director of arts and culture until 1942.[65]

In the end representatives of the former Ulmanis regime proved more successful in seizing power. They gained the upper hand in the Latvian Self-Administration, amounting to a revival of the Ulmanis state. Subsequently, the interior ministry revived the Latvian eugenic program. Although the revival did not lead to implementation of the radical ideas of Puksis, the proponents of eugenics and racial theory in Latvia increasingly adapted the tenets of Nazi racial science. In continuance of prewar eugenic education and scientific research, in December 1941 the newly established TDSVDK came to replace the former TDSPI.[66] The TDSVDK was later incorporated into the health department and, in 1943, into the social department, under the name of the Main Branch for the Advancement of National Vitality. The head of the newly established institution was no longer a scientist like Prīmanis but a mere administrator, economist Džems Raudziņš.[67] The TDSVDK conducted research and promoted eugenics with the purpose of "improving the vitality and quality" of the Latvian people. Having eugenics, "purity of race," and so-called *Ariertum* (Aryan studies) on its official agenda, the TDSVDK was particularly concerned with boosting the birthrate and conducting anthropological research.[68] It had five departments affiliated with the University of Latvia: the eugenic department, under Dr. Teodors Upners; the department of anthropology, under Prīmanis; the department of demography, under V. Saulitis; the department of ideology, under Professor Pēteris Starcs; and the department of genealogy, under Reverend Klavs Siliņš.[69] In order to promote the benefits of "racial purity" among the general public, the TDSVDK wanted to launch a new journal, *Tautietis* (Fellow countrymen). Although the journal never materialized, articles on eugenic issues regularly appeared in *Tēvija*.[70] In practical terms, the TDSVDK organized public lectures (most of the lecturers worked pro bono), coordinated research, and generally tried to exercise control over the Latvian administration as far as race and eugenics were concerned. The contents of the lectures reveal the shift in post-1941 Latvian eugenics from hereditary hygiene to Nazi racism and racial hygiene. The public learned about issues such as race, population, and eugenics, and the lecturers spoke of "innate skills," "enforced isolation of inferior individuals," "natural selection," and "humanitarian absurdities." Under the general heading of eugenics, racial propaganda addressed topics such as "thoughts about the Nordic race," "Aryan strength," "racial hygiene," and "purity of race."[71] The article by Jeruma-Krastiņa mentioned earlier is a good example of the kind of racial propaganda disseminated in Nazi-occupied Latvia. Jeruma-Krastiņa was an official lecturer with the TDSVDK and probably also conducted research for the anthropological department.[72]

In 1941 eugenics was reintroduced as a subject at the University of Latvia in Riga. Thus Dr. Viktors Mühlenbach and Upners lectured on racial hygiene and race theory.[73] The scientific debate on race in Latvia was undoubtedly influenced by Nazi ideology. As an expert serving on the eugenic commission in Riga and head of the TDSVDK eugenic department, Upners wrote articles on public health from "the eugenic perspective" and in 1943 published the first-ever popular scientific book on eugenics in Latvia.[74] In his book *Eugenikas nozīme tautas un valsts dzīvē* (The meaning of eugenics for the life of the nation and the state) he demanded a more radical application of eugenic practices, praised the eugenic laws introduced in Nazi Germany, and advocated the compulsory sterilization of "inferior" people.

The Latvian eugenic project between 1941 and 1944 was supported not only by the old-guard bureaucrats but also by scientists. As mentioned earlier, Prīmanis assumed the position of head of the department of anthropology at TDSVDK. A group of young psychiatrists and neurologists from the Sarkankalns psychiatric hospital became involved, as did Dr. Hermanis Saltups and Dr. Nikolājs Jerums, who, like Upners, belonged to the second generation of Latvian psychiatrists, educated during the 1930s in Latvia and abroad. Perhaps the only exception to the rule, Kraulis was dismissed from both the department of medicine at the University of Latvia and the eugenic commission as an alleged Communist collaborator.[75] Likewise, the eugenic commission resumed its work in "support of the natural selection of the Latvian people," authorized by the director of public welfare, Oskars Sīlis, on 13 January 1942. The 1937 law, translated into German for the occasion, provided the legal foundation for this decision.[76] The commission's structure remained as it was before the war: chair, health department director Dr. Eriks Bušs (later replaced by Dr. Teodors Vankins) and his deputy, Dr. Aleksanders Vītoliņš; a number of medical and legal specialists, Dr. Kristaps Rudzītis, Dr. Augusts Krastiņš, Dr. Kārlis Briedis, Dr. Nicolājs Jerums, and Dr. Upners; and attorney Aleksanders Cīrulis.[77] The commission dealt with the issues of abortion and sterilization but not euthanasia, in contrast to Nazi policies. Yet in the meantime the program had been expanded, and so in addition to the one in Riga, eugenic commissions were established in Latvian cities such as Ventspils, Talsi, Valka, and Ludza. Because the proceedings of the eugenic commissions are available only for Riga, the total number of patients who went through those commissions is difficult to estimate. To give a general idea, from January 1942 to October 1943 approximately 880 abortions and 24 sterilizations were performed in Riga alone. The eugenic commission in Riga met on a weekly basis; nearly half of the patients screened by the commission suffered from tuberculosis and the rest from mental illnesses like epilepsy or schizophrenia. However, once again, sterilizations were carried out on the basis of a diagnosis of "feeblemindedness."[78] In May 1943 the Latvian Self-Administration issued a ban on marriage for persons suffering from mental diseases, in order to "secure the health of families."[79]

The convergence of Latvian eugenics and Nazi racial hygiene occurred not only in terms of terminology and ideology but also in practice. "Racial purity," in the meaning of preventing miscegenation, became an increasingly important issue for the Latvian administration. In 1942 and 1943 the Self-Administration under Dankers enacted several laws that forbade marriage between Latvians on the one hand and Russian prisoners of war, Germans, and other non-Latvian "migrants" on the other.[80] The marriage act that interdicted Latvian-Russian relationships was passed on 23 December 1942. During his interrogation by the Soviet Security Police in 1948, Upners insisted that the German authorities had forced the Latvian administration to pass this discriminatory act.[81] Upners was disingenuous, since in 1942 there was apparently a consensus between Latvian officials and the German authorities concerning marriage bans. Indeed, at a joint conference in November 1942 Latvian and German officials reached an understanding that "marriages between Latvians and Germans should be made impossible or banned completely."[82] The Latvian Ministry of Justice affirmed the law, while the TDSVDK under Raudziņš intended to check every marriage in Latvia against racial-eugenic criteria by means of special heredity applications.[83] Under Nazi occupation, local eugenicists increasingly spoke of the Latvian nation in racial and eugenics terms, more than they ever did before the war. It did not help that the German authorities required from Latvian civil servants proof that they had "Aryan" blood.[84] Furthermore, the agency in charge of handing out proofs of Aryan descent (*Ariernachweise*) was the department of genealogy at TDSVDK.[85]

Teodors Upners, probably the foremost Latvian eugenicist in 1941–44, was particularly close to principles of Nazi racial hygiene in his thinking. As mentioned earlier, he was concerned about "the racial purity of the Aryan Latvians." As a follower of social Darwinism, he believed that natural selection was a product of the struggle for survival.[86] For Upners, eugenics meant racial hygiene in the sense of racial selection and the improvement of the "biological value" of a nation.[87] Therefore, according to him, individuals forsook their personal rights and freedoms in favor of the biological defense of the nation: "Every member of the Nation has to know that he does not belong as an individual . . . but as a part of the whole nation." Consequently, Upners called for compulsory sterilization, because voluntary sterilization (carried out with the informed consent of the patient) in the context of Latvia had only limited success as far as he was concerned.[88] It was probably not a coincidence that Upners looked favorably upon Nazi racial hygiene as he prepared his doctoral thesis at the German Research Institute for Psychiatry in Munich in the 1930s. His supervisor at Munich was Professor Ernst Rüdin, one of the leading Nazi psychiatrists and racial hygienists.[89] After 1941 the scientific exchange between occupied Latvia and Germany intensified markedly and several Latvian scientists spent time at German research institutions. Upners himself, with the support of Prīmanis, spent another three months at Kaiser Wilhelm Institute in Berlin and Munich in 1942. In Munich he stayed two months with Rüdin while preparing his book on eugenics for publication (it eventually came out in 1943).[90] At the same time, in spite of his belief in the major tenets of Nazi racial hygiene, he condemned the murder of the mentally ill and other "inferior" members of society in several of his publications. In a 1943 study, for example, he took a clear stance against mass murder: "Life is our greatest gift. . . . We should not exterminate the inferior parts of our society as the primeval people once did. We all belong to the Latvian people and therefore every Latvian is dear to us."[91] Upners argued that even "inferior" people should be given a chance to grow old.[92]

Euthanasia Killings in Latvia

These and similar statements by Upners can be interpreted as a backlash against the Nazi euthanasia killings in Latvia the previous year, a sentiment reflected in wider public opinion. Altogether, by 1943 resistance against the German authorities had grown considerably in Latvia. Latvians whose relatives had disappeared from psychiatric hospitals started lodging inquiries with the Latvian health authorities.[93] The Latvian health administration had no influence on the euthanasia killings. Generally, they were informed about the executions ex post facto. However, due to the increasing number of inquiries from family and friends of the mental patients, the health administration had to react somehow. It seems that this particular issue was of extreme sensitivity to the Latvian authorities. For example, on 31 December 1943 the head of the health department, Dr. Teodors Vankins, personally addressed the director of the Liepāja city hospital. Vankins inquired about the fate of patients with mental diseases who were rumored to have been delivered to German authorities.[94] Once he realized the extent of the crime—that is, systematic mass murder of the mentally ill in Latvia—health department authorities instantly tried to wash their hands of the affair. As a rule, the health authorities redirected these kinds of requests to the German Security Police.[95]

Somewhat different was the case of those Latvian physicians who were directly implicated in the killing of the mentally ill. The first mass execution took place in the city of Daugavpils as early as July 1941. A local psychiatry ward received an order from the German local headquarters to deliver its inmates to the town of Aglona, where they were later killed. In order to legitimize the mass killing of the mental patients, the German authorities cited the need to secure hospital beds for wounded German soldiers. Later, after all the inmates were murdered, several other psychiatric hospitals were turned into German field hospitals, for instance, the Sarkankalns psychiatric hospital in Riga, which became a hospital of the Waffen-SS in 1942.[96] Even though the planning and implementation of the mass executions of the mentally ill in Latvia was primarily a task assigned to Einsatzgruppe A, high-ranking members of the German Ostland administration (e.g., head of the health department Harry Marnitz or Jelgava district commissar Wilhelm Freiherr von Medem) openly demanded the introduction of a euthanasia program. Thus, at a meeting of physicians in Jelgava, von Medem explained that the killings of the mentally ill had been carried out in order to provide much-needed hospital space for the medical treatment of German soldiers. According to him, mentally ill patients were incapable of performing productive labor and yet were consuming goods required for the army. Physical liquidation was simply inevitable, he concluded.[97] Nevertheless, the killings at the Jelgava Ģintermuiža psychiatric ward did not take place until much later, in January 1942.

After the "liquidation" of the Daugavpils mental facility in July 1941 further killings temporarily ceased, with the exception of Jewish patients who had been removed from the hospitals and executed en masse in the fall of that year as part of the Nazi Final Solution. The non-Jewish inmates suffered the same horrible fate in January 1942. By then, most patients in Latvian psychiatric hospitals, such as Sarkankalns in Riga, Ģintermuiža in Jelgava, and hospitals in Strenči and Liepāja, had been murdered. A majority of the twenty-four hundred victims were executed by members of the German Security Police in Latvia. As regards the involvement of Latvians in the mass murder, it was mainly the local psychiatrists at the hospitals who took part in the process, mainly by selecting those among their own patients who would be executed. As in the case of the so-called Operation T4 in Germany, the collection of statistical information concerning the mentally ill in Latvia commenced at the beginning of 1942. That is when the Latvian health department circulated a letter requesting local hospitals to provide data concerning mentally ill patients, including their diagnosis, length of stay in the institution, and so on.[98] Apparently the order came from the German Security Police in Latvia. In accordance with this order, Latvian physicians had to prepare statistical sheets on their patients. Subsequently, members of the Security Police arrived at the hospitals, distributing lists of patients who were to be executed.

As far as the Latvian personnel were concerned, Dr. Hermanis Saltups, head of the men's insulin department at the Sarkankalns psychiatric hospital, stated during his interrogation by the Soviet Security Police in 1945 that he and his colleagues were forced to prepare a patient list for the German Security Police in January 1942. According to Saltups, the director of the hospital, Hermanis Buduls, who was supposed to forward the list to the Germans, hesitated to do so. When Buduls realized that the patients at Sarkankalns would share the fate of the mentally ill in Daugavpils, he allegedly stopped accepting new patients. Saltups claimed that Buduls even tried to save some of the patients, for instance, by asking their relatives to take them home. Supposedly, Buduls lent some inmates to farmers in exchange for money.[99] Emilija Prostaka, then a nurse at Sarkankalns, corroborated in her testimony that director Buduls was hesitant to surrender his patients to the German Security Police. If true, this kind of behavior was not an exception; according to historian Rudīte Vīksne, the head of the mental institution at Daugavpils tried to save some of the inmates by sterilizing them, upon an oral agreement with German officers.[100]

The case of Buduls is rather peculiar, as he had been a committed eugenicist since his time as a medical student at the University of Yuryev (Tartu) at the turn of the twentieth century. In 1909, while still pursuing an advanced degree, he published the first monograph on eugenics in the Baltic (as well as being the first such publication in Latvian), in which he emphasized the importance of "racial improvement" and, following the German scholar Wilhelm Schallmayer, spoke of "negative instruments of eugenics."[101] In 1913 Buduls graduated with a PhD thesis on "comparative racial psychiatry," and a year later, as assistant at the psychiatric hospital of the University of Yuryev, he traveled to Berlin to study with the professor of psychiatry Karl Bonhöffer, an important figure in German eugenics.[102] During the interwar years Buduls became one of the leading figures in Latvian psychiatry as well as a recognized expert for the Latvian eugenic project in the 1930s. His students and assistants, such as Kraulis and Upners, also achieved the status of leading Latvian eugenicists. The few articles on eugenics that Buduls published in the 1930s did not expose a radical point of view. Thus he questioned the large-scale sterilization program in Nazi Germany by arguing that contemporary genetic research yielded ambiguous results. This skepticism does not mean, though, that Buduls condemned sterilization as such. As a matter of fact, he promoted the sterilization of the "feebleminded," which condition he considered hereditary. Furthermore, he believed that the "feebleminded" tended to have many children, thereby constituting a threat to the Latvian nation.[103]

The resentment of ordinary Latvians toward the mass murder of people with mental disabilities found reflection in Upners's writings, since he was not only a scientist but also a prominent member of the Latvian health administration in his capacity as head of the eugenic department at TDSVDK. His book was commissioned by the TDSVDK and sponsored by Dankers as leader of the Latvian Self-Administration. Yet there might have been other than political reasons for his argumentation against the murder of mentally ill patients. It is likely that his argument against the killings reflected a more widely held fear in the wake of the First World War about the numerical decline of the Latvian population. Ethical and moral considerations, on the other hand, played little, if any, role in Upners's argument. On the contrary, Upners had explicitly stated on numerous occasions that he rejected what he called "overblown humanism."[104] Nonetheless, despite his belief in the urgent need for hereditary hygiene in Latvia, he admitted that the killing of the mentally ill would make the general population feel insecure. In Upners's utopia, every Latvian was expected to actively choose eugenics as a part of his or her daily life; the killing program would jeopardize the eugenic project.[105]

The end of the German occupation spelled the end of the Latvian eugenic project. Many Latvian eugenicists fled to the West, while others stayed put and eventually ended up in the Gulag camps, as, for example, did Teodors Upners. Yet other scholars prospered in postwar Latvia. Scientists such as Pauls Stradiņš, who had applied eugenics in his capacity as a leading physician in Latvia, emerged unscathed. Eventually Stradiņš became one of the most important figures in postwar Latvian medicine. Physical anthropology and genetics experienced a revival in Soviet Latvia following the death of Stalin.

In interwar Latvia, the debate on race had been gravitating toward the "Nordic/Aryan" discourse. Racial and eugenic paradigms in Latvia became increasingly radical in the aftermath of the Nazi invasion of 1941, even if a German health expert at the Commissariat General thought of his Latvian counterparts as backward scientists influenced by Anglo-American research traditions.[106] By the same token, SS-Untersturmführer Hans-Jürgen Bosse considered Professor Jēkabs Prīmanis a researcher who "could not advance beyond exact science" to the "ideological science" of Nazi racial theory.[107] During the Nazi occupation Latvian scientists and health officials increasingly came under the influence of Nazi racial theory and sought to biologically re-create a nation with a higher percentage of Nordic racial traits. Several factors lent currency to racial theory based on biological characteristics in Nazi-occupied Latvia. Regarding the protagonists' probable motivations beyond opportunism, the positions of influence that they secured enabled them to realize the kind of radical projects impossible prior to 1941. In the initial period members of the Pērkonkrusts took the lead in the race debate. Even after the Pērkonkrusts had been banned, some of its members held on to their positions, for example, Džems Raudziņš as the head of the TDSVDK or Gustavs Celmiņš as the senior interpreter in the service of Dankers. As much influence was exercised by eugenicists such as Teodors Upners, Verners Kraulis, or Hermanis Saltups, who had spent time studying or conducting research in Germany and became inspired by Nazi racial hygiene in the process. Moderate bureaucrats from the Ulmanis government had also been increasingly referring to Latvians as "Aryans." This was part of a general strategy to achieve recognition as equal partners with the Germans, admitting that at least some protagonists seem to have been convinced of the Latvians' racial worth as a group. Furthermore, with the assistance of the Latvian Self-Administration, Latvian eugenicists were able to radicalize and enlarge the scope of the eugenic project. The eugenic commissions newly established outside Riga concurred with the intensified application of negative eugenics in the form of sterilization and abortion. Shaping the national body required more than just denying people with mental disabilities the right to reproduce. More than ever before, the biological thrust was exemplified by eugenic marriage control and the ban on interethnic marriages. Thus the Latvian eugenic project attempted to render not just a healthy but an ethnically and racially pure nation. In the process, individual rights and freedoms, even in matrimony, were to be subordinated to the national and racial interest, as was the case in Nazi Germany.

During the period of Nazi occupation Latvian eugenics exhibited a strong element of racial biology. Eugenics was meant to create a "pure Latvian race," superior not only in a hereditary, biological sense but also in the ethnic sense, that is, free of any non-Latvian influences. This kind of ethnocentric racial hygiene was, in fact, a fundamental part of prewar Latvian and Estonian eugenics. While Baltic eugenics on the whole proved less aggressive vis-à-vis ethnic minorities, it nevertheless aimed to exclude them from the eugenic project and consequently from the national body. So successful were local scholars such as Jeruma-Krastiņa in promoting the racial issue that even the reluctant Germans sometimes readily admitted the perceived racial value of the Latvians. As compared with the early statements of Hitler and Himmler, who saw the Latvians as inferior to "Nordic" nations, by 1943 the German Ostland administration was increasingly describing Latvians as a racially valuable stock. Ironically enough, this shift in the German point of view was influenced by the reception of the anthropological research of Prīmanis and his students.[108] Between 1941 and 1944 *race* was not only a commonly used term in Latvia but also a defining element of the nation in its biological sense. Racial and eugenic paradigms had been commonplace in interwar Latvia. Albeit more moderate in their interpretation and application than they were in Hitler's Germany, racial theory and negative eugenics underwent a significant radicalization in Latvia under the conditions of Nazi occupation.

Notes

Abbreviations Used in the Notes

BAB Bundesarchiv Berlin (German Federal Archives, Berlin)
LVA Latvijas Valsts arhīvs (Latvian Historical Archives, Riga)
LVVA Latvijas Valsts vēstures arhīvs (Latvian State Historical Archives, Riga)

1. Lūcija Jeruma-Krastiņa, "Par latviešu rasisko būtību," *Tēvija*, 20 December 1941.
2. See Lūcia Jeruma, "Die Lettin vom anthropologischen Standpunkt" (PhD diss., Anatomisches Institut der Univesität Lettlands, Riga, 1935).
3. Hans F. K. Günther, *Rassenkunde Europas: Mit besonderer Berücksichtigung der Rassengeschichte der Hauptvölker indogermanischer Sprache* (Munich: J. F. Lehmanns Verlag, 1929). See also Jeruma, "Die Lettin" (diss.), 135.
4. Teodors Upners, draft speech, "Tautas dzīvā spēka kvalitates celšana," LVA, 1986/2/P-6831.
5. See, e.g., Peter Weingart et al., *Rasse, Blut und Gene: Geschichte der Eugenik und Rassenhygiene in Deutschland* (Frankfurt: Suhrkamp Verlag, 1988); Magdalena Gawin, "Progressivism and Eugenic Thinking in Poland, 1905–1939," in *"Blood*

and Homeland": Eugenics and Racial Nationalism in Central and Southeast Europe, 1900–1940, ed. Marius Turda and Paul Weindling (Budapest: Central European University Press, 2007), 167–84; Marius Turda, "The First Debates on Eugenics in Hungary 1910–1918," in Turda and Weindling, *"Blood and Homeland,"* 185–222; Marius Turda, "'To End the Degeneration of a Nation': Debates on Eugenic Sterilization in Inter-war Romania," *Medical History* 53, no. 1 (2009): 77–104.

6. Arunas Germanavicius, "Development of Lithuanian Psychiatry in 1918–1940," paper delivered at the conference Eugenics, Race and Psychiatry in the Baltics: A Transnational Perspective, 1900–1945, Riga, Latvia, 7–8 May 2009.
7. Ken Kalling, "Introduction to the History of Estonian Eugenics," *Tartu University History Museum: Annual Report 1998* 3 (1999): 31–42; Kalling, "Professor Juhan Aul and Eugenics," *Papers on Anthropology* 7 (1997): 67–73; Kalling, "The Self-Perception of a Small Nation: The Reception of Eugenics in Interwar Estonia," in Turda and Weindling, *"Blood and Homeland,"* 253–62.
8. Andrew Ezergailis, *The Holocaust in Latvia, 1941–1944: The Missing Center* (Riga: Historical Institute of Latvia, 1996); Ezergailis, *Nazi/Soviet Disinformation about the Holocaust in Nazi-Occupied Latvia; Daugavas Vanagi: Who Are They?—Revisited* (Riga: Latvijas 50 gadu okupācijas muzeja fonds, 2005).
9. Andrew Ezergailis, "Kolaborācija vācu okupētajā Latvijā: Piedāvātā un atraidītā," *Latvijas Vēsture* 54, no. 2 (2004): 42–53.
10. Björn Felder, "Mazvērtīgo samazināšana-eigēnika Latvijā," *Kultūras Diena*, 23 April 2005.
11. Vita Zelče, "Vara, zinātne, veselība, un cilvēki: Eigēnika Latvijā," *Latvijas Arhīvi* 3 (2006): 94–137.
12. Seppo Myllyniemi, *Die Neuordnung der Baltischen Länder 1941–44: Zum nationalsozialistischen Inhalt der deutschen Besatzungspolitik* (Helsinki: Dissertationes Historicae, 1973); Ezergailis, *Holocaust in Latvia*; Katrin Reichelt, "Kollaboration und Holocaust in Lettland 1941–1945," in *Täter im Vernichtungskrieg: Der Überfall auf die Sowjetunion und der Völkermord an den Juden*, ed. Wolf Kaiser (Berlin: Propyläen, 2002), 110–24; Geoffrey Swain, *Between Stalin and Hitler: Class War and Race War on the Dvina, 1940–46* (London: Routledge, 2004); Daina Bleiere et al., *History of Latvia: The 20th Century* (Riga: Jumava, 2006).
13. Following Foucault, these categories are a form of racism; see Philipp Sarasin, "Zweierlei Rassismus? Die Selektion des Femden als Problem bei Michel Foucaults Verbindung von Biopolitik und Rassismus," in *Biopolitik und Rassismus*, ed. Martin Stingelin (Frankfurt: Suhrkamp, 2003), 55–79.
14. For Aul, see Kalling, "Professor Juhan Aul," 67–73; for Prīmanis, see Björn Felder, *Lettland im Zweiten Weltkrieg: Zwischen sowjetischen und deutschen Besatzern 1940–1946* (Paderborn, Germany: Ferdinand Schöning, 2009), 279–85.

15. Kārlis Kangeris, "Nacionālsociālistiskās Vācijas plānotā represīvā politika pret latviešu inteliģenci," in *Totalitārie režīmi un to represijas Latvijā 1940–1956: gadā/ Totalitarian Regimes and Their Repressions Carried out in Latvia in 1940–1956* (Riga: Historical Institute of Latvia, 2001), ed. Irēne Šneidere et al. 7:240–66; Kangeris, "Die nationalsozialistischen Pläne und Propagandamassnahmen im Generalbezirk Lettland 1941–42," in *Collaboration and Resistance during the Holocaust: Belarus, Estonia, Latvia, Lithuania*, ed. David Gaunt et al. (Bern: Peter Lang, 2004), 161–86; Katrin Reichelt, "Latvia and Latvians in the Nazi Race and Settlement Policy: Theoretical Conception and Practical Implementation," in *Latvija Otrajā pasaules karā/Latvia in World War II*, ed. Daina Bleiere et al. (Riga: Historical Institute of Latvia, 2000), 1:266–77.
16. Christian Geulen, *Geschichte des Rassismus* (Bonn: Bundeszentrale für politische Bildung, 2007); Geulen, *Wahlverwandte: Rassendiskurs und Nationalismus im spätern 19. Jahrhundert* (Hamburg: Hamburger Edition, 2004); Stefan Kühl, *The Nazi Connection: Eugenics, American Racism and German National Socialism* (Oxford: Oxford University Press, 1994); Kühl, *Die Internationale der Rassisten: Aufstieg und Niedergang der internationalen Bewegung für Eugenik und Rassenhygiene im 20. Jahrhundert* (Frankfurt: Campus, 1997).
17. Georg von Rauch, *Geschichte der baltischen Staaten* (Munich: dtv, 1977), 132–41.
18. See Geulen, *Geschichte des Rasissmus*, 78.
19. Christian Promitzer, "A Lithuanian in the Balkans: Jonas Basanavičius (1851–1929) and Bulgarian Racial Anthropology," and Maciej Gorny, "First World War and National Characterology in East-Central Europe," papers delivered at the conference Eugenics, Race and Psychiatry in the Baltics: A Transnational Perspective, 1900–1945, Riga, Latvia, 7–8 May 2009.
20. Kalling, "Self-Perception of a Small Nation," 254.
21. Mathew Kott, "Antropologen Gaston Backman och den uppsaliesnsiska rasbiologins spridning i tid och rum," in *Rasen och vetenskapen*, ed. Helmut Müssener and Per Jegebäck (Uppsala, Sweden: Centrum för multietnisk forskning, 2009), 59–82.
22. Prīmanis's university personal file, LVVA, 7427/13/1367.
23. Nicolājs Jerums and Teodors Vītols, "Beiträge zu Anthropologie der Letten," *Latvijas Universitātes raksti* 18 (1928): 273–386; Jēkabs Prīmanis, *Latviešu ķermena uzbūve Latvju Daiņās* (Riga: A-S Golts un Jurjans spiestuve, 1929).
24. Jēkabs Prīmanis, *Latviešu antropoloģiskais raksturojums* (Riga: Valtera un Rapas akc. sab. apgads, 1937).
25. Lūcija Jeruma, "Die Lettin vom anthropologischen Standpunkt," *Latvijas Universitātes raksti, Medicīnas Fakultātes sērija* 2 (1935): 1–151.
26. Lūcija Jeruma-Krastiņa, "Latviešu rasiskās īpašības," in *Jaunais nācionālisms*, ed. Jānis Lapiņš (Riga: Valters un Rapa, 1936), 46–49.

27. Jeruma-Krastiņa used the anthropological textbook of Rudolf Martin, *Lehrbuch der Anthropologie*, 3 vols. (Jena, Germany: G. Fischer, 1928).
28. See Jeruma, "Die Lettin" (*Latvijas Universitātes raksti*), 18.
29. Jeruma-Krastiņa's university personal file, LVVA, 7427/6/410.
30. On the Ulmanis's regime, see Ilgvars Butulis, "Autoritäre Ideologie und Praxis des Ulmanis-Regimes in Lettland 1934–1940," in *Autoritäre Regime in Ostmittel-und Südosteuropa 1919–1944*, ed. Erwin Oberländer (Paderborn, Germany: Ferdinand Schöningh, 2001), 249–98; Inesis Feldmanis, "Umgestaltungsprozesse im Rahmen des Ulmanis-Regimes in Lettland 1934–1949," in Oberländer, *Autoritäre Regime*, 215–48.
31. Kārlis Ducmanis, "Mūsu nācionālo centienu garīgie pamati," in Lapiņš, *Jaunais nācionālisms*, 111–28.
32. Ducmanis, "Mūsu nācionālo centienu garīgie pamati," 124; Ernests Brastinsch, *Volkstumslehre Dienstliche Übersetzung der Publikationsstelle des Preussischen Geheimen Staatsarchivs in Berlin-Dahlem ausgeführt von Dipl. Volkswirt Paul Sender im Auftrage des Instituts für Osteuropäische Wirtschaft* (Berlin: Publikationsstelle Dahlem, 1937), 184.
33. Brastinsch, *Volkstumslehre Dienstliche Übersetzung*, 44–50.
34. Jānis Lapiņš, "Ievadam: Jaunais nācionālisms," in Lapiņš, *Jaunais nācionālisms*, 5–19.
35. Jūlis Vecozols, "Latvijas tautas asiņu kopība," *Latvijas Kareivis*, 19 October 1935.
36. Prīmanis's draft paper, no date, LVVA, 1642/1/38.
37. The new medical law (*ārstniecības likums*) was enacted after Christmas 1937. See *Valdības Vēstnesis*, 28 December 1937.
38. Teodors Upners, *Eugenikas nozīme tautas un valsts dzīvē* (Riga: Latviju Grāmata, 1943), 63.
39. Report on activities of the eugenic commission, 3 January 1938, LVVA, 4578/1/204. For other documents produced by the commission, see LVVA, 4578/1/202, 4578/1/204, 4578/1/205.
40. Protocol no. 1 of the eugenic commission, 12 January 1938, LVVA, 4578/1/202.
41. Proceedings of the eugenic commission, LVVA, 4578/1/202, 4578/1/204, 4578/1/205.
42. See *Valdības Vēstnesis*, 28 December 1937.
43. Report on the activities of the eugenic commission in 1938, LVVA 4578/1/204; report on the activities of the eugenic commission for 1939, reproduced in *Ārsts* 2 (1940): 127.
44. See the protocol of the Society for the Promotion of Health, 25 January 1938, LVVA, 3112/2/9.
45. Annual report of the Society for the Promotion of Health, 1937, LVVA, 3112/1/19.

46. On eugenics in interwar Sweden, see Gunnar Broberg and Mattias Tydèn, "Eugenics in Sweden: Efficient Care," in *Eugenics and the Welfare State: Sterilization Policy in Denmark, Sweden, Norway, and Finland*, ed. Gunnar Broberg and Nils Roll-Hansen (East Lansing: Michigan State University Press, 1996), 77–149.
47. Felder, *Lettland*, 282–83.
48. Verners Kraulis, "Über die Behandlung der Schizophrenie mit protrahiertem Insulinchock," *Zeitschrift für die gesamte Neurologie und Psychatrie* 166 (1939): 36–49.
49. See Björn Felder, "Stalinismus als 'russisch-jüdische Herrschaft': Sowjetische Besatzung und ethnische Mobilisierung im Baltikum 1940 bis 1941," *Zeitschrift für Geschichtswissenschaft* 57, no. 1 (2009): 5–25.
50. Report of Dr. V. Barkāns, executive officer at the Latvian Ministry of Health, 29 August 1941, LVVA, P-1023/1/83.
51. Volonts's NKVD file, LVVA, 1986/2/P-6853. Documentary evidence indicates continuous employment of the personnel at the Ministry of People's Welfare: LVVA, 4578 and P-1023.
52. *Cīņa*, 11 July 1940.
53. Proceedings of the eugenic commission from 1941, LVVA, P-1023/1/40.
54. Czeslaw Madajczyk, ed., *Vom Generalplan Ost zum Generalsiedlungsplan: Dokumente* (Munich: K. G. Saur, 1994).
55. Andrej Angrick, *Besatzungspolitik und Massenmord: Die Einsatzgruppe D in der südlichen Sowjetunion 1941–1943* (Hamburg: Hamburger Edition, 2003).
56. Ezergailis, *Holocaust in Latvia*, 173–202.
57. Rudīte Vīksne, "Garīgi slimo izncināšana Latvijā vācu okupācijas laikā," in *The Issues of the Holocaust Research in Latvia*, ed. Dzintars Ērglis, et al. (Riga: Historical Institute of Latvia, 2003), 8:324–50; Anton Weiss-Wendt, *Murder without Hatred: Estonians and the Holocaust* (Syracuse NY: Syracuse University Press, 2009), 148–49; Felder, *Lettland*, 206, 287, 294–96.
58. Felder, *Lettland*, 266–75.
59. Felder, *Lettland*, 242–43.
60. Björn Felder, "'Die Spreu vom Weizen trennen . . .' Die Lettische Kartei—Pērkonkrusts im SD Lettland 1941–43," in *Sphere of Influence: Yearbook of the Museum of Occupation of Latvia*, ed. Valters Nollendorfs (Riga: Museum of Occupation of Latvia, 2004), 47–68.
61. Report of the German Security Police and the Security Service in Riga, 13 November 1941, BAB, R-92/6.
62. See the manuscript collection "Ideoloģikas sagatavošanas kursos" (Ideology preparation lessons), 19–22 August 1941, BAB, R-92/6.
63. Puksis's draft speech, "Race, People, and State from the Perspective of National Socialism," BAB, R-92/6.

64. Felder, "Die Spreu," 309–11.
65. Survey of the Latvian Ministry of Education and Culture, 1 September 1941, BAB, R-92/71.
66. Bylaws of the TDSVDK, no date, BAB, R-92/581.
67. Draft plan of the TDSVDK and the report of the Main Branch for the Advancement of National Vitality, 8 October 1942, BAB, R-92/581.
68. TDSVDK agenda, BAB, R-92/581; position paper, "Basic Ideas on Future of the Latvian Nation," no date [1942], LVA, 1986/2/P-6831.
69. Draft plan of the TDSVDK, 8 October 1942, BAB, R-92/581.
70. Raudziņš to the Health Department of the Latvian Commissariat General, 9 December 1941; Raudziņš to the Latvian Department of Culture, Commissariat General, 4 and 9 December 1941, BAB, R-92/581.
71. List of TDSVDK lectures, no date, LVVA, 4578/1/2609.
72. List of public lectures on eugenics by the TDSVDK, BAB, R92/581.
73. Upners's interrogation records, 10 August 1948, LVA, 1986/2/P-6831.
74. Upners, *Eugenikas nozīme tautas.*
75. Upners's interrogation records, 10 August 1948, LVA, 1986/2/P-6831.
76. Dr. Bušs to Sīlis, 9 February 1942, LVVA, P-1023/1/54; The German translation of the eugenic law is available at LVVA, P-69/1/20.
77. Protocols of the Riga Eugenic Commission, January 1942–October 1943, LVVA, P-1023/1/54.
78. There exists only fragmentary evidence on local eugenic commissions. See, e.g., the protocol of the Walk Commission, 26 October 1943, LVVA, 4578/4/639. A total of 756 eugenic abortions were performed between January 1942 and October 1943; by the end of that year a further 164 applications for abortion were filed. Documents for the Riga Eugenic Commission, 1941–43, are available at LVVA, P-1023/1/40, 4578/4/639, and 4578/4/640.
79. Act to Secure Healthy Families, May 1943, BAB, R-90/377.
80. Meeting protocol of the Social Department of the German civil administration in Latvia, 19 December 1942, BAB, R-92/580; report of the Social Department, 25 March 1944, BAB, R-90/379.
81. Upners's interrogation records, 10 August 1948, LVA, 1986/2/P-6831.
82. Protocol of the joint German-Latvian conference, 19 November 1942, BAB, R-92/580.
83. Comments of the TDSVDK to the proposal of the Latvian Ministry of Justice concerning marriage laws, 23 September 1942, LVA, 1986/2/P-6831.
84. Commissar General of Estonia Karl Litzmann to the Estonian Ministry of Education, 20 December 1941, Estonian Historical Archives, Tartu, 2100/13/35.

85. Rauziņš to Bosse, 18 December 1941, BAB, R-92/581. See, e.g., the proof of Aryan descent of Lucija Jeruma-Krastiņa, 18 December 1942, LVVA, 7427/13/704.
86. Upners, *Eugenikas nozīme tautas,* 19.
87. Teodors Upners, "Tautas veselība eugeniskā skatījumā," *Ārstnicības Žurnals* 3 (1943): 197–202.
88. Upners, "Tautas veselība eugeniskā skatījumā," 201–2; Upners, *Eugenikas nozīme tautas,* 62–4.
89. Travel permits issued to Upners by the University of Latvia, 15 September 1936 and 19 June 1938, LVA, 1986/2/P-6831.
90. Prīmanis to the president of the University of Latvia, 7 March 1942, LVA, 1986/2/P-6831; Upners's interrogation records, 20 September 1948, LVA, 1986/2/P-6831.
91. Upners, *Eugenikas nozīme tautas,* 25.
92. Upners, "Tautas veselība eugeniskā skatījumā," 201.
93. See, e.g., M. Barkovskis's request regarding the whereabouts of her husband, 19 July 1943, LVVA, P-1023/1/10.
94. Vīksne, "Garīgo slimo iznīcinašana," 340–41.
95. Latvian Health Department to A. Kokons, June 1943, LVVA, P-1023/1/10.
96. Health administration to the hospital director, 18 April 1942, LVVA 2917/3/35.
97. For Marnitz, see Felder, *Lettland,* 287; for von Medem, see Vīksne, "Garīgo slimo iznīcinašana," 332.
98. Vīksne, "Garīgo slimo iznīcinašana," 331.
99. Saltups's interrogation records, 29 November 1944 and 14 August 1945, LVA, 1986/2/P-7280.
100. Vīksne, "Garīgo slimo iznīcinašana," 331, 336.
101. Hermanis Buduls, *Lauliba un zilweka dsihwes mehrķis: Bioloģisks un etisks apzerejums* (Riga: Ed. Sirgela, 1909).
102. Hermanis Buduls, *K sravnitel'noj rassovoj psichiatrii* (PhD diss., University of Yuryev, 1914). See also his student file at the University of Yuryev, Estonian Historical Archives, Tartu, 402/1/3449.
103. Hemanis Buduls, "Rases labidzimtība," *Jaunākās Ziņas,* 6 June 1936.
104. Upners, *Eugenikas nozīme tautas,* 23.
105. Upners, "Tautas veselība eugeniskā skatījumā," 202.
106. Proposal to establish a local institute for population policy, no date [1942], BAB, R-92/578.
107. Bosse's memo, 20 February 1942, BAB, R-92/580.
108. A RuSHA anthropological study of Latvians specifically referred to Prīmanis as a major source. RuSHA report on the racial situtaion in Lativa, Czech National Archives, Prague, VRP/DOD II/ box 57. I am grateful to Mihal Šimůnek for sharing this document with me. See also Anton Weiss-Wendt's chapter in this volume.

13

The Eternal Voice of the Blood

Racial Science and Nazi Ethics

WOLFGANG BIALAS

Walter Gross, from 1934 onward director of the NSDAP (Nationalsozialistische Deutsche Arbeiter Partei, or National Socialist German Workers' Party) Office of Racial Policy (Rassenpolitisches Amt der NSDAP), spelled out the new Nazi ethic at a party rally in 1933: "Compassion for the hereditarily ill contradicts the laws of nature and life, laws that are apathetic to the trivial fate of single individuals, seen as drops in the huge stream of blood that flows eternally through history. . . . Whenever compassion and false humanity help the unhealthy to survive, man sins against the will of the creator who established the laws of life that, brutal as they are, always destroy the sick as soon as the existence of a race is in jeopardy."[1] This early statement exemplifies Nazi racial ethics, which provided that only those whose racial criteria qualified them as human beings deserved to be treated with compassion. In other words, the laws of nature and God himself favored the strong and healthy. The laws of life might be brutal, but they were, after all, humane, contradicting only the false humanism of unconditional, racially indifferent mercy. Left to his own devices, the individual counted for nothing. Only through his surrender to the eternal laws of life, through his commitment to the higher cause of race and history, did he attach meaning to his otherwise meaningless life. Following this line of argumentation, this chapter will discuss Nazi attempts to develop a unique race ethics meant to provide its adherents with a clear conscience while assuming that they were acting from selfless motives.

Nazi Perpetrators with a Clear Conscience: Monsters or "Ordinary Men"?

Who were the Nazi perpetrators? Were they just ordinary men or ordinary Germans? Or were they pathological criminals whose flawed worldview and perverted fantasies of humiliation, torture, and murder could come to fruition only under the protective umbrella of a racist ideology? Scholarly discourse on Nazi criminality has tended to argue that there is no prototypical Nazi perpetrator and that attempting to construct one might miss the point completely.[2] There is no doubt that the genocide of the Jews qualifies as immoral behavior beyond imagination. The Holocaust was indeed a monstrous evil, but does this necessarily imply that it was the product of evil monsters? Adolf Eichmann, for instance, was responsible for "monstrous" deeds, yet he was quite an ordinary man, neither demonic nor monstrous. As Hannah Arendt has argued, "The trouble with Eichmann was precisely that so many were like him, and that the many were neither perverted nor sadistic, that they were, and still are, terribly and terrifyingly normal."[3] Where do such considerations lead us?

Nazi perpetrators differed in their attitudes and characteristics. Those who rose to the top ranks of Nazi institutions were often very eager to act on their own initiative and to take responsibility for their actions. It was not enough for them to just obey orders or to meet the expectations of their superiors. Nazi ideology itself clearly favored the responsible, creative, and dedicated "political soldier" or, as Götz Aly put it, the ambitious activist who is always ready to improvise and take responsibility for his actions.[4] Heinrich Himmler proved just that when in October 1943 he complimented his Schutzstaffel (SS) officers for remaining "decent fellows" while murdering Jews. As he explained, "Most of you will know what it means when one hundred bodies lie together, when there are five hundred, or when there are thousand. And to have seen this through, and—with the exception of human weaknesses—to have remained decent, has made us hard and is a page of glory never mentioned and never to be mentioned."[5] These "decent fellows" retained their feelings for others, including even the Jews, but refused to submit to them when such feelings ran contrary to duty.[6]

Nazi perpetrators were encouraged to develop split personalities that allowed them to act in different capacities in accordance with their own rules, ethics, and rationale.[7] Various forms of splitting became common: of self from task, of partial task from whole task, of means from end, of procedure from its purposes, and of action from its moral implications. The result was the willingness of individuals, as members of a bureaucracy, to commit acts that they would never have committed outside of these administrative structures. Acting in the capacity of death squad members, camp commanders, or racial experts never overlapped with the perpetrators' roles as "decent" husbands, fathers, or friends. Membership in agencies with clearly defined rules and hierarchies eliminated personal responsibility and potential feelings of guilt. The gap between private and professional permitted the creation of a sense of self that would be functional in places like Auschwitz-Birkenau death camp. They had learned no longer to perceive atrocities as such but rather to see them as acts of duty.[8] Many of the worst known Nazi perpetrators are reported to have behaved in their "off-duty" hours as ordinary, decent human beings in accordance with recognized moral standards. Men like Rudolf Hoess, the commandant of Auschwitz, oversaw the killing of Jewish women and children by day but were loving husbands and fathers by night; the only observable difference in their behavior was the refusal to behave decently toward people who belonged to "degenerate" racial groups.

At the same time, the ruthlessness of death squads was linked to Nazism's self-image as a crusade against staid bourgeois conformity. The SS weekly *Das Schwarze Korps*, for example, suggested in many of its articles that such behavior was fitting of the "new man," a revolutionary Germanic warrior who disdained the security and safety of capitalist society.[9] These articles stated that the new man—the dedicated, militant political soldier—took shape in war and in war alone. Simultaneously, they acknowledged that this new man remained an exception to the rule on the "home front." They severely criticized the civil servant's opportunistic, conventionally timid mentality and lack of personal commitment, pouring scorn on old-fashioned, compliant bureaucrats who refused to take responsibility beyond the narrow domain delineated by their professional duties.[10] How can moral behavior be explained? What makes people act morally under conditions that reward immoral behavior such as terrorizing, humiliating, deliberately hurting, and even killing those belonging to the enemy camp of "inferior races" who, according to the prevailing ideology, are undeserving of a decent life and ultimately forsake the very right to live?[11] Hannah Arendt employed the "two-in-one argument" to explain acts of moral behavior under circumstances that promoted immoral behavior. Arendt viewed the conscience as the incarnation of our inner struggle to justify what we are doing, including what we, after having considered the pros and cons of different options, decide not to do. Our conscience forces us to align our current actions with our own criteria of right and wrong.[12] Arendt insisted that, though we might be able to deceive others with respect to how we feel about what we are doing, we cannot deceive ourselves. For her, the conscience operates as an inner dialogue, a thorough, genuine communication, in which we scrutinize our self-understanding from the viewpoint of morality. A potential murderer, to paraphrase her argument, as soon as he realizes what he is about to do, is unable to continue because he suddenly becomes aware of the personal implications of committing murder; the prospect of living as a murderer for the rest of his life is simply too much to bear.[13] Unfortunately, those in the SS who surrendered to their moral reasoning and conscience were few compared to the majority of morally indifferent collaborators and perpetrators with a clear conscience. In regard to the latter group, the question is how to explain their behavior, their reasoning, and their ability to cope with what they were doing. They appear to have replaced moral reasoning with something similarly effectual that allowed them to murder without having to examine their consciences.

Nazi scholarly and ideological discourse on the existence of a unique racial ethics contested the concept of conscience. Nazi theorists blamed Jews not only for the racial contamination of the Germans but also for having introduced ethics into history and specifically the moral institution of conscience into Western ethical discourse. This moral corruption consequently weakened the spirit of the strong, who could no longer rule the weak with a clear conscience. The perceived unconditional Jewish prohibition on killing and the empathy for each human being regardless of race that "Nazi experts on Jewish affairs" identified as the core of Jewish religion were considered especially dangerous to the race-conscious ethics that the Nazis sought to develop. According to the historian Gunnar Heinsohn, Hitler attempted to eradicate Jewish ethics in order to liberate the Nordic peoples from their conscience while killing and conquering. Had this attempt been successful, the Germans would have gained a clear strategic advantage over those who were still conscience-stricken while committing morally dubious acts framed as patriotic duties.[14] However, this was only one line of antisemitic argumentation utilized by the Nazis to stigmatize the Jews as a danger to the Germanic and Nordic cause. They also turned racist arguments on their head to blame the Jews in general and Jewish ethics in particular for discriminating between members of the Jewish race and others. Jews were blamed for granting the status of moral subjects only to members of their own race and conversely treating non-Jews arbitrarily as objects of unlimited cruelty and perversion. In fact, Nazi antisemitic activists accused Jews of doing to "Aryan Germans" what they themselves planned to do, or were already doing, to German Jews. The leading Nazi ideologist, Alfred Rosenberg, and the editor of *Der Stürmer*, Julius Streicher, for example, were notorious for such attacks against the immoral, sexually perverted, cruel Jew.

Medical ethicist Martin Staemmler, meanwhile, insisted that racially conscious Germans should internalize the Nazi ideology through the development of a special Nazi racial conscience.[15] The Nazis considered the human conscience the essence of man's moral life, against which he measured his deeds. Therefore, they attempted to replace conscience based on mutual exchange, trust, and empathy with one's fellow human beings with a racial conscience directed only toward members of one's own race. Nordic morality listened to the inner voice of conscience as the moral guideline of unwritten laws, as Lothar Stengel von Rutkowski, lecturer in the Department of Race Hygiene, Cultural Biology, and Eugenic Philosophy at Jena University, put it. In all moral decisions man had to consult his conscience.[16] The Nazis themselves thought of their ideas as an "ethics of race."[17] As a consequence of the Nazis' attempt to introduce a race ethics, moral standards were no longer applied to all human beings, but only to those who belonged to the superior Nordic race or the racial community of pure-blood Aryans.

In order to understand what drove ordinary Nazis to act the way they did, one must reconstruct the Nazi's moral order. The Nazi worldview clearly had an ethical dimension, rooted in notions of an evolutionary ethic that legitimized the struggle for existence. National Socialist ethics was remarkably syncretic: it was composed of elements and value systems that contradicted each other yet were fused together by racial ideology. Nazi ideology imbued Germans with a sense of high moral purpose and ethnic righteousness. It adhered to a "higher" morality that transformed traditional values regarding human life.

The dignity of life, based on the idea of its unconditional worth, was replaced by the distinction between life worthy of being promoted and life unworthy of being lived.[18]

In essence, Nazi race ethics and morality provided a framework through which Nazi leaders could legitimize mass murder as moral and concurrently induce ordinary Germans to actively participate in, or at least acquiesce to, the violence. Converting ideologically justified mass murder into state politics is impossible without moral justification. It is not enough to justify obvious crimes as lawful through the introduction of appropriate laws. In order to make morally questionable deeds appear morally correct, the perpetrators must also be morally conditioned. Their moral reasoning and capacity to judge can be used to convince them that they merely did what had to be done for higher purposes, in this particular case the well-being and survival of the *Volksgemeinschaft* (racial community).

Nazi Racial Biology and Ethics

The Nazi ideology of racial biological ethics played on the antisemitic sentiments and prejudices of the time. Simultaneously, it converted the existing scientific-technological Zeitgeist of "everything is possible" into pragmatic, ideological politics. Thus the Nazis, like many other authoritarian regimes in the 1930s, aimed to inculcate the population with a new set of moral principles. The idealization of pioneering modernizing processes, irrespective of their moral implications, was part and parcel of the "social engineering" mentality that preceded Nazism.[19] Governments of the era frequently used the imagery of gardening to call for the elimination of the rapidly proliferating "human weeds" in order to stop the "degradation of the human race."[20] Zygmunt Bauman has famously argued that the Holocaust was based upon the technological and organizational achievements of advanced industrial society. "Exterminatory racism" was tied to conceptions of social engineering, to the idea of creating a new order by transforming the present one and eliminating those elements that could not be reshaped. Under Nazi rule, this idea took on a renewed urgency. The Reich Office for Peoples' Growth, Racial Improvement, and Nordicization (Reichsbehörde für Volkswachstum, Aufartung und Aufnordung) promoted the moral reorganization of German society through the creation of a racial people's community. The Nazis not only claimed the moral right to their actions but even considered the purification of the racially contaminated world their moral and patriotic duty. In their view, only racial purification by means of exclusion of racially inferior members would offer the people's community an opportunity to regain its strength and health. This would protect it from racial intermixing, which might place the German nation in peril.[21] A political program of complete extermination became possible because of the collaboration of science, technology, and bureaucracy in modernity.[22] The Nazis considered promoting the laws of life and nature in society their moral duty, claiming that what was perfect by the standards of natural sciences was also morally perfect.[23] Thus they essentially stated that the strong and racially healthy should rule, while the weak and unhealthy, that is, those who were unable to live on their own and care for themselves, were to be eradicated. This juxtaposition was supposedly in accordance with laws of nature and life, while the "dubious" moral and religious sentiments inherent in caring for the needy were not and should be discouraged.

According to Giorgio Agamben, National Socialism subscribed to a totalitarian concept of biopolitics, that is, the growing inclusion of human life in the rationale and the mechanisms of power.[24] The Nazis' "politically applied biology" conceptualized society as a body similar to a biological organism.[25] In a similar way, Hitler was praised for having created a "biological state" aimed at bridging the gap between politics and the biological laws of life.[26] Indeed, Bernhard Rust, the Prussian minister of science, arts, and people's education, proclaimed on 15 September 1933, "The knowledge of basic biological facts is the necessary condition for the renewal of the German people."[27] The director of the Office of Racial Policy, Walter Gross, meanwhile, was entrusted with the racial education of Germans, who should from that point on be encouraged to develop biased biological sentiments.

Nazi racial policy was intrinsically defined by racial eugenics, both positive and negative. Whereas positive eugenics was designed to promote and increase the number of racially pure "Aryans," negative eugenics attempted to prevent the procreation of racially impure and degenerate persons. Since negative eugenics alone would not change the value of the German race, Hitler's regime introduced a program of positive eugenics—deliberate breeding that would promote the stronger, healthier, and more competent individuals. The leading Nazi feminist, Sophie Rogge-Börner, insisted that *Aufnordung* and *Aufartung* (i.e., Nordicization and improving the race through racial hygienic measures) would simultaneously reduce the number of less valuable elements in society while increasing the number of the racially superior.[28]

The Nazi program of euthanasia was the most radical aspect of this campaign of race improvement. In its early stages it mainly involved the killing of the mentally ill and unfit. Gradually, though, the pool of individuals slated for destruction grew to include the socially unproductive, the ideologically dangerous, and the racially unfit. Emil Abderhalden, editor of the monthly *Ethik*, tried to ground ethics in biology even before the Nazis came to power, but he also discussed the limits of a purely biological understanding of ethics.[29] His hesitation in fully subscribing to Nazi racial biological ethics highlighted an awareness of the dangers of a Nazi type of ethics solely based on biology. He was convinced that replacing humanistic principles with biological standards would negatively affect the weak and needy, who were unable to care for themselves.[30]

Abderhalden foresaw dangerous consequences from a future Nazi euthanasia program. In contrast to the popular scientific view, according to which the eradication of racially inferior people served the interests of both a healthy national community and the racially degenerate, who would no longer suffer or live at the expense of the community, he wrote,

> A purely biological ethics can live with exterminating the weak. The idea of helping the weak and providing them with special care is something novel from a biological point of view. Some doubt that charity for the weak is reasonable. Some even fear that spending so much money on sheltering and maintaining mentally retarded or chronically ill individuals will lead to a lack of resources for the physically and mentally healthy. Instead, from a biological point of view it is possible to justify an ethics that promotes the killing of such unhealthy people for the sake of the healthy. Our entire inner conscience is repelled by such an ethics.[31]

As it turned out, Abderhalden's warning proved indeed correct. The Nazis believed that superior races had not only the right but also the obligation to subdue and even exterminate inferior races. They also believed that this struggle of races was consistent with the laws of nature and that the eradication of weaker races was the most effective means of safeguarding the purity of the Germanic "race." Unlimited humanity, by contrast, would lead to racial degeneration and ultimately become an economic burden. False humanity and exaggerated compassion had caused the disproportionate breeding of the racially inferior and unfit, with the consequence that ever-growing resources had to be spent keeping human beings alive who would otherwise never be able to survive on their own. While "healthy" Germans ready to work roamed the streets penniless, the state wasted huge sums of money on the hereditarily ill—individuals unworthy of life. Nature, free from artificial interference through irrational charity, would mercilessly exterminate them because it always supported the healthy and strong. This would bring the laws of human society back into agreement with the higher laws of nature and life.[32]

The German *völkisch* right blamed the Weimar Republic for having weakened the German "Aryan" race by tolerating procreation among people whom the Nazis and other far-right German parties considered genetically degenerate: people with physical and mental disabilities; criminals and social "misfits," including the homeless; allegedly promiscuous women; and alcoholics.[33] Nazi racial politics claimed to be in harmony with the laws of life and nature. Opposed to human intervention against the apparent cruelty of natural selection, racial politics was meant to eliminate the weak, those unable to prevail in the struggle for life. It considered the real barbarity to be the unnatural preservation of the weak. This supposed misbalance needed to be corrected through a responsible racial demographic policy that would return natural selection to its proper place. The once again unhindered laws of life would help the strong and healthy to succeed over the weak, unhealthy, and racially inferior, who would be exterminated by natural selection.[34] Against a culture guided by humanism and charity that had irresponsibly diminished nature's ability to eradicate the inferior, Nazi racial politics hoped to ensure that the basic laws of biology would have an appropriate cultural impact, favoring the racially valuable and destroying the racially inferior. Modern technology and medicine would protect those unfit for life from the consequences of natural selection.[35]

To this end, leading Nazi ideologues sought to cultivate the moral education of ordinary Germans, centered on biology. Particularly interesting in this respect is the Nazi racial-medical discourse of *Volksgesundheit* (people's health) and eugenics. The center of moral attention shifted from the individual body to the biologized body politic. The medical profession took the lead in the ideological struggle for cultural influence by shifting its focus from the health and well-being of the individual to the health of the "national body." Now German physicians were instructed by the Nazi Party that caring for an individual human counted for less than caring for the future of the German people. One of the key principles of medical practice in Germany was the distinction between "care" and "preventive care." While care (*Fürsorge*) referred to genuine concern for the individual sick person, preventive care (*Vorsorge*) referred to preventive medicine for the benefit of the *Volksgemeinschaft*, which Nazi medicine clearly prioritized. In 1932 the National Socialist Doctors' Association (Nationalsozialistischer Ärztebund) officially announced this change of focus. In fact, the foundations of a transformation in medical thinking had been set in motion before the Nazis came to power. In 1933 Dr. Gerhard Wagner, who soon became the leader of the entire German medical profession, contrasted the Nazi medical ideal with that of its predecessors. He emphasized health leadership rather than health care, preventive rather than curative medicine, racial rather than individual hygiene. Furthermore, he insisted that health leadership implied distinguishing between valuable life and life "not worth living."[36]

Eugenics attempted to create a more prosperous society by maximizing health and efficiency. Eugenic projects entailed the "normalization" of the individual, his physical characteristics and behaviors, while approaching aberrations as pathological differences in need of containment or elimination.[37] Nazi eugenicists combined this guiding principle with notions of racial purification. For them, the radical separation of superior and inferior races was the only possible way to prevent racial intermixing, which they considered contagious. The *internal* threat lurked in intermarriage between "Aryan" Germans and members of inherently inferior races: the offspring of these marriages supposedly diluted the superior characteristics of German blood, thus weakening the race in its struggle for survival. For the Nazis, the survival of the race depended on its ability to reproduce, multiply, and maintain the purity of its blood. They instructed Germans, that is, members of the Nordic race, to preserve the unique "racial" characteristics with which "nature" had equipped them in order to succeed in the struggle for survival.[38] The aim of creating racially pure Aryans eventually took precedence over the most fundamental ethical issues in medicine and the medical profession. Thus the racially fit were encouraged to marry and have as many children as possible, and couples who planned to get married were advised to exchange health certificates with each other to ensure their racial compatibility.

The individual was seen as a temporary vessel for storing racial value.[39] According to racial thinking, the value of human beings depended upon the racial category they belonged to. As mere individuals, however, they were considered worthless.[40] Nazi racial thinking established that humans' racial makeup determined their chances of developing and making a useful contribution to society. This composition could be suppressed or promoted but never exchanged by or replaced for another set of racial characteristics. Politics would, therefore, work with the respective racial composition at hand.[41]

Racial belonging determined an individual's life chances. Belonging to the superior Nordic race by definition qualified one for life, whereas membership in the supposedly inferior and potentially dangerous Jewish race doomed one to humiliation and destruction. According to Ernst Lehmann's comprehensive survey of medical practice in the new National Socialist Germany, written in 1934, Germans were no longer willing to spend disproportionately more on the racially inferior and degenerate than they did on the healthy and racially valued.[42] Maintaining race purity was important, according to Hitler and other Nazi leaders, because mixing with other races was a degeneration of the race and would inevitably lead to its demise.[43] Hitler believed that Germans were members of a superior "Aryan race" within a hierarchy of unequal races, a race threatened by dissolution from within and without. In defense of the race, Hitler and the Nazis utilized a secular conception of original sin that they specified as sin against blood and race.[44] The dissolution of the people's community was seen as the result of Jewish contamination. Exterminating them as the carriers of this disease, therefore, appeared as the only possible solution. Thus the Nazi Party saw itself as a "movement of convalescence against the unhealthy," to quote racial psychologist Erich Rudolf Jaensch.[45]

In fact, the idea of the Jews as a threatening "other" contaminating the "racial purity" of the German nation was a defining theme of Nazi antisemitic discourse. Jews were considered a medical problem to be eradicated by the German nation and were commonly referred to as parasites and bearers of numerous diseases that might spread to the rest of society.[46] As Nazi Minister of Propaganda Joseph Goebbels stated in a speech to the party faithful in June 1943, "The disappearance of all Jews from Europe is not an ethical matter, but one of state security. The Jew will always act the way his inner essence and racial instinct tell him to. He simply cannot help but act this way. Just as the potato beetle destroys the potato fields, the Jew destroys states and peoples. There is only one remedy: the radical extinction of the danger itself."[47] In dealing with the Jews, those who were put in charge of the Final Solution of the Jewish

Question were instructed not to apply Aryan humane standards. Jewish food rations were to be based on their efficiency and performance and no effective distinction was to be made between their treatment and that of farm animals.[48]

Social Darwinism versus Christian Charity: Nazism as a Racial Religion

Social Darwinism was a part of German public discourse long before the Nazis came to power; it can actually be considered the moral antecedent of Nazi genocide.[49] Social Darwinism suggested that natural selection and the struggle for the survival of the fittest were the basic forces that governed human society. Humans would remain subject to the laws of nature, and the struggle for life and existence would ultimately lead to the extermination of "inferior" races. The assumption that everyone with a human face was equal would contradict experience and biology.[50] Instead of ignoring basic differences between human beings, Nazism would highlight that human rights, duties, and responsibilities depended on hereditary characteristics.[51]

Social Darwinism blamed the Judeo-Christian moral tradition for standing in the way of a world ruled by the strongest and fittest. Christianity's instruction to love one's enemy was blamed for having destroyed the biological instinct for racial health and morality. Under Christianity's influence, mankind had developed institutions that contradicted race, instinct, and reason. As the Nazi theoretician Karl Pintschovius argued in 1940, Christian ethics and charity were to be blamed for having suppressed nature while undermining natural instincts through feelings of guilt and shame. According to him, unconditional care for the weak violated the natural order in which only the strong were preordained to survive.[52] Similarly, in *Mein Kampf* Hitler wrote, "A stronger race will drive out the weak, for the vital urge in its ultimate form will, time and again, burst all the absurd fetters of a so-called humanity of individuals, in order to replace it with the humanity of Nature that destroys the weak to give his place to the strong."[53]

In particular, Nazi theoreticians argued that the universal Jewish ethics of mutual love and empathy was a baleful influence and hindrance to the development of a master morality. The Nazis were eager to replace "racially ignorant" universal ethics with the selective ethics of racial superiority and inferiority. They valued belonging to the Aryan race at the expense of belonging to the human race. New SS recruits were informed that granting human rights to everybody would amount to the irresponsible promotion of the weak at the expense of the strong. Instead, they were trained to ignore, deny, and overcome their personal feelings if these feelings contradicted the attitudes they were expected to develop in order to fulfill their duties as "political soldiers" or "racial warriors."[54] Similarly, in order to give his men the impression that he was a role model who had unconditionally fulfilled his duty, Rudolf Hoess insisted that he had to suppress all humane feelings of mercy when, for instance, confronted with mothers and their crying children about to go to the gas chambers.[55]

Paradoxically, although the Nazis attempted to replace traditional Christian ethics with a secular ethics, they continuously saw themselves as a part of the Christian tradition. While Nazi ideologists opposed what they perceived as Jewish-influenced Christianity, they also attempted to establish Nazism as the German-Aryan version of Christianity.[56] German youth were taught to believe in the Führer as the new God, exemplified in the "invocations" that they were instructed to recite before meals:

> Führer, my Führer, bequeathed to me by the Lord,
> Protect and preserve me as long as I live!
> Thou hast rescued Germany from deepest distress,
> I thank thee today for my daily bread.
> Abideth thou long with me, forsake me not,
> Führer, my Führer, my faith and my light![57]

Based on their belief that the Jews posed a threat to Germany, the Nazis saw the persecution of the Jews as an act of self-defense. Only if they struck preemptively before they themselves became the target of a supposed Jewish aggression would they be able to win this struggle and survive as a race. Viewed in this light, Himmler's moral code could be legitimized as justifiable killing in self-defense, striking first, before the Nordic race became the target of aggression and extermination. Nazi perpetrators claimed to have acted free from selfish motives. The Nazis insisted that their anti-Jewish politics served the interests of the German people. They gave priority to communal duties and not to individual rights, to the common good and not to individual interests, and they used this prioritization as proof that their politics were morally justified. The common public interest was supposed to precede self-interest and duties were to come before rights.[58] Service to the people's community was considered not only a duty but a unique privilege of belonging to the Nordic race. Those few entitled to membership in the master race were not to take their membership for granted but rather to prove their worth through their attitudes and deeds.[59]

Thus ordinary Germans were turned into "agents of history" as part of a world historical enterprise. They were offered a stage on which to act and perform in a way that would supposedly allow them to transcend their earthly limitations and become part of the eternal flow of history.

Belonging to the supposedly racially superior Aryan German *Volksgemeinschaft* elevated its members regardless of their individual capabilities. Members of the Nordic race were told that they were part of a hereditary chain that they should not break through racial misbehavior. At the same time, however, their membership in the Nordic master race was considered a probation that had to be passed and could also be forfeited. Pure-blooded Aryans had to prove through racially conscious behavior that they were worthy of belonging; inappropriate behavior or a visible lack of enthusiasm for the Nazi cause could turn them into unreliable and devious fellow citizens.

Nazi ethics worked within a concept of "higher morality" that prescribed followers to dedicate their lives to a "higher cause" and to turn this dedication into a personal commitment. Due to their ideological conditioning, they took their belonging to the Nordic race more seriously than their individuality. Hans Globke and Wilhelm Stuckart, who wrote the commentary on the Law for the Protection of the German Blood (1936) thus stated, "National Socialism has put the people directly into the center of thought, faith, and will of creativity and life. . . . National Socialism does not recognize a separate individual sphere which, apart from the community, is to be painstakingly protected from any interference by the state. The moral personality can prove itself only within the community. . . . [T]he position of the individual is no longer determined in terms of the person as such, but in terms of the community."[60]

But personal interests were not completely absent from Nazi perpetrators' mindsets. They also learned that it would be ignorant, irresponsible, and simply stupid not to act as law-abiding citizens, good Germans, and members of families for whose well-being they were responsible. The message that Nazi ideology tried to convey was that those Germans who acted in accordance with racial laws, supported the fatherland, and cared for their families would be acknowledged and rewarded as reliable people's comrades (*Volksgenossen*). It offered not-to-be-missed opportunities and advantages, such as a communal sense of belonging, career opportunities, access to Jewish property, and a feeling of superiority.[61] Many ordinary Germans who were neither Nazis nor held prominent political or military positions knew, or at least suspected, what was being done to the Jews, though few protested or tried to intervene. Why so many Germans acquiesced to, looked the other way, or sometimes eagerly participated in the persecution of the Jews remains the subject of sophisticated scholarly discourse with a variety of explanations and answers offered.[62] Nazi morality effectively meant a war to the death between the races. The Nazis' racial version of the "survival of the fittest" theory categorically rejected the fundamental idea of human coexistence. In effect, well-established biological laws of life were used to inform state policies concerned with family planning and to address the supposed burdens imposed on society by those with "lives not worth living."[63]

At the very root of Hitler's worldview was the glorification of "nature" and the idea that progress could be attained only through a ceaseless and merciless struggle for survival among humans. In the spirit of social Darwinism, Hitler wrote in *Mein Kampf*, "The battle among humans arises less from inner aversion than from hunger and love. In both cases, Nature looks on calmly, with satisfaction, in fact. In the struggle for daily bread, all those who are weak and sickly or mediocre succumb, while the struggle of the males for the females grants the right or opportunity to propagate only to the healthiest. And struggle is always a means for improving a species' health and power of resistance and, therefore, a cause of its higher development."[64] Acting free from the interference of any religious or moral considerations was deemed moral and noble. The weak were no longer to be protected but rather left to their fate as determined by the laws of nature and God's will.

This ideology was introduced as the will of an infallible leader who supposedly acted in line with destiny and divine providence. It asked for an unconditional belief in both the Nazi cause and Hitler, who embodied this moral crusade. It gave its followers the impression that they were not acting from selfish motives but rather doing their duty or even fulfilling God's eternal will. The Führer was frequently presented as the incarnation of all that defined Aryan German virtues: unconditional reliability, steadfastness, strength, fidelity, and selfless dedication to the *Volksgemeinschaft* and the common good.[65] Moral and religious concerns with respect to intervention in the rights of the hereditarily ill were considered superfluous because their weakness merely confirmed the sacred law of natural selection.[66] According to this law, the extermination of those unfit, and therefore unworthy of life, was in the interest of life and humankind.[67] Promoting this law would necessarily fulfill the will of the Creator, who was supposedly in favor of healthy humans, whereas the empty rhetoric of acting in the name of humanity would have betrayed his will. For the Nazis, nature was truly the only source of morality.

Yet the Nazis' racial thinking also exposed a strong religious component, even as it outlawed mercy and proscribed empathy. Instead of subscribing to a set of universal ethics that granted all humans equality before God and the entitlement to empathy, care, and mercy, the religious elements of Nazi ideology singled out the German people as an elite, members of the Nordic race who should be proud of belonging to the "community of the chosen." In this respect, Nazis saw themselves as the inheritors, not the opponents, of the Christian tradition. While claiming to be the chosen people or acting in line with God's will, they sought to religiously justify their race politics. After coming to power, the Nazi Party created a separate church, the German Christian Church, which combined aspects of Christian tradition with paganism. For members of this church, the Germans were a new chosen people of God as well as a community of blood, and Christ was a Nordic pagan martyr.[68] Guidelines for the "German Christians" put it as follows: "Through God's creation we have been put directly into the community of blood and fate of the German people and as the bearers of this fate we are responsible for its future."[69] Hitler himself made a point of stressing that he had been chosen by destiny to fulfill God's eternal will, as can be seen at the end of a messianic speech he gave on 10 February 1933 in the Berlin Sportpalast, emphasizing the divine destiny of the German national community: "I cannot free myself from belief in my people, can't get away from the conviction that this nation will once again arise, can't distance myself from the love of this, my people, and hold as firm as a rock to the conviction that sometime the hour will come when the millions who today hate us will stand behind us and, with us, will welcome what has been created together, struggled for under difficulty, attained at cost: the new German Reich of greatness and honor and strength and glory and justice. Amen."[70]

Nazi Perpetrators in Hindsight: Justification and Defenses

Many Nazi perpetrators denied that they could be held morally accountable and legally responsible for their actions, committed in an environment in which their moral judgment had clearly been distorted. With few exceptions, those who had to face trial after the defeat of Nazism did not appear to have suffered from a guilty conscience. Instead, they protested their innocence on account of their limited knowledge of the atrocities, the camps, the medical experiments, and the extermination program against the European Jews. Some of them justified the medical and euthanasia experiments conducted on humans in the name of scientific progress.[71] While shifting responsibility to higher ranks, anonymous structures, and institutions, many defendants claimed to have only been following orders and performing their duty to the fatherland. Whether they truly believed such rationalizations or not, many Nazi perpetrators retreated into the safe haven of the narrowly defined concept of responsibility and professional duty.

The defendants typically claimed mitigating circumstances, insisting that they had personally harmed nobody, that they were free from antisemitic hatred and racial prejudice, and that they did not care about politics and ideology. On the contrary, they reasoned that because of their lower standing in the Nazi ranking system they only had limited knowledge and, therefore, could not be held accountable for what went beyond their expertise and duty. Many of them also claimed that they were not in a position to question orders, to consult their conscience, or to concern themselves with moral considerations since they knew that they were acting within the framework of German law and jurisdiction. Ideological indoctrination might have been the starting point for moral indifference and a marked lack of empathy for the victims of racial hatred and prejudice, but so, too, was indifference toward ideology and politics.

Most Nazi physicians believed their actions were morally defensible. Rather than apologize, they attempted instead to explain and justify their actions, often in explicitly moral terms. Though gestures of repentance and pity—however insincere—toward their victims might have lessened their sentences, many of the accused Nazi perpetrators were unable to take this step. Instead they insisted that what they did was just and right. This moral stance suggests that for the accused, escaping the inner dialogue with their own conscience took precedence over all other concerns, including their own lives. Until the very end, when the death sentences were being handed down, the most important aim for some of the perpetrators was to live at peace with themselves. For example, Karl Brandt, who was in charge of both the euthanasia program and later medical experiments in the camps, refused to condemn his actions for as long as possible, in spite of his apparent knowledge and approval of the extermination program. He did not plead ignorance of ethics as such but did constantly allude to medical ethics and the Hippocratic tradition in defense of his actions. Brandt constructed an image of himself as a dutiful doctor who acted in the best interest of his patients by sparing them unnecessary, unbearable pain and suffering. A more rigorous defense might have carried with it an implication of culpability, thereby shattering the delusion that he was not responsible for the brutal murders and that he had not broken his professional oath. Ironically, in his obituary he was referred to as a doctor driven by respect for life.[72] Although Brandt accepted personal responsibility for the euthanasia program, he denied that it was linked to the extermination of Jews in Poland.[73] He insisted that "the only rationale for the euthanasia program had been to free handicapped and incurably ill patients from suffering." He continued, "I do not feel that I am being incriminated. I am convinced that I can bear responsibility for what I did . . . before my conscience. I was motivated by utterly humane considerations. I never had any other intention. I just believed that those miserable creatures, the painful lives of these creatures, were to be shortened."[74]

Brandt and perpetrators like him also attempted to defend themselves by redefining the criminal deeds that they committed or oversaw not as murder but as a measure of racial hygiene implemented to save the fatherland from racial decline. Consequently, they were able to portray themselves as reliable professionals who fulfilled their patriotic duty. Indeed, Nazi ideology had introduced the policy of racial extermination as population policy founded on the individual's duty to his community. This nexus suggests that at least some Nazi perpetrators were sincerely committed to working in the best interest of the national community, as they saw it.

Nevertheless, the basic question remains as to whether Nazi perpetrators who participated in the mass murder of the Jews comprehended that their actions were criminal and immoral.[75] Seemingly, some of them were aware that they had crossed a line by committing murder on a mass scale for ideological reasons. However, the process of rationalization they were engaged in allowed them to live out their alleged roles as actors on the stage of racial history while remaining decent human beings in their capacities as husbands, fathers, friends, and colleagues.

Thus Nazi ideology spared its followers the trouble of having to make decisions for which they would eventually be held accountable; they were merely asked to do as they were told. No justification was necessary as long as there was reason to believe that they would be absolved of responsibility for their actions. They believed that either Germany would win the war—in which case, as usual, the winner would not have to face difficult questions—or that nobody would find out about the death camps, as there would be no survivors or eyewitnesses to testify. The use of euphemistic language when discussing the organizational, practical, and economic aspects of the Final Solution of the Jewish Question, even among the architects of the Holocaust, demonstrates intent on the part of the Nazis to keep the genocide secret.[76] Even the minutes of the notorious Wannsee Conference in January 1942 did not overtly refer to the extermination of the Jews, speaking instead of the "evacuation of the Jews to the East": "In the course of the Final Solution, the Jews should be brought under appropriate direction in a suitable manner to the East for labor utilization. Separated by sex, the Jews capable of work will be organized in large labor columns to build roads, in the course of which a majority would undoubtedly fall away through natural decline. The residual element, which undoubtedly represents the hard core of Jewry, would have to be treated accordingly, since these Jews have been shown in the light of history to be the dangerous Jews, the people who could rebuild Jewish life."[77]

Nazi ideology succeeded in mobilizing a section of the German population by effectively utilizing the notion of society as a diseased organism that could be turned into a healthy body through racial biopolitics and the promotion of a multifaceted antisemitism based on pseudoscientific racial and ethical arguments. All that was left to do was to turn this racial discourse into state politics and treat the Jews and other groups whom the Nazis considered inferior as subhumans unworthy of life.

Biological ethics did not grant each person the same rights but rather subscribed to the principle *to each his own*. It considered the assumption that all human beings were equal a violation of nature's laws: there was no mankind but only unique human beings with various backgrounds defined by natural laws of blood and race. Humanity and race were seen as incompatible universal value systems. Nazi ethics accepted the principle of humanity only insofar as it did not diminish the quality of the race. Consequently, human rights were not universal but could only be granted to members of the race community; people were not entitled to human rights by birth, as claimed by liberal humanism, but only by biological high value.[78] The well-being of one's own race would clearly come before the well-being of mankind. There was a clear-cut distinction between those who did not belong and those who could expect to be treated morally, to be cared for when in need, and to be looked on with empathy as members of the German racial community.

Social Darwinist ethics penetrated German culture to such an extent that the Judeo-Christian moral tradition was effectively superseded. In line with contemporary eugenics, moral relativism redefined the value of life and death. In Nazi Germany the cross-cultural moral principle of humanism was restricted to members of the Germanic racial community. It shifted from an unconditional, all-embracing rule to a selective principle used to justify the unequal treatment of human beings based on their racial belonging.

The Nazis claimed that the Jewish conception of morality differed from that of non-Jews. They valued belonging to the Aryan race at the expense of belonging to the human race. Therefore, the Jews could not expect to be treated by the same moral standards as the German people were. The Nazis were very clear that there would be no space for Jews in the future new world order dominated by Aryans. Their antibourgeois attitude not only was directed against unconditional empathy and charity toward all human beings—no matter what their cultural or racial characteristics—but also militated against exaggerated pity for the weak and mentally or terminally ill. The Final Solution was, ultimately, the logical conclusion of the Nazis' moral philosophy.

Notes

1. Walter Gross, "Politik und Rassenfrage," *Ziel und Weg*, 3, no. 14 (1933): 413. On Walter Gross, see Roger Uhle, "Neues Volk und reine Rasse: Walter Gross und das Rassenpolitische Amt der NSDAP" (PhD diss., University of Aachen, 1999).
2. For the recent debate on Nazi perpetrators, see Donald Bloxham, "Organized Mass Murder: Structure, Participation, and Motivation in Comparative Perspective," *Holocaust Genocide Studies* 22, no. 2 (Fall 2008): 203–45; Michael Mann, "Were the Perpetrators of Genocide 'Ordinary Men' or 'Real Nazis'? Results from Fifteen Hundred Biographies," *Holocaust Genocide Studies* 14, no. 3 (Winter 2000): 331–66; Harald Welzer, *Täter: Wie aus ganz normalen Menschen Massenmörder werden* (Frankfurt: Fischer Verlag, 2005); Leon A. Jick, "Method in Madness: An Examination of the Motivations for Nazi Mass Murder," *Modern Judaism* 18, no. 2 (May 1998): 153–72.

3. Hannah Arendt, "Eichmann in Jerusalem," in *The Portable Hannah Arendt* (New York: Penguin Books, 2000), 373. For a summary of the critical debate on Arendt's Eichmann book, see Garrath Williams, ed., *Hannah Arendt: Critical Assessments of Leading Political Philosophers* (London: Routledge, 2006), 2:244–72, 299–328; David Cesarani, *Becoming Eichmann: Rethinking the Life, Crimes, and Trial of a "Desk Murderer"* (Cambridge MA: Da Capo Press, 2006), 343–56.
4. Götz Aly, *Hitlers Volksstaat: Raub, Rassenkrieg und nationaler Sozialismus* (Bonn: Bundeszentrale für politische Bildung, 2005), 14, 22. On Nazi ideology and the "political soldier," see Frank Werner, "'Hart müssen wir hier draußen sein': Soldatische Männlichkeit im Vernichtungskrieg 1941–1944," *Geschichte und Gesellschaft* 34, no. 1 (2008): 5–40; Mark Mazower, "Military Violence and National Socialist Values: The Wehrmacht in Greece 1941–1944," *Past & Present* 134, no. 1 (February 1992): 129–58.
5. Heinrich Himmler's Posen speech of 4 October 1943, available at the Holocaust History Project, accessed 20 September 2010, http://www.holocaust-history.org/himmler-poznan/speech-text.shtml.
6. Bernd Wegner, *The Waffen-SS: Organization, Ideology and Function* (London: Blackwell, 1990).
7. With respect to the split self ("the Auschwitz-Self") of the Nazi doctors, see Robert Jay Lifton, *The Nazi Doctors: Medical Killing and the Psychology of Genocide* (New York: Basic Books, 1986), 418–65; Leni Yahil, "The Double Consciousness of the Nazi Mind and Practice," in *Probing the Depths of Antisemitism: German Society and the Persecution of the Jews, 1933–1941*, ed. David Bankier (New York: Berghahn Books, 2001), 36–53.
8. Christopher Browning, *Ordinary Men: Reserve Battalion 101 and the Final Solution in Poland* (New York: Harper Collins, 1992); George C. Browder, "Perpetrator Character and Motivation: An Emerging Consensus?" *Holocaust Genocide Studies* 17, no. 3 (Winter 2003): 480–97; Jürgen Matthäus, "Controlled Escalation: Himmler's Men in the Summer of 1941 and the Holocaust in the Occupied Soviet Territories," *Holocaust Genocide Studies* 21, no. 2 (Fall 2007): 218–42.
9. Mario Zeck, *Das schwarze Korps: Geschichte und Gestalt des Organs der Reichsführung SS* (Tübingen, Germany: Max Niemeyer Verlag, 2002).
10. "Mass und Wert," *Das Schwarze Korps*, 16 December 1943.
11. Hermann Langbein, *People in Auschwitz* (Chapel Hill: University of North Carolina Press, 2004); Primo Levi, *If This Is a Man: Remembering Auschwitz* (New York: Summit Books 1986); Robert Antelme, *The Human Race* (Evanston IL: Marlboro Press, 1998).
12. David H. Jones, "Freud's Theory of Moral Conscience," *Philosophy* 41, no. 155 (1966): 34–57.

13. See Hannah Arendt, "From the Life of the Mind," in *The Portable Hannah Arendt*, 408–13.
14. Gunnar Heinsohn, *Warum Auschwitz? Hitlers Plan und die Ratlosigkeit der Nachwelt* (Reinbek bei Hamburg: Rowohlt, 1995), 18. In his book, Heinsohn provided a list of forty-two conceptual approaches to the Holocaust. For a similar list, see Gavriel D. Rosenfeld, "The Politics of Uniqueness: Reflections on the Recent Polemical Turn in Holocaust and Genocide Scholarship," *Holocaust and Genocide Studies* 13, no. 1 (Spring 1999): 28–61.
15. Martin Staemmler, "Aufgaben und Ziele der Rassenpflege," *Ziel und Weg* 3, no. 14 (1933): 415–22.
16. Lothar Stengel von Rutkowski, *Von Allmacht und Ordnung des Lebens* (Berlin: Nordland Verlag, 1942), 18; "Moral-kritisch betrachtet," *Das Schwarze Korps*, 31 August 1944.
17. One of the most influential "racial thinkers" behind the Nazi movement was Fritz Lenz. His treatise "The Renewal of Ethics"—first published in 1917 and reprinted in 1933 under the title "Race as a Principle of Value"—regarded the prosperity of the Aryan race as the ultimate moral goal. Fritz Lenz, *Die Rasse als Wertprinzip: Zur Erneuerung der Ethik* (Munich: Lehmann 1933).
18. Peter Fritzsche, *Life and Death in the Third Reich* (Cambridge MA: Belknap Press of Harvard University Press, 2008).
19. Jeffrey Herf, "The Engineer as Ideologue: Reactionary Modernists in Weimar and Nazi Germany," *Journal of Contemporary History* 19, no. 4 (October 1984): 631–48; Richard L. Rubenstein, *The Cunning of History* (New York: Harper Collins, 1987).
20. F. C. S. Schiller, "Die Eugenik als sittliches Ideal," *Archiv für Rassen-und Gesellschaftsbiologie* 24 (1930): 342.
21. Alexandra Przyrembel, *"Rassenschande": Reinheitsmythos und Vernichtungslegitimation im Nationalsozialismus* (Göttingen, Germany: Vandenhoeck & Ruprecht, 2003).
22. Zygmunt Bauman, *Modernity and the Holocaust* (Ithaca NY: Cornell University Press, 1989).
23. Viktor Franz, "Aufsteigende Entwicklung," *Rasse* 2, no. 5 (1936): 175–76.
24. Giorgio Agamben, *Homo sacer: Die souveräne Macht und das Leben* (Frankfurt: Suhrkamp, 2002), 127. For biopolitics and Nazism, see also Simona Forti, "The Biopolitics of Souls: Racism, Nazism, and Plato," *Political Theory* 34, no. 1 (2006): 9–32.
25. Alfred Strassburg, "Volksstaat, Weltstaat und nationalsozialistische Weltanschauung," *Ethik* 10 (May–June 1934): 311.
26. G. Timmer, "Die Berufung des Arztes," *Ziel und Weg* 5, no. 6 (1935): 139.

27. Ministry of Science, Arts, and People's Education, proclamation, 15 September 1933, reprinted in *Der Biologe* 3, no. 14 (1933): 344.
28. Sophie Rogge-Börner, *Nordischer Gedanke und Verantwortung* (Leipzig: Klein, 1930), 81. On Rogge-Börner, see Liliane Crips, "'National-feministische' Utopien: Sophie Rogge-Börner und die 'Die deutsche Kämpferin' 1933–1937," *Feministische Studien* 8, no. 1 (May 1990): 128–37.
29. On Abderhalden, see Andreas Frewer, *Medizin und Moral in Weimarer Republik und Nationalsozialismus: Die Zeitschrift "Ethik" unter Emil Abderhalden* (Frankfurt: Campus, 2000).
30. Emil Abderhalden, "Sind ethische Grundzüge wandelbar?" *Ethik* 5 (May 1929): 413–14.
31. Abderhalden, "Sind ethische Grundzüge wandelbar?," 413.
32. Walter Gross, "Die ewige Stimme des Blutes im Strome deutscher Geschichte," *Ziel und Weg* 3, no. 10 (1933): 257–60.
33. Elof Axel Carlson, *The Unfit: A History of a Bad Idea* (Woodbury NY: Cold Spring Harbor Laboratory Press, 2001). For a Nazi perspective, see Heinrich Wilhelm Kranz, *Die Gemeinschaftsunfähigen* (Giessen, Germany: K. Christ, 1940).
34. Alfred Mjøen, "Die biologische Lebensauffassung und Sippenpflege," in *Kultur und Rasse: Otto Reche zum 60. Geburtstag*, ed. Michael Hesch and Günther Spannaus (Munich: J. F. Lehmann, 1939), 131–39.
35. Walther Brunk, "Nationalsozialistische Erbpflege, Blutmaterialismus oder göttliches Naturgesetz?" *Der Schulungsbrief* 6, no. 3 (1939): 356.
36. Robert N. Proctor, *Racial Hygiene: Medicine under the Nazis* (Cambridge MA: Harvard University Press, 1988), 73.
37. Edward Ross Dickinson, "Biopolitics, Fascism, Democracy: Some Reflections on Our Discourse about 'Modernity,'" *Central European History* 37, no. 1 (2004): 4.
38. "SS-Vollblut?" *Das Schwarze Korps*, 25 March 1937.
39. H. Finck, "Volksgesundheit und Liebesleben," *Ziel und Weg* 8 (1934): 287–94.
40. Finck, "Volksgesundheit und Liebesleben," 292.
41. Walter Gross, "Rasse und Weltanschauung," *Der Weltkampf* 171 (1938): 102.
42. Ernst Lehmann, *Biologischer Wille: Wege und Ziele biologischer Arbeit im neuen Reich* (Munich: J. F. Lehmann, 1934), 35.
43. Otto Reche, "Sippenschande," *Rasse* 8, no. 7 (1941): 296.
44. Adolf Hitler, *Mein Kampf* (Boston: Houghton Mifflin, 1971), 249.
45. Erich Rudolf Jaensch, *Der Gegentypus.:Psychologisch-anthropologische Grundlagen deutscher Kulturphilosophie ausgehend von dem, was wir überwinden wollen* (Leipzig: Barth, 1938), xxxii.
46. David Norman Smith, "The Social Construction of Enemies: Jews and the Representation of Evil," *Sociological Theory* 14, no. 3 (November 1996): 203–40.

47. Joseph Goebbels, *Der steile Aufstieg: Reden und Aufsätze aus den Jahren 1942/43* (Munich: Franz Eher, 1944), 301.
48. Christopher R. Browning, *Der Weg zur "Endlösung": Entscheidungen und Täter* (Bonn: Dietz Verlag, 1998), 48.
49. Richard Weikart, *From Darwin to Hitler: Evolutionary Ethics, Eugenics, and Racism in Germany* (Basingstoke UK: Palgrave Macmillan, 2006).
50. Walter Gross, "Der Rassegedanke des Nationalsozialismus," *Der Schulungsbrief* 2, no. 2 (1934): 9.
51. Gross, "Der Rassegedanke des Nationalsozialismus," 16.
52. Karl Pintschovius, "Die Wiedergeburts des Instinkts," *Das Reich* 13 (1940): 17–18.
53. Hitler, *Mein Kampf*, 132. See also Richard Weikart, *Hitler's Ethic: The Nazi Pursuit of Evolutionary Progress* (Basingstoke UK: Palgrave Macmillan, 2009).
54. Kameradschaft, *SS-Leitheft* 5, no. 9b (February 1940): 1–3; Claudia Koonz, *The Nazi Conscience* (Cambridge, MA: Harvard University Press, 2005), 221–52.
55. Martin Broszat, ed., *Kommandant in Auschwitz: Autobiographische Aufzeichnungen des Rudolf Höss* (Munich: Deutscher Taschenbuch Verlag, 1996), 198.
56. Kevin P. Spicer, ed., *Antisemitism, Christian Ambivalence, and the Holocaust* (Bloomington: Indiana University Press, 2007).
57. Quoted in George L. Mosse, ed., *Nazi Culture: A Documentary History* (New York: Schocken Books, 1966), 241.
58. Emil Aberhalden, "Gemeinnutz geht vor Eigennutz," *Ethik* 12 (September–October 1935): 1.
59. Otto Eberhard, "Die bindende Norm im politischen Sein des Menschen," *Ethik* 10 (May–June 1934): 296.
60. Wilhelm Stuckart and Hans Globke, "Civil Rights and the Natural Inequality of Man," in Mosse, *Nazi Culture*, 328, 330.
61. Aly, *Hitlers Volksstaat*.
62. Robert Gellately, *Backing Hitler: Consent and Coercion in Nazi Germany* (Oxford: Oxford University Press, 2002); Peter Longerich, *"Davon haben wir nichts gewusst!" Die Deutschen und die Judenverfolgung 1933–1945* (Munich: Bertelsmann Verlag, 2007).
63. Gross, "Die ewige Stimme des Blutes."
64. Hitler, *Mein Kampf*, 285.
65. Thomas Schirrmacher, *Hitlers Kriegsreligion*, 2 vols. (Bonn: Culture and Science, 2007).
66. Gerhard Wagner, "Rasse und Volksgesundheit," *Ziel und Weg* 4, no. 18 (1934): 683.
67. Wagner, "Rasse und Volksgesundheit."
68. Susannah Heschel, *The Aryan Jesus: Christian Theologians and the Bible in Nazi*

Germany (Princeton NJ: Princeton University Press, 2008); Richard Steigmann-Gall, *The Holy Reich: Nazi Conceptions of Christianity, 1919–1945* (Cambridge: Cambridge University Press, 2004).

69. Quoted in Mosse, *Nazi Culture*, 241.
70. Ian Kershaw, *Hitler, 1889–1936: Hubris* (New York: W. W. Norton, 2000), 454.
71. Michael S. Bryant, *Confronting the "Good Death": Nazi Euthanasia on Trial, 1945–1953* (Boulder: University Press of Colorado, 2005); Paul J. Weindling, *Nazi Medicine and the Nuremberg Trials: From Medical War Crimes to Informed Consent* (Basingstoke UK: Palgrave Macmillan, 2004).
72. Ulf Schmidt, *Karl Brandt: The Nazi Doctor, Medicine, and Power in the Third Reich* (London: Continuum, 2007), 398.
73. Schmidt, *Karl Brandt*, 365.
74. Interrogation of Karl Brandt, 1 October 1945, quoted in Schmidt, *Karl Brandt*, 370.
75. Raul Hilberg, *Perpetrators, Victims, Bystanders: Jewish Catastrophe, 1933–1945* (New York: HarperCollins, 1993).
76. Wannsee Conference House, *The Wannsee Conference and the Genocide of the European Jews: Guide and Reader* (Berlin: Wannsee Conference House, 2007), 106, 116–18.
77. I have used Raul Hilberg's translation of the original Wannsee Conference documents. See Raul Hilberg, *The Destruction of the European Jews*, (New York: Holmes & Meier, 1985), 2:405.
78. Daniel Levy and Natan Sznaider, "The Institutionalization of Cosmopolitan Morality: The Holocaust and Human Rights," *Journal of Human Rights* 2 (2004): 143–57.

Contributors

WOLFGANG BIALAS is a research fellow at the Hannah Arendt-Institute in Dresden, Germany, where he has just completed a project on Nazi ethics and ideology. His intellectual interests include the history of Nazism and the Holocaust, political philosophy, and nineteenth- and twentieth-century European intellectual history. He is the author of *Politischer Humanismus und "verspätete Nation": Helmuth Plessners Auseinandersetzung mit Deutschland und dem Nationalsozialismus* (2010) and the coeditor of *Nazi Germany and the Humanities* (2007). The collection of essays he is currently coediting with Lothar Fritze, *Ideology and Morality in Nazism*, will be published shortly by Vandenhoeck & Ruprecht in Göttingen.

AMY CARNEY received her PhD in German history from Florida State University. Currently she is an assistant professor of European history at Pennsylvania State University, the Behrend College. Her research focuses on fatherhood and the family community in the German SS. She is currently revising her dissertation, "Victory in the Cradle: Fatherhood and the Family Community in the Nazi Schutzstaffel," for publication.

TERJE EMBERLAND is a senior researcher at the Center for the Study of the Holocaust and Religious Minorities in Oslo, Norway. He is the author of *Religion og rase: Nyhedenskap og nazisme i Norge 1933–1945* (2003) and the coauthor of *Det ariske idol: Forfatteren, eventyreren og nazisten Per Imerslund* (2004). Together with Jorunn S. Fure he coedited a collection of essays on Himmler's Ahnenerbe, *Jakten på Germania: Fra nordensvermeri til SS-arkeologi* (2009). His most recent book (together with Matthew Kott), *Himmlers Norge: Nordmenn I det storgermanske prosjekt*, was published last year by Aschehoug & Co. in Oslo.

BJÖRN M. FELDER is a researcher in the Seminar for Medieval and Modern History at the Georg August University of Göttingen, Germany. He is the author of *Lettland im Zweiten Weltkrieg: Zwischen sowjetischen und deutschen Besatzern 1940–46* (2009). His current research project examines eugenics and race in the Baltic States and Russia, focusing on interconnections between scientific debates on the one hand and national identity and politics on the other.

ISABEL HEINEMANN is an assistant professor of modern and contemporary history at Münster University, Germany. She is currently leading a junior research group on "Family Values and Social Change in the Twentieth-Century United States," funded by the German Science Foundation. She is the author of *"Rasse, Siedlung, deutsches Blut": Das Rasse-und Siedlungshauptamt der SS und die rassenpolitische Neuordnung Europas* (2003), the coeditor of *Wissenschaft, Planung, Vertreibung: Neuordnungskonzepte und Umsiedlungspolitik im 20. Jahrhundert* (2006), and the editor of *Inventing the Modern American Family: Family Values and Social Change in 20th Century United States* (2012).

GERALDIEN VON FRIJTAG DRABBE KÜNZEL is an assistant professor of history at the University of Utrecht in the Netherlands. She is the author of *Het geval Calmeyer* (2008), which deals with the Holocaust in the occupied Netherlands and, more specifically, the application of racial laws and regulations in that country. Her current research project examines Dutch participation in the Germanization project in the occupied East.

THOMAS MAYER is a research associate in the Department of Contemporary History at the University of Vienna, Austria. He has published numerous articles and chapters on the history of eugenics in Austria, and he served as the coeditor of *Eugenik in Österreich: Biopolitische Strukturen von 1900 bis 1945* (2007). He is currently completing his doctoral thesis, which examines the history of the Department of Racial Biology at the University of Vienna between 1938 and 1945.

VLADIMIR SOLONARI is an associate professor of history at the University of Central Florida. His book *Purifying the Nation: Population Exchange and Ethnic Cleansing in Nazi-Allied Romania* was published by Woodrow Wilson Center Press in cooperation with the Johns Hopkins University Press in 2010. He is the author of a number of articles on Romanian, Moldovan, and Soviet history. His current research focuses on the social history of southwestern Ukraine under the Romanian occupation during the Second World War.

MARIUS TURDA is a reader in Central and Eastern European biomedicine at Oxford Brookes University. His current research focuses on the history of eugenics, race, and anthropology, particularly in Hungary and Romania. His recent publications include *Crafting Humans: From Genesis and Eugenics and Beyond* (2013) and *Modernisme et eugénisme* (2011). His new monograph on the history of eugenics in Hungary is forthcoming from Palgrave Macmillan.

ANTON WEISS-WENDT heads the research department at the Center for the Study of the Holocaust and Religious Minorities in Oslo, Norway. He is the author of *Murder without Hatred: Estonians and the Holocaust* (2009) and *Small-Town Russia: Childhood Memories of the Final Soviet Decade* (2010) and the editor of *Eradicating Differences: The Treatment of Minorities in Nazi-Dominated Europe* (2010) and *The Nazi Genocide of the Roma: Reassessment and Commemoration* (2013). He is currently writing a book tentatively titled "The Soviet Union and the Genocide Convention: Exercise in Cold War Politics."

STEFFEN WERTHER is an assistant professor of history at Södertörn University. His doctoral thesis examines SS ideology and the German minority in South Denmark. Previously he held research fellowships at the universities of Kiel, Copenhagen, Århus, and southern Denmark. His is primarily interested in nineteenth- and twentieth-century German and Scandinavian history, with a particular focus on nationalism, German-Scandinavian relations, racial theory, and Nazi ideology. He is the author of *SS-Vision und Grenzland Realität* (2012).

ELISABETTA CASSINA WOLFF is an associate professor of history at the University of Oslo, Norway. Previously she worked as a political analyst at Europa-programmet, a research institute in Oslo. She is the author of *Det latinske Europa: Frankrike, Italia, Portugal og Spania* (1999), *L'inchiostro dei vinti: Stampa e ideologia neofascista 1945–1953* (2011), and numerous articles and book chapters about Italian fascist ideology before and after the Second World War and social and political developments in Italy during the Berlusconi era. She is currently writing a book on Italy's political history.

RORY YEOMANS is a senior international research analyst at the UK Ministry of Justice. He is the author of numerous articles and book chapters about eugenics, cultural politics, and social experimentation in the Independent State of Croatia as well as in interwar and socialist Yugoslavia. He is the author of *Visions of Annihilation: The Ustasha Regime and the Cultural Politics of Fascism, 1941–1945* (2013) and is currently writing a book about racial politics and social mobility in fascist Croatia.

Index

In the Critical Studies in the History of Anthropology series

Invisible Genealogies: A History of Americanist Anthropology
Regna Darnell

The Shaping of American Ethnography: The Wilkes Exploring Expedition, 1838–1842
Barry Alan Joyce

Ruth Landes: A Life in Anthropology
Sally Cole

Melville J. Herskovits and the Racial Politics of Knowledge
Jerry Gershenhorn

Leslie A. White: Evolution and Revolution in Anthropology
William J. Peace

Rolling in Ditches with Shamans: Jaime de Angulo and the Professionalization of American Anthropology
Wendy Leeds-Hurwitz

Irregular Connections: A History of Anthropology and Sexuality
Andrew P. Lyons and Harriet D. Lyons

Ephraim George Squier and the Development of American Anthropology
Terry A. Barnhart

Ruth Benedict: Beyond Relativity, Beyond Pattern
Virginia Heyer Young

Looking through Taiwan: American Anthropologists' Collusion with Ethnic Domination
Keelung Hong and Stephen O. Murray

Visionary Observers: Anthropological Inquiry and Education
Jill B. R. Cherneff and Eve Hochwald
Foreword by Sydel Silverman

Anthropology Goes to the Fair: The 1904 Louisiana Purchase Exposition
Nancy J. Parezo and Don D. Fowler

The Meskwaki and Anthropologists: Action Anthropology Reconsidered
Judith M. Daubenmier

The 1904 Anthropology Days and Olympic Games: Sport, Race, and American Imperialism
Edited by Susan Brownell

Lev Shternberg: Anthropologist, Russian Socialist, Jewish Activist
Sergei Kan

Contributions to Ojibwe Studies: Essays, 1934–1972
A. Irving Hallowell
Edited and with introductions by Jennifer S. H. Brown and Susan Elaine Gray

Excavating Nauvoo: The Mormons and the Rise of Historical Archaeology in America
Benjamin C. Pykles
Foreword by Robert L. Schuyler

Cultural Negotiations: The Role of Women in the Founding of Americanist Archaeology
David L. Browman

Homo Imperii: A History of Physical Anthropology in Russia
Marina Mogilner

American Anthropology and Company: Historical Explorations
Stephen O. Murray

Racial Science in Hitler's New Europe, 1938–1945
Edited by Anton Weiss-Wendt and Rory Yeomans

www.ingramcontent.com/pod-product-compliance
Lightning Source LLC
Chambersburg PA
CBHW020331270326
41926CB00007B/143

9780803245075